FROM GALILEO TO THE NUCLEAR AGE

GALILEO GALILEI (1564-1642)

FROM GALILEO TO THE NUCLEAR AGE

AN INTRODUCTION TO PHYSICS

BY

HARVEY BRACE LEMON

PHOTOGRAPHS BY THE AUTHOR
DRAWINGS BY CHICHI LASLEY

THE UNIVERSITY OF CHICAGO PRESS
CHICAGO & LONDON

The University of Chicago Press, Chicago & London
The University of Toronto Press, Toronto 5, Canada

Copyright 1934 and 1946 by The University of Chicago. All rights reserved. Published 1934 under the title From Galileo to Cosmic Rays. *Second Edition 1946. Seventh Impression 1965. Printed in the United States of America*

FEB 3 1967

TO THE UNDERGRADUATE STUDENTS
IN THE INTRODUCTORY GENERAL COURSE IN THE PHYSICAL SCIENCES
AT THE UNIVERSITY OF CHICAGO AND TO THOSE STUDENTS AT LARGE,
WHETHER THEY BE IN COLLEGE OR NOT, WHOSE NEEDS AND WHOSE
INTEREST IN THESE MATTERS HAVE FURNISHED ITS INSPIRATION
THIS BOOK IS DEDICATED

PREFACE TO FIRST EDITION

SINCE the adoption of the new undergraduate curriculum at the University of Chicago four years ago, it has become increasingly evident, as the annual revision of its four introductory general courses has been undertaken, that in addition to comprehensive syllabi an entirely new type of book is needed at least in certain fields of work.

In the physical sciences, an examination on the content of the Introductory General Course is required of all students. The majority of the students have no background in this field and little or no previous training in any of its departments. Nearly all of them are majoring in other fields. This is due to the fact that students entering the University with good preparation in mathematics, physics, or chemistry, or in any two of these subjects, are advised to read for their physical science examinations, or to review and prepare for them in some manner that requires less time than a full year's attendance in this course. Furthermore, a few of the best students in other fields find it possible to do creditable work and save much time by intensive individual study, guided by conferences with their instructors and advisers, and supplemented by attendance only for certain periods of the course.

THERE are many excellent textbooks on physics. The present syllabus is replete with references to the two texts provided for our students in their rental sets. Abundant testimony from students of all types, however, has taught us that selected paragraphs, no matter how clear in their context in a well-organized body of material, become obscure to one who has almost no background in the subject. These books, furthermore, have been designed for full-year courses in physics. They cannot be used in entirety in a year's general course in the physical sciences that includes astronomy, chemistry, some mathematics, three months of geology, and some geography, in addition to physics.

For these reasons it has seemed desirable to put in the hands of our students, while physics is the subject of their study, a book with continuity, designed for reading from cover to cover within a reasonable time, stressing the source material, phenomena, and giving interpretations in non-technical style and by homely analogy. It should also have something of that human interest that attaches to great personalities, but most of all it should try to show explicitly and implicitly how a great scientific field of thought has come into being. Moreover, it should indicate the satisfaction enjoyed by those who participate in such a development of thought, as well as its practical advantages to mankind. A book of this kind is needed not only for physics but also for each of the other sciences listed above. These have been our reasons for taking our integrated lecture notes of the past three years and, in the same spirit of those lectures, and in the first person, attempting to set down in type and halftone a story to which no man can do justice.

THAT a book designed for the purposes outlined above may also meet the interests of the adult reading public, is not perhaps too much to be hoped. The latter comprise a great body of students at large—students whose years and experience in general are much greater than those of our undergraduate body but whose lack of background is almost identical with that of those in attendance upon our formal courses. Furthermore, where background does exist in the case of this older group, it has been, for the most part, acquired before the days when modern physics started on its meteoric career.

Adult lectures in physics, covering material identical to that given to college sophomores, draw capacity crowds, especially if source material in the way of demonstrated phenomena is presented. Our experience is that the attendance is quite as good when only the simpler undebatable material of the subject is presented as when the more modern controversial aspects are given. In the latter case it may be that the personal contact with, or at least the hearing of men recognized as experts in their subject, constitutes the major source of satisfaction to the listeners. For it is doubtful if the force of

PREFACE

the arguments presented is fully appreciated by the audiences either at the time of the lecture or later.

The same thing may be said with respect to the reading of books. There have been, and there are being continually, offered to the reading public popular expositions of science. Many of these deal with subjects that are so recondite that they are still matters of controversy among scholars who have devoted a lifetime of study and experience to them. In regard to such books it must be the talent of the author as a writer, or a sense of personal contact with him through his writing, that remains afterward with the reader, rather than any real understanding or comprehensive knowledge of a subject that is not yet understood or fully comprehended by anyone.

THERE is in physical science, on the other hand, a large body of material, universally accepted by all well-trained and critical scholars, with which the layman, even if he has been formerly "exposed" to it in conventional courses for the brief time of a crowded academic year, is quite as unfamiliar as he is with the arguments for and against a pulsating universe. This material is the solid substance of science, proved, checked, cross-checked. It is not the tenuous stratosphere of its speculations. To do it justice no specially gifted pen is needed, no rhetoric is necessary. For the story enthralls if only it can be woven into the pattern of our everyday experiences and be told by words, rather than by equations, and by simple drawings that speak for themselves, rather than by cuts, inadequately explained and of too technical a character. The well-known psychological technique of alternation between intense concentration and relaxation is seldom used in books on the more difficult subjects, where it is most needed.

These are the devices we have tried to use to make readable a story about things largely unseen and entirely analytical in character. The critical physicist who looks at our pages may be somewhat shocked at times, but he will recognize that every effort has been taken to make our narrative authentic and without exaggeration or distortion with respect to its factual content.

PART I deals with the fundamental aspects of *Mechanics*, upon which all the rest of the physical sciences depend. In connection with falling bodies, there is discussion of several matters, such as terminal velocity, which, though they are of common knowledge and interest, seldom seem to be given any place except in more technical and advanced works. For integration with the sister-science of astronomy, such matters as speculations about tidal evolution have been included, chiefly to show the broad scope of application of fundamental principles but partly to provide relief after discussions that require concentrated attention. Energy, the central character in the entire story, is given four chapters, in the hope, perhaps vain, that the reader may acquire by manifold repetition an enduring impression of how it is defined and what it means.

Part II first presents the phenomena of *Heat* entirely from the experimental point of view and stresses the difference between heat and temperature; then it gives the interpretation of the kinetic theory of heat. In this way when the theory is taken up some desirable repetition as to the facts is made necessary. Furthermore, the method by which a hypothesis becomes acceptable and is made the reference point for all future work can thus be set off by itself and be illustrated by one of the most magnificent cases in the history of the development of science.

Part III, on *Electricity and Magnetism,* takes up aspects of the subject in which it begins to recede far from the realm of daily experience. The historical approach is used; and the experiments are associated, as far as possible, with the experimenters. A very definite attempt is made from the outset to induce the reader to formulate various interpretations for himself and then, following through their implications, to recapitulate in his own thinking what has been the racial development in regard to them.

Part IV, dealing with *Electricity and Matter*, introduces modern physics. The events of the twentieth century cast their shadows ahead of them out of the nineteenth; and from such early indications as the discoveries of conductivity in gases, the photoelectric and thermionic effects, of the latter part of that century, and those of X-rays and radioactivity that ushered in the next one, the story is put together.

Part V, on *Waves and Radiation*, combines in a very unconventional association a great variety of things: mechanics again, and sound and light. Emphasis on wave characteristics is certainly demanded in an age that attempts the solution of every unknown problem or the interpretation of any new set of data by the writing of the wave-equation. Sins of omission are bound to be numerous in any attempt to cover so vast a field. That our account may at times, at least on a first reading, submerge the reader beyond his depth is almost inevitable. We can but hope that the more general aspects of the subject will be impressed upon him and that some basis for future reading may be laid.

ACKNOWLEDGMENTS are due to so many who have co-operated in the making of the book that we hardly know where to begin the list. To Mr. Egbert G. Jacobson, for time and unrequited effort in planning and design, as well as to Mrs. Chichi Lasley, the artist illustrator, goes all credit for such artistic merit as the volume possesses. Mr. Donald Bean, the Manager of the University of Chicago Press, and his able staff have been inspired and sagacious collaborators as to methods, ways, and means. Professor A. H. Compton has read helpfully that section of the manuscript that deals with cosmic rays, and Professor Walter Bartky has done the same for the section on mechanics. Mrs. Ardis T. Monk and Dr. Reginald J. Stephenson, of our staff, who have participated in the design, administration, and teaching of small group sections of the General Course, have read the manuscript with great care and have made many valuable suggestions. The same has been done by Dr. W. K. Chandler, of the Department of English. To the Director and staff of the Museum of Science and Industry of Chicago, especially to Dr. A. M. MacMahon, we wish to express our appreciation of their co-operation in the development of a physics museum used by our students in these courses. Apparatus provided by them has furnished us with no small amount of material for our illustrations. For the rest, Mr. Stanley Anderson, our assistant in preparing demonstration experiments, has furnished and set up special settings for the cameraman. To Erpi Picture Consultants, New York City, our collaborators in educational sound pictures, our acknowledgments are due for prints and films for reproduction. To Professor Paul Kirkpatrick, of Stanford University, for valuable suggestions based on careful class use of the first printing received in time to be incorporated in subsequent impressions. Last but not least, our esteemed friend, the other co-director of this course, Professor Hermann I. Schlesinger, of the Department of Chemistry, has this year generously assumed more than his share of responsibilities in that work to free the author for the purpose of writing this book. Moreover, on those sections of the volume where physics and chemistry are not to be differentiated his suggestions have been of the utmost value.

HARVEY B. LEMON

RYERSON PHYSICAL LABORATORY
UNIVERSITY OF CHICAGO
July 1, 1934

PREFACE TO SECOND EDITION

IN THE twelve years that have elapsed since the first edition of this book was offered to nonprofessional students and to the public, the boundaries of physical knowledge have been vastly extended. Items in the first edition, like the neutron tentatively mentioned in a footnote, have become firmly established in atomic architecture; contrariwise, nuclear structures, believed in 1934 to be composed of electrons and protons, have been entirely abandoned in view of new evidence; and proton-neutron composition has taken their place. Cosmic rays are no longer the last word in the present nuclear age, although still most suggestive by their recent disclosure of energies much greater, per particle, than those released in the sadly misnamed "atomic" bombs, of such grave moment to the entire world.

Educational experimenting likewise has produced curricula, at least in some institutions, as altered from those of twelve years ago as are details of nuclear structure; whether they are as sound only time and experience in this swiftly changing world can decide.

Fortunately for those intensely concerned in both aspects of our subject, there are solid and unchanging foundations on which the science rests and at least a few ideas with respect to its pedagogy that still find increasingly wide acceptance.

Thus our title has been brought up to date—a simple matter, at least for the author. The later chapters that come close to the border line of knowledge have been advanced with it, largely by re-writing and considerable additions of things undreamed of even seven years ago. Here and there in earlier chapters, where the longest shadows of coming events reached back, minor alterations have been made—the most difficult part for the author.

Finally, lest educational changes of emphasis and later experiments may cause earlier ones to be forgotten—since here we inevitably are less objective and build more on sand and less on rock—this second edition carries without alterations the Preface to the first—for the record, ill or good.

Harvey B. Lemon

University of Chicago
October 1, 1946

APROPOS OF OUR THREE-DIMENSIONAL ILLUSTRATIONS

ONE feature rarely found in books has been included in this one. Many of our photographs may be seen in solid relief, STEREO-SCOPICALLY, as it is called, being double in the reproduction. These double pictures are not two identical prints from the same negative, as many persons suppose, but are taken by two lenses on two negatives, in most cases simultaneously.

These two lenses are separated by approximately the distance of our two eyes. Thus the angle of view is slightly different in the two photographs, as a careful comparison of them will show. Technically we give this shift the name "parallax," and we might describe what it is by saying that objects (or parts of objects) near the observer change their relative positions with respect to objects (or other parts of the same object) more remote as we look at them from different points of view, namely, those occupied by the pupils of our left and our right eye, respectively. You can observe this effect yourself by holding the head still and looking at some objects in front of you, first with one eye closed and then with the other closed.

When both eyes are open, the sensations of both superimpose and we see things in relief, in three dimensions, provided our eyesight is normal and each eye has approximately the same degree of vision. To make our two photographs so superimpose, we use two prismatic lenses* as we look at these

---

* A prismatic lens is one that combines deviation of the rays through it, here necessary to superimpose the pictures, with some degree of convergence of the rays, which magnifies the picture.

pictures. They must be held so that the line between their centers is parallel to the line between the centers of the two photographs; and the line between the centers of our eyes must also have the same direction, i.e., pictures, lenses, and eyes must be lined up. The appropriate distance from the page, which should lie as flat as possible, is about six inches but varies somewhat with different individuals and is soon found by experience.

BY A little practice it is possible for many persons to learn to see such illustrations in relief *without* any prismatic lenses. One does this by placing the two photographs two or three feet away at first and holding a long strip of cardboard at right angles to them and lined up on the line between them. One then places the tip of his nose upon the cardboard, which thus prevents either eye from seeing the picture in front of the other eye.* On looking "far away" then, through the pictures as it were, you will see but one, and that one will be in relief. As this facility is acquired, the length of the cardboard strip may be reduced until the pictures may be brought eight to ten inches from the eyes when they are larger and details are more clear. Finally with practice the cardboard may be dispensed with entirely. Without the cardboard but with the eyes properly aligned, three pictures will always be seen; the center one is in relief and appears "solid." The advantage of such illustrations is obvious in bringing out details and relationships of different parts of a piece of apparatus. One unfamiliar with the object photographed thus actually sees many more details of it.

IN A few illustrations where *a small change takes place* in some experiment most parts of which are unchanged, we have made one of our photographs before the change occurred and the other afterward. In addition to saving space and reproductions, the slight change produced in the experiment becomes extremely conspicuous. These two changed portions do not blend, and the two eyes pick out the difference much more quickly than does one eye alone.

Astronomers search for unknown variable stars in fields that contain thousands of stars by this method, by

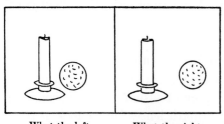

What the left eye sees   What the right eye sees

* As is being done by the young lady in the drawing on p. xiii.

THREE-DIMENSIONAL ILLUSTRATIONS

combining or by "blinking" first one eye and then the other at two identical pictures of the star field taken at different times. In the interval between the exposures a star of variable intensity will have changed and its two images will not be alike. The eyes at once detect such differences when used in conjunction in this manner.

Stereoscopic photography has a multitude of uses. By taking the two photographs from distances separated by many feet instead of by the distance between the two eyes, when distant objects are involved, the third dimension (that from the observer to the object) is greatly magnified in comparison with the other two. Such "giant vision" views reveal aspects of topography that may be seen in no other way. Microscopic photographs, X-ray photographs, and the photography of cloud-forms all find increased possibilities of delineation by the uses of the stereoscopic principle. For other illustrations we refer to Judge, *Stereoscopic Photography* (London: Chapman and Hall, Ltd., 1926).

CONTENTS

PART I. MECHANICS

CHAPTER	PAGE
1. Galileo and the Principle of Inertia	3
2. Newton's First Two Laws of Motion	9
3. Falling Bodies	13
4. Terminal Velocity	20
5. Gravitational Attraction	26
6. Paths of Projectiles	32
7. Balanced Forces and Newton's Third Law of Motion	39
8. Applications of the Law of Conservation of Momentum	49
9. Work	57
10. Energy, Potential and Kinetic	65
11. Work and Heat: Conservation of Energy	77
12. Power	83
13. Our Two Oceans—Atmosphere and Hydrosphere	91

PART II. HEAT

14. Temperature and Expansion	109
15. Thermometers	116
16. Temperatures of Various Places	125
17. Quantity of Heat and Temperature	135
18. The Kinetic Theory of Heat	149

PART III. ELECTRICITY AND MAGNETISM

19. Electroscopes: Pith Balls and Gold Leaves	177
20. Conduction, Induction	183
21. Thunderstorms	191
22. Artificial Production of Charge	197

CHAPTER	PAGE
23. MAGNETISM	205
24. FIELDS OF FORCE	214
25. THE ENERGY OF ELECTRIC AND MAGNETIC FIELDS	222
26. CHARGES IN MOTION	231
27. MAGNETIC FIELDS IN MOTION	246

PART IV. ELECTRICITY AND MATTER

28. CONDUCTORS, SOLID AND LIQUID	265
29. THE CONDUCTION OF ELECTRICITY BY GASES	275
30. PHOTOELECTRIC AND THERMIONIC EFFECTS	283
31. CATHODE RAYS	293
32. ELECTRONS: THEIR MASS AND VELOCITY	301
33. POSITIVE RAYS, PROTONS, NEUTRONS, AND ISOTOPES	311
34. RADIOACTIVITY AND TRANSMUTATION	316

PART V. WAVES AND RADIATION

35. ABOUT WAVE PHENOMENA IN GENERAL	333
36. THE SPECIAL ATTRIBUTES OF WAVES	347
37. ON SOUND	372
38. ELECTROMAGNETIC WAVES	385
39. VISIBLE LIGHT	398
40. RADIATION AND ATOMIC STRUCTURE	415
INDEX	443

PART I
MECHANICS

CHAPTER 1

GALILEO AND THE PRINCIPLE OF INERTIA

MODERN science, that has done so much in the last one hundred and fifty years to transform the conditions under which mankind lives, has in large part adopted a new manner of attack upon the unknown. To a limited degree this approach was hesitatingly attempted by men before the sixteenth century. Galileo (1564–1642), (Plate 1), however, was the first to use it openly, consistently, and effectively. As a result, to him more than to any other single individual goes the credit of being the father of modern science.

His method was to ask, not *why* things happen as they do, but rather what it is that happens. He searched for accurate descriptions of events and interrelations between various aspects of them. He did not attempt to find in his own consciousness reasons why things were as they seemed; rather, he inquired whether they really were what they seemed to be.

The scientific method before Galileo

Galileo performed experiments

One great field of Galileo's endeavor had to do with the motions of inanimate objects. He was an ardent student of astronomy. He investigated what it was that happened to terrestrial objects that caused their motion. Thus, in a world where everything from the skies above to the whispering winds below spoke of movement both seen and unseen, his studies were most fundamental. The analyses he was the first to make with respect to the motions of objects, celestial or terrestrial, hold good to this day. He began with the study of pendulums and inclined planes. It has since been found that the analyses he used not only apply as well to the basic aspects of modern intricate machines but also are fundamental in wider ranges including stars and galaxies.

If to his contributions be added those of Isaac Newton (1643–1727), (Plate 2), then it can be said that more recent times, up to the twentieth century at least, have done little but amplify and add detail, then adapt and apply to a myriad of uses. In every connection there has been found new confirmation of the correctness of Galileo's and Newton's ideas, principles, and laws. Not until the twentieth century were new fields in the microcosmos entered where new genius was, and still is, needed for unerring guide.

WITH respect to the motions of ordinary material bodies, such as a stone that is thrown, or a rock that becomes detached and crashes down a mountain side, the ancients had what seem to us curious, naïve, and conflicting ideas. They held a doctrine about "natural places"—levels, as it were—toward which fire, air, water, and the different materials of the earth showed a "natural" tendency to settle or to rise. Each substance sought to return to its own special plane when for any reason it was disturbed and moved elsewhere. A ball thrown by the hand, according to one view, was kept moving onward in its journey because it displaced some air in its passage. The moving-in of the air behind it kept pressing it forward and thus supplied the motive power. Elasticity also was appealed to. When one bent a short bit of spring between finger and thumb and snapped it forward, its elasticity furnished in some fashion the driving mechanism to keep it going after it had left one's hand.

PLATE 2 SIR ISAAC NEWTON (1643-1727)

LET us not think ourselves so far beyond these ancients, however. No event of scientific and of news value ever occurs today but that the many letters of inquiry, of criticism, of denial, of what is universally accepted as correct by competent judges, clearly reveal that there are still very many people in all walks of life who are, in the main, quite as far from understanding natural processes as were the men and women of the sixteenth century. Fakers and quacks, either ignorant or unscrupulous, still continue to exploit great masses of people who, if they but had a mere bowing acquaintance with the vast store of knowledge available to our generation, could never be deceived by such obvious fallacies and trickery.

WITH respect to moving objects, Galileo contributed a new point of view. *Why* things kept moving was to him not the important thing. Indeed, to this day *we do not know why things set in motion keep on going*. We know, of course, that they *do* keep on moving. Galileo placed the emphasis where it belonged: the tendency to keep going admitted, how is it that moving objects are stopped? With respect to objects at rest, likewise, there was a precisely similar situation, too obvious almost to be noticed. The fact is that all objects remain at rest, they "stay put," if undisturbed. To move them by setting them in motion requires effort just as effort is required to stop a body once it has been put in motion.

Equally important is the fact that objects moving freely move along a straight line. Nothing can be found to draw a straighter line than a moving object entirely free from all constraint. This straightness is not seen when something is hurled through the air, since it is then impelled to fall to the earth. But, let the object be smooth and round and rolled along a level surface, then not even a ray of light can go more directly. Intuitively, also, we recognize that were it not for disturbing and extraneous forces, especially friction, this constant speed in a straight line might be maintained forever.

That this by now is common knowledge possessed by all normal adults and even by many children is abundantly attested by any device that appears to function otherwise. If in a show window there be an aluminum ring that suddenly jumps about from one place to another, a crowd will at once gather, attracted by something obviously unnatural

GALILEO AND THE PRINCIPLE OF INERTIA

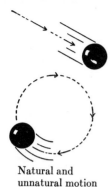

Natural and unnatural motion

and mechanically unorthodox. "What is the hidden mechanism that does this?" they inquire. Everyone knows there is a hidden mechanism. Bets may be made as to its nature, bets difficult to settle, since often neither side is right. The *rectilinear property* of unconstrained motion, likewise, by some arrangement that appears to deny it, makes an excellent advertising device. It commands instant attention from almost any passer-by. Imagine a ball rolling round and round a circle, or in any other curved path for that matter. To keep it upon its curved path, quite as much as to speed it up or to slow it down or to start it or to stop it, means *interference* with its natural behavior.

THIS natural behavior is a manifestation of a general property exhibited by all material objects, the property of INERTIA. This property, by its name, connotes a universal truth about all matter, which is that it *does* remain motionless, if at rest, and that it *does* continue to move with a constant speed in a straight line, if already in motion—unless something outside and beyond itself interferes.

ALREADY, alas, before the first section of our exposition is completed, you are beginning to become conscious of an inevitable preciseness of language that is always found in scientific writing. Of course that is the only way that it can be scientific. We genuinely intend to emphasize this scientific terminology as little as possible, but now and then, in order that you may understand what we say, we must explain the meanings of our words.

By "speed" we mean nothing more than the word connotes to you—motion—fast or slow, along a path curved or straight—so many inches in so many hours, so many miles in so many seconds—up, down, east, west, or wandering about in any old direction. The idea of speed takes no cognizance of direction. The word VELOCITY does. When we say "constant velocity," we mean "constant *speed in a straight line*." The latter phrase has no ambiguity about it; and, from now on, the former should not have, either. The former is more concise since in it one word takes the place of five.

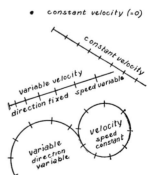

If a body is at rest and starts to move, its velocity is changed. So is its speed. If it is moving in a straight line and if its motion along this line is slowed down or "speeded" up, its velocity is likewise changed. But suppose now that it moves with a constant "speed" along a curving path. Then its velocity is changing, although its speed is not, just as truly as its velocity was changing in the other cases. Velocity, by definition of the word, is varied either if the speed changes *or if the direction changes* (and, of course, if a simultaneous change of both occurs).

THE property of inertia, which we described above as the one involved in the natural tendency of objects, left alone, to maintain a constant speed along a straight line, can now be more succinctly stated. *By the "property of inertia" we mean the property of all objects to maintain a constant velocity.*

CHAPTER 2

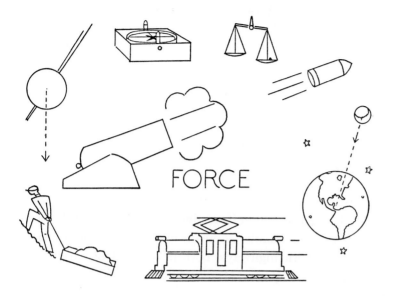

NEWTON'S FIRST TWO LAWS OF MOTION

ANY change in velocity, exhibited by inanimate matter, we have seen is of signal importance, since it is indicative of interference in some way with the natural *status quo*. We call rate of change in velocity by the name ACCELERATION. The accelerator in an automobile is what one steps on to increase the velocity. You have probably heard this little lever called the "exhilarator" by persons more learned in the vocabulary of thrills than in that of dynamics.

Speeds of motor cars are usually indicated in miles* per hour. They have *speed*ometers, not velocity meters. When the *direction* of the velocity is to be changed, you yourself must make the effort upon the steering wheel. You are well aware of how the inertia of the car tends to straighten out its path after it rounds a sharp corner.

An accelerator

* For brevity we use the following abbreviations: miles, mi.; feet, ft.; seconds, sec.; millimeters, mm.; centimeters, cm.; meters, m.; kilometers, km.; pounds, lb.; grams, gm.; kilograms, kg.; square centimeters, sq. cm,; cubic centimeters, cc.; liter, l.; degrees Centigrade, °C.; degrees Fahrenheit, °F.; degrees Absolute, °K.; per, /.

Acceleration measurement

ALONG a straight road the acceleration, indicative of "pick-up" or "get-away," of modern machines would naturally be expressed in miles per hour per second; for example, a recent advertisement of a "well-known car" states that it will pick up speed from 7 mi. per hr. to 75 mi. per hr. in 37 sec. Admitting this claim for the sake of argument, this gain in speed of 68 mi. per hr. in this interval, if uniform, would mean an increase in speed of 1.84 mi. per hr. every second.*

THE cause here responsible for the acceleration of the aforesaid automobile is, obviously, to be found in its engine. Acceleration, in the opposite sense of slowing down, is available by means of the brakes. Fortunately for all concerned along modern highways, both in and out of cars, brakes have been improved as well as engines. Modern brakes are still several times more effective in producing acceleration in the negative sense than are the engines in a positive one.

From 7 to 75 mi. per hr. in 37 sec., + acceleration

From 75 mi. per hr. to rest, in 12 sec.; retardation (i.e. −acceleration)

The same engine, in the same state of effectiveness, in the same automobile, and—we must add—with the same driver, always produces the same acceleration along a level road.

In a more massive car the pick-up would be sacrificed. Much of the recent trend toward liveliness of behavior has been due, not to increasing sizes of motors, but rather to diminishing the massiveness, and thus the inertia, of the bodies, of the chassis, and even of the motors themselves.

---

* Scientific procedure abhors mixed units—as here, hour and second. Feet per minute would be given by $1.84 \times 5{,}280/60 = 162$ ft. per min. per sec. acceleration; but this is open to the same objection. Since 162 ft. per min. is 2.7 ft. per sec., the advertiser interested only in an appeal to scientific men would claim a "pick-up" of 2.7 ft. per sec. per sec., or, abbreviating further, 2.7 ft./sec.^{2}!

NEWTON'S FIRST TWO LAWS OF MOTION

Lively behavior may be maintained with lighter cars and correspondingly less powerful motors, which means correspondingly fewer gallons of gasoline to buy.

SINCE the effort to speed up or slow down any object (or to swing it out of a straight path) depends on the magnitude of its inertia and upon the amount of acceleration desired, it would seem logical to measure the effort by these two items "conjointly," an ancient term which means "by the product of the two." This extremely fundamental quantity today is called the FORCE. Both Galileo and Newton used this term in its present meaning. What we have discussed above is the essence of Newton's so-called Laws of Motion, the first two of them specifically.

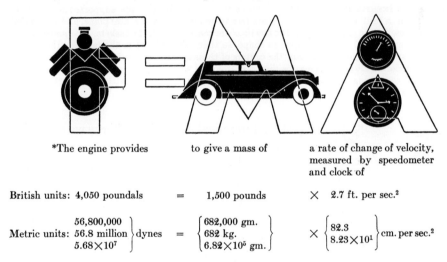

*The engine provides	to give a mass of	a rate of change of velocity, measured by speedometer and clock of
British units: 4,050 poundals =	1,500 pounds	× 2.7 ft. per sec.²
Metric units: $\begin{Bmatrix} 56{,}800{,}000 \\ 56.8 \text{ million} \\ 5.68 \times 10^7 \end{Bmatrix}$ dynes =	$\begin{Bmatrix} 682{,}000 \text{ gm.} \\ 682 \text{ kg.} \\ 6.82 \times 10^5 \text{ gm.} \end{Bmatrix}$	× $\begin{Bmatrix} 82.3 \\ 8.23 \times 10^1 \end{Bmatrix}$ cm. per sec.²

Newton's First Law restates Galileo's principle of inertia, to the effect that: **Any body persists in its state of rest or uniform motion in a straight line, unless acted on by some external force.**

Newton's Second Law amplifies and gives the exact relation implied by the first, to the effect that: **The measure of the force which has acted on a body to alter the state of its motion is the product of the latter's inertia (often called MASS) by the acceleration produced.**†

---

* It is suggested that the reader simply let the picture sink in at the moment, and return to comtemplation of figures after reading the chapter.

† To give unit mass (defined in next paragraph) an acceleration of 1 cm/sec² requires unit force, called 1 dyne.

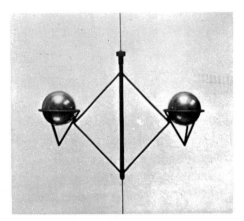 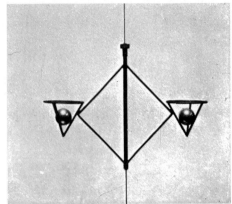

PLATE 3 THE INERTIA BALANCE. In this device two conical baskets are supported horizontally from a taut vertical wire. They are free to oscillate in a horizontal plane, and the period of oscillation depends upon the inertia of the masses held in the baskets. The masses being compared should be spherical in shape, so that, when placed in the baskets, the conical shape of the latter results in the centers of these spheres, irrespective of their size, being supported always at the same distances from the wire. Symmetry and balance are served by comparing masses that have been divided into two approximately equal parts. If t_1 be the period with mass m_1, and t_2 that with mass m_2, then $m_1/m_2 = t_1^2/t_2^2$.

THE dynamical definition of force we have just outlined involves, of course, a selection of some definite standard of inertia, some reference mass. This by international agreement is the GRAM, a mass 1/1,000 that of a certain block of platinum-iridium known as the International Prototype Kilogram, deposited at the International Bureau of Weights and Measures at Sèvres, France. Dull as it looks, utterly uninteresting to tourists, it is guarded with great care, and its exhibition is attended with great ceremony. Its mass was intended to be, and is very nearly, that of 1,000 cc. (1 liter)* of water at 4° C. Copies of this standard are possessed by various governments, and with them all other subsidiary standards may be compared. Although this comparison is usually made by the process of weighing (see p. 17), inertias may be compared directly by devices such as the inertia balance (Plate 3), in which the force of the earth's attraction for the masses being compared plays no rôle whatever.

* About a quart.

CHAPTER 3

FALLING BODIES

THAT part of Galileo's work best known because of the tremendous row it produced had to do with certain of his observations on falling bodies dropped from the overhanging side of the Leaning Tower in the plaza of the cathedral at Pisa. These experiments, however, were not the primary ones. They were offered as confirmation of deductions from many other experiments that had gone before and of which one seldom hears any mention.

When an object is dropped, it seems to go faster and faster—at least if it be large and massive. On the other hand, a light object like a feather gently settles to the ground with a constant speed not difficult to observe. Raindrops and snowflakes also are good examples of the latter case.

The difference in types of motion exhibited by light and heavy objects in falling produced the utmost confusion in the minds of the earlier philosophers. Modern philosophers tend to leave such matters to physicists. Democritus (460–370 B.C.) had argued that in a vacuum (which he held to be conceivable) heavier objects would fall faster than light ones. Aristotle (384–322 B.C.) held that they would all have the same velocity in a vacuum, but that, since it was inconceivable for falling bodies to have all the same velocity, the idea of any such concept as a vacuum was absurd.

THIS sort of inquiry seemed to Galileo to be getting nowhere; consequently, he attempted to measure the actual motion of falling objects, that of heavier and more massive ones especially. This being difficult to do directly because of the swiftness of the motion, he ingeniously devised a simple way of diluting this motion, so to speak, without changing its general character.

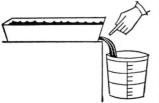
Galileo's water-clock

In sliding downhill, objects go faster and faster, just as they do when falling freely. The pull of the earth, however, does not have its full force along an inclined plane; hence it cannot produce the same degree of acceleration that it normally does. Furthermore, by changing the slope of the incline, in going from the vertical to the horizontal, all degrees of diminution of the effect of gravity from its full value to its complete elimination are possible.

In those days there were no clocks. Galileo used a broad, shallow tank of water with a hole that could be opened and closed with the finger. Measurement of the amount of water that ran out during an interval of time was taken as the measure of the interval.

Suppose, however, that now we use a clock and observe the motion of an object sliding with as little friction as possible* down an inclined plane and note first its positions at the end of equal consecutive intervals of time, indicated by tick and tock. We will observe, as did Galileo, that the distances covered between consecutive ticks of the clock increase. Furthermore, if we could be enticed to take a ruler and compare these various distances with that covered during the *first* interval, taking this as 1, we will find they vary as 1, 3, 5, 7, 9, etc. This is true no matter how steep the slope: the actual lengths covered during each interval increase as the slope increases, but the relations of these distances to each other remain unaltered. Remembering our argument that the motion along the inclined plane which we have actually observed approaches that of a freely falling body as the plane becomes vertical, we might have the

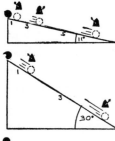
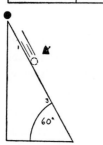
Distances down an inclined plane in consecutive intervals

---

* The object is best mounted on very light wheels having ball or roller bearings.

FALLING BODIES

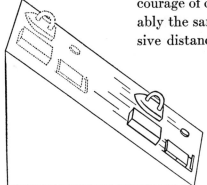

courage of our convictions sufficiently to say that probably the same relations that we found between successive distances along the plane would hold for a body falling freely. If we are not sure, however, any physics laboratory has an apparatus that will demonstrate this directly.

Still a further statement can be made from observations on the plane. If various objects of different size, shape, and

Objects slide together down a *smooth* incline irrespective of their masses

material, which have therefore quite different amounts of inertia, slide down a very smooth incline, they descend at almost the same rate. The motion of any one of them differs so little from that of any of the others that the discrepancy seems unworthy of mention. Since natural common sense would prevent our trying to slide either a feather or a 10-ton truck down our experimental plane, any difficulties that might arise from such foolishness we might properly think were hardly pertinent.

To obtain the *total distance* traversed by any of these objects in any given interval of time, one has only to add up the distances covered in *each one of the successive intervals*. Taking the distance during the first interval as 1, the total distance

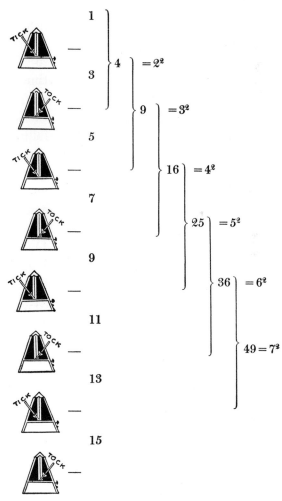

Time-distance relations of a freely falling body

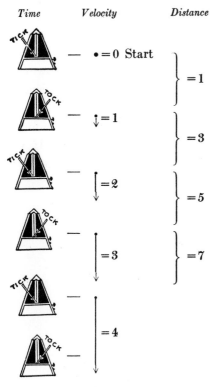

Time-velocity-distance relations of a freely falling body

during the first two intervals is $1+3=4$; during the first 3, is 9; during the first 4, is 16; etc.; thus we see that the *total distance* covered by a body sliding down an incline (or falling freely, for that matter) is proportional to the *square of the time* elapsed.

When Galileo first attacked this problem, he was, of course, aware that the *velocity* of the sliding or falling objects increased, but he did not know whether this increase was proportional to the distance traversed or to the time elapsed. The observations outlined above, however, are consistent only with the latter alternative. The *velocity* of a falling body *increases exactly as does the time* that it has been falling. It gains equal amounts of velocity in equal intervals of time. This is what is called a constant, or a *uniform, acceleration*.

From what has been said above (p. 11) about force (Force = Mass × Acceleration), we see that, since the mass and the acceleration of a falling body are both constant, their product, which is the force of the earth's attraction upon the body, is also a constant thing.*

Having found by experiment, therefore, that irrespective of their masses, bodies under the action of the earth's constant pull always exhibit one and the same acceleration, Galileo announced the fact that all bodies dropped simultaneously fell with the same speed at any subsequent moment, i.e., that a heavy one or a

Galileo's experiment on the Leaning Tower

* This is strictly true only for one particular location, and there only within a range of a few score feet of elevation. The acceleration of falling bodies varies from about 978 cm./sec.^{2} at the equator to 983 cm./sec.^{2} near the poles (since the earth rotates). It also varies slightly at different altitudes above sea-level (since these are at different distances from the center of the earth). These differences are detectable only by apparatus of the utmost refinement. You will learn more about this in chapter 5, p. 26.

FALLING BODIES

light one should be observed to fall together. The story is that he demonstrated this to be the case by pushing a small ball and a large ball off the top of the overhanging side of the Leaning Tower. The university was invited to witness that both objects hit the ground at the same time. We may scoff at the fact that many of the audience refused to believe the testimony of their own eyes, since it contradicted the authority of Aristotle. But do we not, at the present day, behave similarly with respect to new ideas in fields where we still cling to past tradition or bow to vested interest or to authority?

One additional point requires emphasis, and if you haven't given it all up by now, take courage; it's easier further on!

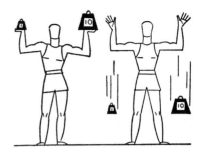

The weight-acceleration-mass argument

We have already noted the fact that larger, more massive objects have the larger inertia, resisting more effectively attempts to change their motion. It is equally obvious as we lift various objects from the ground that the pull of the earth for the larger and more massive is greater also. When various objects are allowed to fall, however, all of them fall at the same rate; the more massive ones have more inertia to be overcome, but the earth pulls upon them correspondingly harder, and exactly so, since all accelerate equally in falling. Thus we see that the ratio of the *inertias* of different masses is experimentally demonstrated to be the same as the ratio between the forces with which they are attracted by the earth, i.e., their *weights*. It is for this reason that we usually compare masses (inertias) by weighing rather than by using those very queer inertia balances (p. 12) that are never seen in butcher shops. Furthermore, balances designed for comparisons of weight can be made almost unbelievably sensitive and delicate. Even when so made, they are far more convenient to use than those that compare inertias directly.

The numerical relations between time elapsed, distance covered in consecutive intervals, total distance, velocity acquired at the end of each interval, as well as velocity gained during each interval, are summarized

FROM GALILEO TO THE NUCLEAR AGE

for the first 5 seconds of fall (and in the absence of any frictional resistance) in table below.

TIME	DISTANCE						VELOCITY			
	In Consecutive Units of Time		Ratios	Total		Ratios	At End of Each Second		Gained during Each Second	
In Seconds	Ft.	Cm.		Ft.	Cm.		Ft./Sec.	Cm./Sec.	Ft./Sec.	Cm./Sec.
0				0	0		0	0		
1	16	490	1	16	490	1	32	980	32	980
2	48	1,470	3	64	1,960	4	64	1,960	32	980
3	80	2,450	5	144	Meters 44.1	9	96	Meters/Sec. 29.40	32	980
4	112	Meters 34.30	7	256	78.4	16	128	39.20	32	980
5	144	44.10	9	400	122.5	25	160	49.00	32	980
6	*									
7	*									
8										

*A conventional book on this subject would here suggest that the student fill out the remaining blanks with the appropriate numbers. We do not request this.

Probably you skipped this table. One usually does skip anything that even remotely resembles a table of numbers. It must be remembered, however, that natural phenomena are not entirely understood, nor is their significance fully comprehended, nor the business of their recording exact, unless they are described, not in words whose meaning is often ambiguous, but by means of tables of numbers. Does not the baker, the butcher, the candlestick maker, as well as your banker, submit to you at the end of each month a table of numbers that represents your current financial status? Do not you carefully compare debits and credits to discover whether your color is to be red or black? There is, nevertheless, another way. Instead of numbers one might use symbols for numbers—algebra instead of arithmetic. Table I

FALLING BODIES

would then take on this appearance:

$$S_t - S_{t-1} = \tfrac{1}{2} g\{t^2 - (t-1)^2\} \; ; \quad S_t = \tfrac{1}{2} gt^2 \; ; \quad V_t = gt \; ; \quad V_t - V_{t-1} = g \; .$$

where S_t is the distance fallen in t sec.
S_{t-1} is the distance fallen in $(t-1)$ sec.
t is the number of seconds
g is the acceleration of gravity
V_t is the velocity at the end of t sec.
V_{t-1} is the velocity at the end of $(t-1)$ sec.

and the body is supposed to start from rest, when $t=0$.

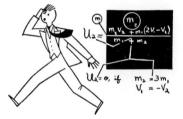

Perhaps this appeals to you more than the table. It certainly has two great advantages: first, it is much briefer—indeed, it represents the preceding table indefinitely extended downward and not stopping with the fifth time unit; furthermore, it does not require separate columns for British and metric systems of units. However, we suspect the nature of your reaction and will continue to use words as far as possible, tables of figures as little as possible, and equations almost not at all.

CHAPTER 4

TERMINAL VELOCITY

SOME pages earlier we mentioned the falling of a feather. Its motion does not seem to be as complicated as that of falling bodies that are heavier. Its motion is simpler and yet the same. If a heavy object, such as a man, fell out of an airplane a couple of miles high, his velocity as he neared the earth would be quite as *constant* as is that of the feather gently settling downward. It would be quite as constant, but it would not be as leisurely; and the bump would do him infinitely more damage. His velocity, although constant, would be rather large—perhaps over a hundred miles an hour, instead of only a few inches a second. Of course, if the man had a parachute to open and "catch the air," he would fall at a constant and much slower speed—a few feet per second—but not as slowly as a feather.

IT IS all a matter of air resistance. You may think we have been rather fussy in some of our preceding statements about what might have appeared to be rather simple facts. Unfortunately, to us as physicists, they are not quite as simple as they appear. We have a dread, more than anyone else who is not a physicist can imagine, of being caught up on slightly incorrect, inexact, and therefore somewhat untrue statements. A feather and a brick

TERMINAL VELOCITY

obey throughout their downward path exactly the same laws. The fluffy and finely divided material of the feather offers vastly more surface for a given weight than does the densely packed material of the brick. For an almost infinitesimal moment after being released the feather falls with an accelerated motion. Before it is dropped, there can be no frictional resistance of the air to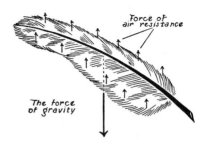

The forces on a falling feather

its falling. This, at least, seems obvious. Now as it falls, friction begins to work; and the faster the feather falls the greater become these frictional effects. The structure of the feather helps to make the utmost of the friction it encounters; so that, before one notices it, friction has become so effective that all acceleration has disappeared. Friction, of course, cannot bring it to rest, since the amount of the friction depends on its speed. Should it stop, all friction would disappear. In this way a balance is obtained between the pull of the earth urging it to speed up, and a promptly increasing friction to slow it down as it yields to the gravity urge.

The case of the brick is just the same. It starts to fall with a swiftly increasing velocity and with a constant acceleration. Air friction starts to work to slow it down the moment it begins to move, but the air has no such hold on the brick as it has on the feather's myriad little sails. For such hold as it has, the air's grip gets tighter the faster the brick goes. Indeed, as often as the velocity doubles, the grip of the air is quadrupled. Ultimately the air catches up; the result is a draw; and then, as in the case of the feather, the brick falls with a constant speed—the force of the earth's attraction just balanced by the force of the air's resistance. When two forces are equal and opposite, the result is just as if there were no force at all—there is no acceleration and the velocity remains unaltered. The brick acquires a swift velocity before this happens. It then continues with that velocity unaltered. It thus acquires what is called its TERMINAL *or* LIMITING *velocity*.

A NEWSPAPER story recently told of the adventure of a parachute-jumper who "bailed out" about 5 mi. aloft and deliberately waited to pull the release ring of his 'chute until only a few hundred feet above the ground. How he or anyone else measured the velocity with which he fell was not explained, but the statement was made that he at first acquired a veloc-

Air density

ity well up in the vicinity of 150 mi. per hour, then ceased accelerating, and subsequently *slowed down* to about 100 mi. an hour before the earth became perilously close. His courage and coolness were no more perfect than was the soundness of the physics underlying the facts reported. The air near the surface of the earth is much denser than it is 5 mi. up—or even 3 mi. up, for that matter. It serves as a more effective brake the denser it becomes. A given object that would acquire a terminal velocity of 150 mi. an hour in thin air could easily have that terminal velocity reduced to two-thirds or even to one-half that value as the air became more dense below.

We might imagine some other planet with an atmosphere so composed that near the surface of the "ground" the air was some score times as dense as in our world. Plane-jumpers or even window-jumpers there might find their terminal velocities reduced to so small a value that from whatever height they leaped they would land as on a bed of down—a down exquisitely soft and gentle for the breaking of their fall. They would acquire a very low terminal velocity and land as lightly as do feathers in this world.

Look at the other extreme. Take our feather and put it in an atmosphere like that of Mars, which is far thinner than ours. If you have no way of making this trip or cannot spare the time, any long closed tube will illustrate what you would learn. You can pump out nearly all the air and have an artificial atmosphere, thin like that of Mars or even negligible like that of the moon. You can then place the feather and a small rock in the tube from which the air in very large part has thus been removed. The feather will fall "like a rock" down the inverted tube, so fast you can see it only a little better than the stone. The stone over this small distance falls very much in the same manner as it did before. Had you a long length of tube at your disposal, you would find that both rock and feather still possessed terminal velocities, both far greater and both requiring considerably longer time for their attainment than before.

TERMINAL VELOCITY

THE case of falling rain is interesting. It is intermediate in behavior between the feather and the brick. It has other aspects in addition because it is a liquid rather than a solid object. Very minute drops of water fall with a very small terminal velocity. Clouds and fog consist of such very small particles that they give no appearance

Rain and mist

of falling at all. This is in part due to the fact that currents of air stir them around with velocities much greater than that with which they slowly settle earthward. In the case of clouds, moreover, these particles form in regions where the air is cooler; and, if they do settle downward into warmer strata, the drops evaporate again, their liquid contents returning to the gaseous condition.

Larger drops of water fall with greater, but nevertheless with constant, speeds. A drop whose diameter is 1/50 mm. has a terminal velocity of only 12 mm. per second. If we consider larger drops, their speed of fall increases very rapidly as the drop size increases. Drops ten times as large, i.e., 1/5 mm. diameter, fall one hundred times as fast, 1,200 mm. (or 1.2 m.) per sec.*
Very much larger drops fall so much more rapidly that they lose their spherical form because of the rushing of the air past them. They flatten out like a saucer and are then broken up into smaller drops. Under normal air density at sea-level drops cannot exceed 4.5 mm. in size without breaking up into

smaller ones as they reach the terminal velocity of 8 m. per sec. that corresponds to this diameter of drop. At a height of 3 km. (1.86 mi.) above the earth, where atmospheric density is about 0.7 of its surface value, water drops of the maximum size have, of course, a somewhat higher terminal velocity, 9.4 m. per sec. The very large drops that sometimes fall during thunder showers do not come from any very considerable heights. Waterfalls plunging over high cliffs, like the Bridal Veil in Yosemite Valley, are completely broken up into a mist of small drops before they strike the ground.

Falling water

---

* The relation between terminal velocity V, of a drop of a liquid, and its radius (R) was given by Sir George Stokes (1819–1903) as $V = \frac{2}{9} \frac{g(\sigma - \rho)}{\eta} R^2$, where g is the acceleration of gravity, σ and ρ the densities of the drop and of the air, respectively, η is a coefficient of air friction called the "viscosity," and equal to 0.00018.

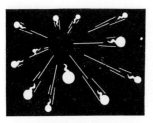

Meteor shower

Another case of falling bodies is not without interest. Who has not seen the spectacle of a "falling star"—on rare occasions an awesome sight, and always productive of queries by the uninitiated. Are they really stars that fall? How fast do they come? From where do they come? What are they anyway? Not all of these questions can be answered with complete assurance, but some of them may.

They are not stars. Heaven help our tiny earth if a star should fall upon it! Wandering bits of matter they are, minute fragments of disintegrated comets sometimes—no one knows whence they ultimately come. Few of countless thousands of these meteorites that strike our atmosphere ever penetrate it far. The speed with which they come is terrific; and the friction of the air, which gets denser and denser, as through this protecting blanket they approach the earth's surface, swiftly heats them to incandescence. Most of them burn up in the outer regions of the atmosphere.

Now and then a chunk more massive than the average succeeds in plowing its way through. Most of these fall hissing into the sea that covers the greater part of our globe. Out in the deserts of northern Arizona is the

Courtesy of the Studio Grand, Winslow, Arizona

PLATE 4 An Enormous Crater Made by Meteors near Winslow, Arizona.

TERMINAL VELOCITY

record of a monster meteorite or a colossal group of smaller ones that hit the earth in some past far distant age. The force of its impact moved millions of tons of earth, pushing up circular crater-like walls several hundreds of feet high and nearly a mile in diameter (Plate 4).

Some years ago (June 30, 1908) a much smaller one fell in the remote wilds of Siberia. Stories slowly filtered out of forests felled for many miles around by the terrific shock to earth and air, of a roar and a concussion heard 600 mi. away, of scorching heat 50 mi. distant, of 1,500 reindeer killed in the neighboring area. Fortunately, such events are very rare. The velocities with which these objects fall can be calculated from certain very probable assumptions. Details are rather beyond our scope; but one circumstance, hitherto unmentioned in connection with the earth's attraction, will be emphasized in the next chapter.

CHAPTER 5

GRAVITATIONAL ATTRACTION

THE acceleration of a body near the surface of the earth is, as we have seen (p. 18), 980 cm. per sec. every second. This is not true when one goes to great distances from the surface. The acceleration then begins to show its dependence upon distance from the center of the earth. At the earth's surface we are approximately 4,000 mi. from the center, and the force of gravity has its well-known value. If we could ascend 4,000 mi. above the surface, thus being 8,000 mi. from the center, we would find that bodies weighed only one-fourth as much as before and (possessing still the same inertia) would now fall with only one-fourth the acceleration to which we surface-dwellers have always been accustomed.

The fact that the earth's gravitational attraction diminishes as the square of the distance from the center increases was a hypothesis first put forth by Newton. It was based on astronomical observations of the planets made by his predecessor, Kepler, and it can be confirmed by direct terrestrial experiment. (Cf. Plate 5.)

It is this law of diminishing attraction that renders the calculation of the velocity of a falling meteorite somewhat intricate. Of course, we do not

GRAVITATIONAL ATTRACTION

know whence the meteorite came. If we assume, however, that it originated in very distant regions of space we will not be far wrong. Falling, then, from regions inconceivably remote,* it gathers speed at first very slowly. After a long time it gets appreciably nearer, and the acceleration increases. Out in the depths of space there is no atmosphere to produce any resistance. Every bit of velocity gradually acquired is retained, and more is added as time goes on. The nearer the body gets to the earth, the more quickly it speeds up. By the time our atmosphere is reached, it has acquired almost the full maximum of velocity any object that starts from rest and falls toward the earth can have. This amounts to nearly 7 mi. per sec.! Striking our atmosphere at this velocity, the frictional forces are terrific. The meteor is almost instantly heated to incandescence—sometimes it bursts into smaller pieces, and, except on very rare occasions, is totally consumed.

Conversely, you may ask, suppose we could shoot a projectile away from our planet with a sufficient speed so that it could escape through the retarding atmosphere and depart thereafter with a velocity just a little more than 7 mi. per sec. Would it never, then, fall back to earth? It would be receding so swiftly that the earth's attraction, ever weakening as the distance from the earth increased, could never overcome the velocity remaining. Even when it got infinitely far away, a slight excess of velocity would be left. It must be mentioned that to get this result the great force of the attraction of the sun has been neglected, and only that of the earth included. Actually to be banished from the solar system, a projectile would have to be given a considerably greater velocity than the foregoing.

Effect of distance on the force of gravity

W ITH respect to the Law of Gravitation, Newton's assumption was that the force of attraction between two bodies depended on the mass of each, as well as being inversely proportional to the square of the distance

---

* We assume that this velocity is due solely to the earth's attraction, and that it started out with no initial velocity at all.

PLATE 5

THE CAVENDISH EXPERIMENT

FOR WEIGHING THE EARTH

APPARATUS FOR MEASURING THE FORCE OF GRAVITATIONAL ATTRACTION BETWEEN OBJECTS OF KNOWN MASS, AND HENCE FOR DETERMINING THE MASS OF THE EARTH.

A long fiber of quartz, so fine as to be almost invisible, supports a needle at whose ends are small silver balls about $\frac{3}{4}$ gm. in mass. The long vertical tube protects the fiber from air currents, and the needle is likewise served by a horizontal rectangular box, seen between the two large lead balls in the illustration.

These lead balls, 2,800 gm. in mass, sliding on horizontal rods on opposite sides of the needle, may be placed at opposite ends quite near the silver balls.

Thus, when in the position photographed above, the gravitational force between silver and lead swings the needle in a clockwise direction, seen from above, and, when the lead balls are reversed, in a counterclockwise direction.

Light from a lamp, focused on a tiny mirror on the suspended system and reflected to a distant scale, enables the swing to be measured. The amount of swing, together with knowledge of the stiffness of the fiber, and distances between lead and silver balls, and their masses, makes possible the calculation of G, the gravitational constant, referred to in the text. As explained therein, if G be known, the mass of the earth may be found from knowledge of its size and the acceleration of freely falling bodies.

GRAVITATIONAL ATTRACTION

between their centers. That is to say, the force with which either attracted the other (cf. Third Law of Newton, pp. 45 ff.) was given by

$$f = G\frac{m_1 \times m_2}{r^2}.*$$

To help us gather still further the meaning of this formula, let us go and dwell on the moon *in our imagination*. We use our imagination only because we know no other way of getting there. The Interplanetary Society† may some day build a rocket to make the journey. We may watch its flight through powerful telescopes, but it probably will not carry any passengers. Shot off at just the right instant in just the right direction, the rocket would travel so as to intercept the moon. In this case, the initial speed would not have to be quite as high as that needed to shoot the projectile to the outermost ends of space. The difference amounts only to about 300 ft. per sec. (More precisely, 6.90 mi. per sec. to reach the moon as compared with 6.95 mi. per sec. to travel to infinity.)‡ On arriving at the moon, there would be, of course, quite a jar. To fall from infinity to the moon's surface, a speed of only $1\frac{1}{2}$ mi. per sec. is acquired, since  the moon is much less massive than the earth. Should we travel as passengers in this rocket, we should probably not survive the landing. No atmosphere would be here at our new abode to cushion our fall; no medium en route to entangle and consume the hundreds of thousands of meteorites shooting through this region with velocities much greater than any artillery can

---

* The factor G is a constant depending on the units used, and these were established when we defined "unit force" (1 dyne) as that which gives 1 gm. of mass an acceleration of 1 cm. per sec.^{2}

The earth imparts to m gm. (or to any other number of grams, of course) an acceleration of 980 cm. per sec.^{2} (g). Thus, the force of the earth's attraction, which we call the "weight," W, equals $m\,g$.

From Newton's relation the attraction of the earth on m gm. at its surface is, if we call M the earth's mass and R its radius, $G\dfrac{m\,M}{R^2}$.

Thus we have the equations $$W = m\,g = G\frac{m\,M}{R^2}.$$

The value of the factor G is obtained from the Cavendish Experiment (Plate 5), g from falling bodies or pendulum experiments, R from astronomical observations, so that the only unknown is M, the earth's mass, which can be solved thus:

$$M = \frac{R^2 g}{G}.$$

† This organization actually exists.

‡ Cf. MacMillan, *Theoretical Mechanics* (McGraw-Hill, 1927), p. 221.

If we were constrained only by the moon's surface gravity

produce. To encounter one of these, no larger than our fist, would be a major catastrophe to a passenger-carrying rocket. The journey between the earth and its satellite would take over 4 days —quite time enough for us to have been punctured at least once. In the face of such vicissitudes, we agree that it is much more safe to go to the moon by imagination than by rocket; much safer also to wander about its surface, likewise, in our thoughts, physically breathing terrestrial air rather than gasping vainly in a lunar landscape where no air exists.

Since the moon is much less massive than our globe (mass moon = 1/81.56 that of earth), we would not weigh so much there. Of course it is much smaller than the earth (radius moon = 0.273 that of the earth), so we would be much nearer its center from which gravitational attraction must be reckoned. Taking both factors into account, we find that the moon's surface "gravity" is only about one-sixth $\left(= \frac{1}{81.56} \times \frac{1}{(0.273)^2}\right)$ that which we enjoy here. A big man weighing 180 lb. here would weigh only 30 lb. there: he could be carried by a child. A ball-player who could throw a ball 400 ft. on earth could hurl it on the moon nearly half a mile, since there it would fall so much more slowly. What jumpers we would be! The effort required to clear a 6-foot bar on earth would on the moon be more than enough to send us soaring up over the roof of a two-story house. We would land after such a jump just as lightly as we do after the 6-ft. leap at home.

WE HERE have arrived at the cause of the moon's being without an atmosphere. Later (chap. 18) we shall discuss the fact and give the reasons why we know that the atmosphere, like all other gases, consists of numerous small particles, called molecules, all of which are in continuous, very rapid, random motion. At ordinary temperatures very few of these molecules in our terrestrial atmosphere possess the sufficiently high speed of 7 mi. per sec. necessary to escape from the attraction of the earth. On the moon, however, a speed of only 1.5 mi. per sec. is necessary. Many molecules possess this much; and thus, in the ages that have passed, all trace of any

GRAVITATIONAL ATTRACTION

atmosphere on the moon has vanished. In a subsequent chapter (p. 123) we shall mention another aspect of this fact that would seem very strange to earthdwellers could they explore our nearest neighbor in space.

Some of the other planets in our sun's family are not very dis-

	Diameter	Mass	Surface Gravity
Earth	1.00	1.00	1.00
Moon	0.273	0.012	0.165
Mercury	0.39	0.04	0.27
Venus	0.97	0.81	0.85
Mars	0.53	0.108	0.38
Jupiter	10.95	317.0	2.64
Saturn	9.02	94.0	1.17
Uranus	4.00	14.7	0.92
Neptune	3.92	17.6	1.12
Sun	109.1	331,950.0	27.9

Sizes, Masses, and Surface Gravities in the Solar Family (as Compared to the Earth)

similar to the earth with respect to surface gravity and the possession of an atmosphere. The surface gravities of Mars and Venus are only slightly smaller, and both have atmospheres not dissimilar to ours. Surface gravity on Mars is about one-third that of the earth, and its atmosphere is rather thin. We should doubtless feel quite giddy there for some time until we got acclimated. Indeed, it is very doubtful if we could survive without artificial aids to respiration. On Venus, were it not for its very torrid humid climate, we should probably feel quite at home. Of course, it probably rains there all the time, for never once have its skies cleared to give us, looking at it from afar through telescopes, any glimpse of its surface characteristics. Jupiter is a giant world. A frail youth of 100 lb. weight here would there tip the scales at 264. He could barely drag himself around.

The sun is almost inconceivably larger and more massive even than Jupiter. If our earth were at its center with our tiny moon revolving about it only a quarter of million miles away, the moon would still be only slightly more than half the distance out to the surface of the sun. There is no question of anything solid or liquid or living surviving on that incandescent surface whose temperature is above 10,000° F. Even were it cool enough for habitation, structures like our bodies with frail bony frameworks could not even crawl upon it. A 145-lb. man would weigh there about 2 *tons*. He would collapse, crushed by his own frightful weight.

As stars go, our sun is of rather less than average size and of about average mass. Many there are that exceed it in size as much as it exceeds our moon; many also, ten to fifty times as massive. Perhaps, rather than face some of the almost unimaginable facts about them at this moment, we had best return to earth and to some simpler things in human history.

CHAPTER 6

PATHS OF PROJECTILES

THE early philosophers, because of their lack of understanding of the principle of inertia, quite naturally had entirely erroneous conceptions of the behavior of objects on which several forces were acting in succession or simultaneously. If a stone were dropped from the top of the tall mast of a sailing vessel, they argued that, since the vessel advanced along its course after the stone was released, it would strike the deck some distance aft of the base of the mast. Indeed, if the boat had a sufficiently high velocity, it might not fall aboard at all, but into the water behind the ship. Of course, no one ever observed that this was true, but of observations these early naturalists made little.

Galileo, on the other hand, with the clear conception of inertia derived from his experimental work, saw such matters in their true light. The stone, held by the man on the masthead, partakes of the forward velocity of the ship. When released, its inertia causes this velocity to be maintained unaltered. Except for possible air resistance—here quite negligible—there is present nothing either to accelerate or to retard this horizontal velocity which it enjoyed as ship's passenger. When it is released, a powerful vertical

* The details of motion of telescope falling from the masthead of moving ship are given in footnote on page 41, and in the drawing at the top of this same page.

PATHS OF PROJECTILES

force is given free play upon it. It accelerates; but the velocity thus acquired in no way interferes with whatever other velocity it previously possessed. At one and the same time, then, it continues not only to travel forward with the ship but also to fall downward. It strikes a spot on deck directly below the point from which it was dropped.

On the other hand, if the ship suddenly altered its course, stopped, or reversed its motion after the stone left the sailor's hand, it would then strike quite a different spot on the deck, or indeed might fall overboard. It could "know" nothing of any changes in the ship's motion after being cut loose from it. It would retain only that motion of the ship in which it was participating immediately before its release.

WE SEE, then, that the motion of the stone while falling from the masthead of a moving ship consists of two component parts, or COMPONENTS. One of these is the uniform horizontal velocity imparted to it by the ship; the other is the swiftly accelerating motion of a body falling vertically. As we shall see later (chap. 7), the actual velocity of the body as it falls—seen, for example, by an observer on shore—is the geometrical sum of these two vector quantities. Do not try at the moment to puzzle out our meaning in this last sentence if it is not altogether clear. It will be clear to you later if you will come back and read it again after finishing page 41, of chapter 7. We will put a footnote there to remind you.

THERE are very interesting corollaries to this principle, some of them almost unbelievable to the uninitiated. Imagine a rifle bullet, fired horizontally from the edge of the parapet on the top of a high building. Suppose that at the instant the bullet leaves the gun, another bullet is dropped over the edge of the parapet. It is hard to believe that both bullets will strike the ground at the same instant, and yet this is a fact. The situation is quite parallel to the one just discussed. Since the attraction of gravity acts at right angles to direction initially imparted to the bullet by the gun, it can neither accelerate nor retard that initial horizontal component

of velocity. It can only impart another component, that of a falling body, to the velocity of the speeding projectile. This vertically accelerated component of the motion carries the bullet earthward at precisely the same rate as it carries the bullet that is dropped over the parapet, so that both objects fall together. The fact that the horizontal component of velocity is unchanging gives the projectile a range equal to the product of this horizontal speed times the time taken to fall from the muzzle of the gun to the ground.*

THIS principle of independence of motions is even more vividly seen if the projected object be fired upward at an angle. In the apparatus illustrated in Plate 6A, two balls are used instead of bullets. One of these, A, is projected forward, and the other, B, is dropped at the same moment. The ball C on the right is rolled down the incline. It strikes A and projects it forward in a direction that is not necessarily horizontal. This direction, at the instant of projection, is exactly in line with the ball B which up to that moment had been hanging from the small electromagnet. As ball A leaves the apparatus, it breaks the circuit through the magnet and ball B is instantly released. If gravity were not operative, the projected ball A, because of its inertia, would travel straight toward the suspended ball B, which of course would not fall in the absence of any gravitational pull. With the attraction of the earth in operation, however (since both balls start to fall at the same instant and are originally aimed so as to collide), ultimately they must still collide, and indeed they always do collide in mid-air just as soon as the horizontal part (component) of the motion of the projected ball brings it to the vertical line along which the suspended ball is falling to the ground. Plate 6B is reproduced from a photograph taken of a series of electric sparks while the two balls were falling. The high-light reflection of each spark from the polished surface of the balls has left on the plate a record of their paths. Where the balls collide is clearly shown, as the path of each suddenly changes its direction.

ANOTHER beautiful experiment that shows the paths of falling projectiles can be performed with a jet of water, directed along the edge of a stick (Plate 7). Hanging from this stick are several little pendulums, equally spaced and of lengths equal to the relative total distances covered by a

---

* This time is given by $t=\sqrt{\dfrac{2h}{g}}$, h being the height of the muzzle above the ground and g the acceleration of gravity. See equation $S_t=\tfrac{1}{2}gt^2$ on page 19.

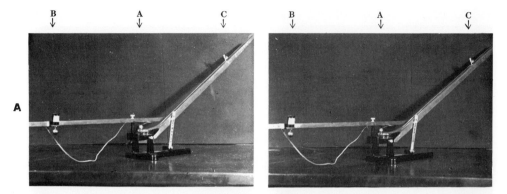

Stereophotograph of device that throws one ball forward and drops another one at the same instant

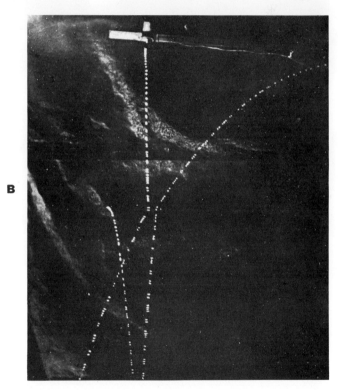

PLATE 6

THE INDEPENDENCE

OF VELOCITIES

Electric-spark photograph of the trail of the two balls, one dropped and the other thrown forward at the same instant by the apparatus illustrated above. They collide, in mid-air, since they fall at the same rate. Furthermore, it is seen they exchange velocities approximately on collision (see p. 52). The loss of energy on rebound of one from the stone slab beneath is also shown (cf. footnote on p. 51).

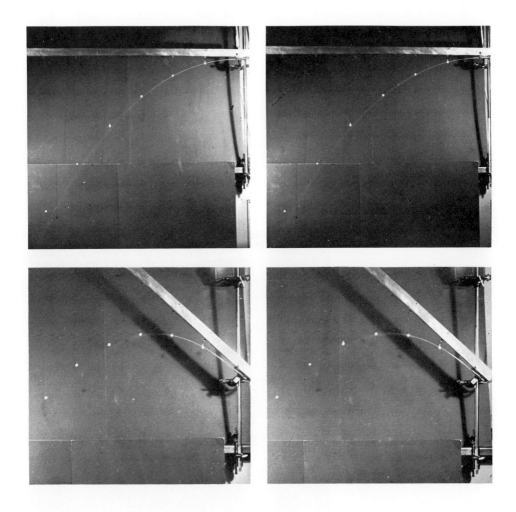

PLATE 7

A JET OF FREELY FALLING WATER

Projected parallel to the edge of the stick the jet at all angles lines itself up with the row of pendulum balls. These have been hung at equal distances along the stick and with lengths proportional to 1, 4, 9, 16, 25, 36, respectively, from right to left. Air friction interferes somewhat, as it breaks the jet into drops. (Stereophotographs.)

PATHS OF PROJECTILES

falling body in consecutive units of time (cf. col. 7 of table, p. 18), i.e., 1, 4, 9, 16, 25. With proper adjustment of the velocity, the water jet will be found to pass by each one of the suspended pendulum balls; and this will be true for whatever angle the stick makes with the horizontal.

Here, however, it will be noticed that the water jet falls a little short of the balls, and the more so the further it is from the nozzle. We have commented on the effects of air resistance in slowing down the motion of light objects like drops of water. Here we see it made visible in its affecting the curve of the water jet and causing it to depart from the ideal curve that would be taken by a jet *in vacuo*.

THE curved paths taken by bullets, called trajectories, show to a much greater degree the effect of air resistance. This is due to the very great velocity of bullets, since, as we have said before, the frictional forces of the air increase very rapidly with the speed of an object through it. In the absence of any friction the greatest range of a projectile is obtained when the

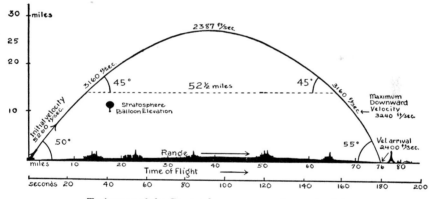

Trajectory of the German long-range gun (approximate)

angle of fire makes an angle of 45° with the horizontal; but because of the falling short of this ideal path when air resistance is encountered, the elevation of guns for their maximum range must be considerably greater than 45°. How much greater depends on the size and shape of the projectiles and upon atmospheric conditions. This very intricate problem is a major field of study by artillery officers responsible for such matters.

During the World War the German army developed a gun of the enormous range of 75 mi., and dropped shells upon Paris by means of it. A sketch of the details of this trajectory, approximately to scale, is reproduced herewith. The nearly mile-a-second initial velocity of this $8\frac{1}{4}$-in. shell was the greatest ever attained in ordnance at that time. This speed enabled the shell to be quickly sent to such a very great height that air resistance became largely negligible. The far greater portion of its path lay in a region where air density was less than one-tenth that at the ground-level. How closely this portion conforms to a trajectory *in vacuo* is clearly seen at all elevations above 13 mi. Above this height the path conformed to the ideal condition for maximum range. It made an initial angle here of 45° with the horizontal. Thus a horizontal distance of $52\frac{1}{2}$ mi. was covered before the shell descended again to levels where air resistance began to slow it up and shorten the range seriously. Its maximum speed along the downward path, amounting to 3,240 ft./sec., was attained shortly after the completion of this low-resistance portion of the path. Air resistance, then increasing rapidly, actually slowed it as it fell to a final velocity of only 2,400 ft./sec.* The maximum height attained was nearly two and a half times as great as that reached in a balloon by Commander Settle on November 20, 1933.

* Note here the confirmation of results of air friction reported in case of plane-jumper on page 22.

CHAPTER 7

BALANCED FORCES AND NEWTON'S THIRD LAW OF MOTION

LET us now once again come back to earth and to our experimental laboratories to study a little more in detail those important cases where, in spite of the applications of several simultaneous forces to an object, no change in motion whatsoever is produced.

The only example of such a result that we have yet mentioned is the case of an object that has acquired its terminal velocity* under the combined action of gravity, tending to accelerate it, and of air resistance tending to retard it. In the cases of falling bodies, feathers, snowflakes, or rocks, just as soon as the action of these two forces becomes equal, the body continues in the state of uniform motion with a constant velocity in a straight line which is also characteristic of motion due to inertia alone, and where no disturbing forces whatsoever are present (Newton's First Law, p. 11).

It is quite obvious that in a tug-of-war, if the groups at each end of the rope were exactly matched in strength, neither side would be able to

* Cf. pp. 21–23.

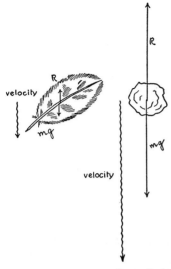

Balanced forces on, and terminal velocities of, a feather and a rock

move the other, and yet no one would attempt to deny that force—and a large amount of it—was being exerted. The case is that of equal forces in exactly opposite directions balancing one another. It is precisely the same as the case of the falling body that has reached its terminal velocity.

When two different forces are applied to one and the same object in different directions, the object will always be given a change of velocity; but now it will move in a direction quite different from that of either of the forces and with an acceleration also different from that which would be produced by either force alone. If a mule pulls on a canal boat, and a man with a pole pushes it away from the bank, the boat moves neither in the direction of the pull nor of the push. Now it so happens that there is a very simple method of finding out the resulting motion

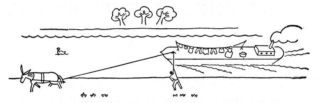

One reason why a canal boat needs a rudder

without having to resort to experiment in every case that arises. It must be understood, of course, that many experiments had to be done before the method of calculation could be established as generally valid.

A FORCE, push or pull, is always uniquely described if three things are specified: a definite direction, a definite magnitude or amount, and a definite point of application; i.e., its effect results from its being applied to some definite part of a material object. Now these three simple elements can be pictorially represented by directed sections of straight lines called "vectors." Their lengths can correspond (on any arbitrary scale) to the magnitudes of the forces they represent; their directions can be the same as (or

The canal-boat vector diagram

BALANCED FORCES

parallel to) the directions of their forces. Furthermore, the object involved can be drawn, and these line segment pictures of forces can be attached to the proper parts of the object in the picture.

This does not, perhaps, seem to contain anything remarkable, as we are quite commonly in the habit of looking at drawings of objects even when highly simplified as cartoons. The remarkable thing, however, is that if we take these arrows representing the system of forces in any actual physical case (e.g., mule's pull, man's push, on canal boat above) and add them all together, *head to tail*, and in any order we please, maintaining, of course, the exact relations in direction and length of each, we can at once find a directed line segment in the picture that on the same scale will represent uniquely that one single force which if applied to the body will cause the same identical motion, i.e., the motion of the boat that results from the several actual forces simultaneously applied. This vector is called the "resultant" of the vectors that represent the several separate forces actually involved. It is found by drawing a line in the picture from the tail of the first vector to the head of the last when they are added together as described above.*

The idea will be clearer if we now apply it to the case already considered, that of a falling body that has reached its terminal velocity. Suppose it to be a round, smooth ball. We can then attach our force vectors to the center. There will be but two, the resisting force and that of gravity. Now, if to a separate picture we transpose these two and add them, head to tail, since they are equal in length and opposite in direction, the tail of the second now occupies the position of the head of the first, and the "resultant" is nothing at all, zero.

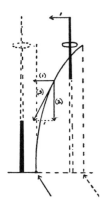

Vectors representing velocity of a body falling from the mast of a moving ship

$$\left(R\bigg|+\bigg|mg\right) = 0$$

Vector diagram of forces where a terminal velocity has been reached

Suppose, now, another case, where a force of 3 lb. acts on some point in a body and at the same time a force of 4 lb. acts at right angles to the first. We show first the

---

* Conversely, the separate vectors which add together to make this resultant are called "components" of it. On page 32, where we spoke of an object dropped from the masthead of a moving ship, this object had two components of velocity: one horizontal and constant (1), the other vertical and accelerated (2). Cf. drawing above to the left. The velocity (as seen from shore) was the resultant or sum of these two (3). The path of fall, seen from the ship, is vertical; seen from shore, is curved.

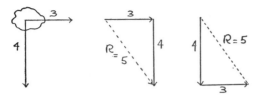

Case of an unfortunate rock that is pulled in two different directions at the same time

picture, then transpose the arrows to another drawing, adding them head to tail, in either order. The result here is not zero. Drawing a line from the tail of the first arrow put down to the head of the last one gives us the single force that will produce the same effect on the body as the simultaneous action of the other two. If you remember anything at all from your early struggles with geometry, you will recall that the "square on the hypotenuse of a right triangle equals the sum of the squares on the other two sides." The resultant, then, represented by letters, could be written $r^2 = a^2 + b^2$; here $r^2 = 3^2 + 4^2 = 9 + 16 = 25$; whence $r = 5$.

Thus we have found the curious result that a force of 3 plus a force of 4 at right angles to it equals a force of 5. "$3 + 4 = 5$" is very bad arithmetic, but it is very good geometry. If forces 3 and 4 have the same direction, then arithmetic and geometry agree that $3 + 4 = 7$; and if forces 3 and 4 should have opposite directions, geometry and arithmetic agree that they equal 1, and here the direction of the 1 is that of the 4. In other instances, however, aspects of the geometry of triangles always enable us to calculate the direction, as well as the magnitude, of the resultant, in any possible case.

Even a complicated case—e.g., of seven forces—yields readily, at least to a graphical solution, often exact enough for ordinary purposes. Here is one such. Crude measurement with ruler and protractor gives the answer to be 2.3* and the direction one that makes with the first force an angle of 120°.

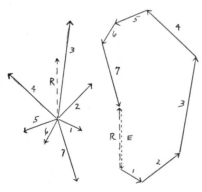

Graphical solution of a complicated vector problem

---

* The number represents the ratio of the length of vector R to vector 1 in the original drawing from which this cut was made. Subsequent change of size of course leaves the ratios of $R:1$, or $R/1$ unaltered.

BALANCED FORCES

ONE final comment with respect to vector pictures of forces should be made. Suppose we draw a vector E, exactly equal to our resultant R, but oppositely directed. It will exactly neutralize the effect of R. The sum of these two is zero. But R, as we have seen, represents the net result of all seven vectors whose sum it is. This new vector E then must exactly neutralize the other seven vectors. To put it in another way, if a force $E\ (=-R)$ be added to the other seven forces as an eighth force, the sum will be zero and no change of motion of the point of application will result. Adding the vector E to the vector diagram, we obtain a closed figure.

Thus we have what might be called by mathematicians a theorem about these matters to the effect that **Whenever the net result of several forces acting on an object is zero, if we add the vectors head to tail in any order, the vector diagram representing this physical situation will be found to consist of a closed polygon.**

WE ARE now in a better position to understand some matters about which very intelligent people often become confused. Questions like the following are often debated: "Is it true that when a horse pulls forward on a wagon the wagon pulls backward to the same extent upon the horse?" In a tug-of-war it would seem, at least from the point of view of the rope, that this was being pulled with equal forces from each end. If there had been inserted, near the middle of the rope, a spring balance, certainly it would  register a very considerable number of pounds tension in the rope as the struggle between the two teams went on. Furthermore, even if one team gradually became weakened and began to yield to the other, one would not expect the balance to show any very great decline in the pull in the rope. It might, indeed, be conceivable that, stimulated by recognition of the beginnings of failure on the part of their opponents, the victorious team might put forth a final spurt and actually increase the reading of the balance as victory appeared to be within their grasp. Furthermore, it is certainly true that in any situation the reading on this balance would not depend on how it was tied into the rope—whether the end with the hook was toward or away from the victorious team.

The tension in the rope and that in the spring of the balance which has been introduced simply to measure it, is continuous throughout its length,

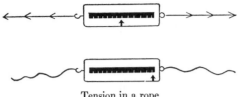

Tension in a rope

and a pull must be exerted simultaneously from both ends for there to be any tension at all. If it is hard for you to see this, just imagine that in the tug-of-war either one of the two sides suddenly lets go. The rope on the instant suddenly goes limp, the spring balance reads nothing at all, and the opposing team collapses in a heap on the ground.

If a pull is transmitted along a rope, then it must be transmitted from both ends. Whether the rope be stretched taut and the two ends held by opposing teams, or between one team and a rigid building, or between two rigid walls, or between a horse at one end and a wagon at the other, there is but one tension, and its effect at each end is a force. The forces at the two ends are equal and opposite; and, as far as the rope is concerned, they always balance, add to zero, and thus can produce no motion.*

When we consider the question of the MOTION of the two athletic teams, or the two rigid walls, or the horse and wagon at opposite ends of this tension-balanced rope, we must consider other factors besides the rope. If the tug-of-war were staged on perfectly smooth ice, the two teams, for all their prowess, might not be able, by pulling at opposite ends of the rope, even to lift its long length off the ground. Indeed, under the conditions now specified they could not pull at all. Granting only perfection in the smoothness of the ice, they could possibly stand erect, but only with difficulty maintain their balance. Having fallen down, they could not even crawl by pushing or pulling in any way, and so move themselves along the ice.†
Indeed, the only way they could ever have come out on the ice would have been to have jumped off upon it from the shore. Had they done this, they could only, by most elaborate arrangements, have stopped themselves from sliding right on across it to the opposite bank. To organize two teams at opposite ends of the rope would have been well-nigh impossible. Anyone who skates knows how important it is for his skate with

Motion picture of action on glare ice

* If horse-wagon system is accelerating and mass of rope is not negligible, a slight difference in tension at two ends exists, owing to rope's acceleration.

† If they had something to throw, shoes or skates, they could, by the reaction to the throwing, acquire momentum, and so move off in the opposite direction.

BALANCED FORCES

each stride to get a good "bite" on the ice. Now perhaps we begin to see why the horse is able to move the wagon even though it pulls back on him just as hard as he pulls forward on it. He has traction on the ground; the wagon has none, or at least very little. By pushing backward on the ground, he is able to create tension in the harness; and the wagon, having little or no attachment to Mother Earth, yields and accelerates forward. Let the wagon sink into mire and get as good a foothold in it as the horse has, and the horse cannot budge it.

The essence of all this is a third Law of Motion stated by Newton, to the effect: **For every action there is an equal and opposite reaction.**

IN THE taut rope and stretched spring let us watch all of the force vectors grow as the two teams, equally matched, begin to put forth their efforts.

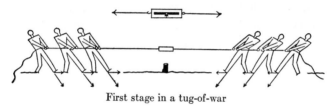

First stage in a tug-of-war

Now one team begins to weaken. It cannot retain its grip on the earth quite as well—one man slips;

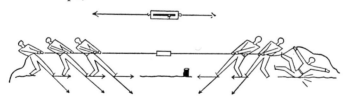

Second stage in a tug-of-war

another's back begins to weaken; pushing less strongly on the earth, he pulls just that much less on the rope.*

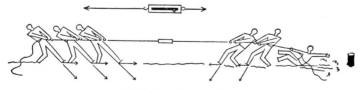

Third stage in a tug-of-war

* The vectors applied to the drawing of the spring balance, transposed there from the men's feet, do not tell the complete story of the pull, tension in the rope. What has been left out?

He and the rest of his companions now begin to move. Adjusting themselves to the motion, they find it becomes increasingly difficult to find new stances; they dig in their heels in desperation, to no avail. The faster they are pulled, moreover, the harder it is to stop. Even Dame Nature is against them now. Their inertia, now they are under way, tends to keep them under way. The round is quickly finished as they are pulled around the lot.

RETURNING to Newton's Third Law for a moment, you may ask if action anywhere is, of necessity, always accompanied by a reaction somewhere? Does not this mean that forces always occur in pairs? The question is shrewd; and its answer, "Yes." We often speak of a single force, but then we are neglecting the reaction on the earth. The Earth's bosom is broad. Our puny efforts disturb her little. This disturbance may often be neglected, but it nevertheless exists even for the slightest effort we put forth.

A train starts to go forward as the engineer opens the throttle on the locomotive. He overdoes it, and the driving wheels lose their grip and spin on the rails. He tries again, and gradually the train gathers momentum.* However, the rails were pushed backward with just the same force by the expanding steam as that which urged the train forward. Spiked to heavy ties, and set in a heavy rock-ballasted roadbed, the rails are securely anchored to the earth. They carry the whole earth with them as they recoil in the opposite direction. How large is this effect you ask—surely not enough to measure? No, not enough to measure, since the mass of the earth is very great.

If a train is mounted, however, on a heavy track which is, nevertheless, not fastened to the earth and is free to move, measurements may be made. They show that, for every quantity of momentum gathered by the train in a forward direction, the track gathers an equal amount backward at the same time.

An experimental device (illustrated in Plate 8) shows how we give the track a chance to move without dragging the whole earth along with it, so that we can see how truly action and reaction accompany each other in this case. The circular track forms the circumference of a large horizontal

---

* By the "momentum" of a body we mean two things taken together, of course—its mass and its velocity. A baby cab and an automobile might both be moving 10 mi. per hr. We would much prefer stopping the momentum of the baby cab if we had to do it with our body. This would be true even if it were going quite a bit faster than the motor car. However, if the baby cab were to be speeded up to 100 mi. per hr., we should probably prefer to take our chances with the 5-mi. per hr. automobile.

BALANCED FORCES

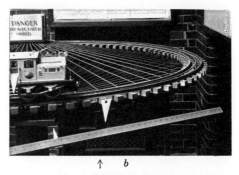

↑   a   ↑   b

PLATE 8   RECOILING RAILS. (*a*) Track and car at rest. (*b*) Car has gained momentum to the left and track has recoiled, gaining an equal amount of momentum to the right.

wheel mounted on a vertical axis with very little friction. Current for the electric locomotive, supplied from the rails, is led in to the latter by light sliding contacts at the center of the wheel; so one is freed from the nuisance of dangling wires. When the switch is closed outside, and the locomotive starts forward, the track recoils backward. Part *a* of the plate was taken before the car was started. Note the position of markers on car and track with reference to the fixed point on the yardstick, in picture *a*. Next note the relative positions of these markers in picture *b*, photographed a few seconds after the car was started.* A comparison of the amount of forward motion of the car and backward recoil of the rails can easily be made.

If the inertia of the car and of the track were exactly the same, it would be found that the forward velocity given the car was exactly equal to the backward velocity given the track. By action and reaction then we must mean forward and backward momentum. Further experiment confirms this. If the track has ten times the inertia of the car, the velocity imparted to the track will be only one-tenth that imparted to the car. When the track is held firmly to the earth, the backward momentum given the latter is still equal to the forward momentum of the car. The inertia of the earth, however, is so vast that the velocity it has imparted to it to make up momentum is to the same degree excessively small, i.e., quite unnoticeable and entirely unmeasurable.

* The marker on the *yardstick* (Plate 8) appears to have been moved slightly to the left in picture *b*. This is due to a slight shift in position of the camera lens in taking the second photograph. In all of the stereoscopic photographs in this book such a lens-shift occurs; the three-dimensional character of the picture depends upon this shift, discussed in an introductory section that deals with this type of photograph, pp. xiii–xv.

FROM GALILEO TO THE NUCLEAR AGE

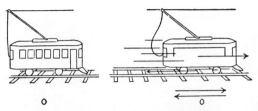

Momentum vectors when the rails recoil

Momentum, like force, is a directed quantity. Equal and opposite momenta can be pictured by equal and opposite arrows. Their sum is zero. Before the car is started, the total momentum is zero. After the car and track are in motion, the sum of the momenta is still zero. This indicates that in this system, or, indeed, in any system of objects, entirely self-contained and subject to no pushes or pulls from without, the momentum never changes. The various parts of such a system can push and pull one another about to any extent, but they can never change their total momentum. To this experimental fact no exception has ever been found, and hence it has been given an important place in the records of science and raised to the dignity of a law, the Law of Conservation of Momentum. It is the embodiment of Newton's three Laws of Motion.

CHAPTER 8

APPLICATIONS OF THE LAW OF CONSERVATION OF MOMENTUM

THE application of the Law of Conservation of Momentum to our solar system—particularly to the three bodies: earth, sun, and moon—and the consideration of the major effects of the tides generated on the earth by our two luminaries bring forth important and astonishing conclusions about our planet's history. Four thousand million years ago, the day may have been only about 5 hr. in length, and the month only a little longer. The moon then was only about nine thousand miles away from the earth, instead of a quarter of a million, as at present. Ever so slowly our days are still becoming longer; and when day and month again become equal, they will be as long as 47 days are now. The moon will then be much farther away—a small dot in the sky—and, unchanging in position with respect to the earth, visible only from one hemisphere. As the moon now shows but one face always to earth-dwellers, so then will the earth, as well, present one unchanging hemisphere to the lunar observers, if there be any there to watch. Many thousand times a thousand million years will elapse, however, ere this takes place.

Far beyond that time even, the law implies that, through the agency of the weaker tides raised by the sun upon our little planet, there will come an age inconceivably more remote when the day will be longer than the month, when the moon will gradually be brought back closer and closer to the earth until possibly it may actually be shattered by tidal forces, and our planet will acquire a ring of fragments such as we see surrounding Saturn at the present time.*

APPLICATION of the Law of Conservation of Momentum may be even more ambitious. It may even presume to inquire into the origin of our entire solar system. Our solar system, with respect to its momentum, is dynamically a very curious affair. Most of its mass resides in the sun, but most of its rotational momentum resides elsewhere—in the major planets. Such a system cannot be imagined to have come into such a situation from dynamical changes within itself, but only by means of some exterior agency. Had our sun, long eons ago, been a simple single body, planetless and solitary, and had it encountered in its wandering through space some other star near which it passed, the forces thus engendered could have drawn out of it prodigious tidal masses of its material. Some of this might have been lost to it forever; but part of it, being swung to one side by the passing of the neighboring body, might, nevertheless, have remained circling round the parent-mass.

Such an encounter, estimated to have taken place about ten billion years ago, would produce something very like the distribution of momentum that our solar system now possesses, i.e., a large amount of velocity associated with an almost negligible part of the mass. (Actually, the planets contain more than 97 per cent of the angular momentum of the solar system, although their combined mass is less than one tenth of 1 per cent of that of the sun.) Various other details of our solar system otherwise inexplicable find interpretation on the basis of such a hypothesis, proposed first by Professors T. C. Chamberlin and F. R. Moulton, of the University of Chicago, about 1904. Satisfactory as this theory may be in many respects, it is far from being provable. Some details of existing fact seem difficult to reconcile with it. But certainly it must be admitted that fascinating, indeed, become the implications of some of these simplest laws when our imagination follows their operations forward and backward in long ages of time.

Unfortunately, there are innumerable other agencies at work in the cosmos about the nature of which we are not so well informed; and, most

* For a somewhat more extended discussion, and the details of the reasoning that leads to these statements, see "Tidal Evolution," p. 302 of *Astronomy* by Russell, Dugan, Stewart (Ginn & Co., 1926).

THE LAW OF CONSERVATION OF MOMENTUM

likely, there are still others of whose existence we at present have no suspicion. What conflicts with our present theories these uncertain elements or totally unknown processes may play in Time's long drama we know not; hence our vision may be incomplete, imperfect, or even utterly at variance with the truth. The preceding discussion has, furthermore, definitely neglected certain small quantities in lunar theory that should be included. Therefore, such matters must still be regarded only in the light of speculation.

TO SUCH far reaches of space and time we can travel only in our minds. Let us return now to simpler objects that we can touch and handle, experiment upon, and measure. Perhaps here also we may be able to discover some implications of the Law of Conservation of Momentum that will render the behavior of common objects more intelligible to us.

LET us consider the collision of two highly elastic* steel balls of the same mass. To make the problem simple, let us suppose that one of them is initially at rest when the other hits it. Now no matter what the circumstances may be, the total momentum of the system of two balls remains unchanged. This is the law to which no exception has ever been discovered, and which we thus take with confidence as our primary guiding principle. Before the impact, all the momentum resides in the moving ball; after the impact, this same amount must be found in one or both of them.

Now, if this momentum be found in both of them, and equally divided between them after impact, they would each be moving along together with *half* the speed that the original ball possessed before impact. But this would mean that upon impact they did not rebound at all from one another but remained in contact. Such a collision is perfectly *in*elastic,

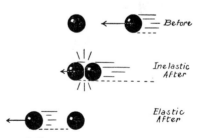

Collision of 2 balls of equal mass

rather than elastic, and to be expected of lumps of putty but not of two balls of steel.

---

* By the term "elastic" in this connection, we do not mean something like a garter, easily stretchable. "Resilient" is more nearly a synonym for our meaning. A perfectly elastic ball, as we use the word, is one that would rebound from a floor of the same quality back to the height from which it was dropped. That there are no materials that will do this need not trouble us. Steel and ivory approximate such behavior and will illustrate this definition sufficiently well.

If, however, the ball that was moving before impact came totally to rest afterward, and the one that was at rest initially took on all the motion, then not only would all the momentum still move forward but the balls would also recede from one another at the same rate as they approached. Furthermore, we can see that no other arrangement of velocities after collision can be found that will do this.

If, for example, the first ball "follows through" to any extent after the impact, the momentum it retains leaves just that much less for the other ball to have. But unless the other ball gets all of the momentum, it cannot satisfy the elastic requirement and provide that the two recede after impact at the same rate as that with which they approached.

If, on the other hand, the first ball bounced backward after collision, this would introduce some negative momentum, which would have to be compensated for by the second ball's rebounding with a greater velocity than the first one possessed. This is even more unthinkable, since it would mean that the steel was more than perfectly elastic, i.e., contained some hidden springs of liveliness that would result, if you dropped it upon another object like itself, in its rebounding higher and higher forever!

THUS it appears that when two elastic balls of the same mass collide, they *exchange velocities* with one another. We have followed only the one case where one ball was initially at rest. But even if both were moving, the same statement can be derived.

Plate 9 (1) shows two snapshots of such a case: (A) was taken just at the moment of release of the first ball; (B), taken at the position of greatest swing of the second ball, shows it has reached a position symmetrical to that from which the first was dropped. The other ball is seen to be at rest; so the fact is evident that the velocities have been exchanged.

Given an entire row of such balls, they exhibit uncanny behavior. Pull out the end one and drop it against the row, and one pops off the other end (Plate 9 [2]). All along the row each has exchanged velocity with its adjacent neighbor on either side. Faster than the eye can follow, the impact has been passed along to the other end.

If two are pulled aside and dropped together (Plate 9 [3]), two pop out from the other end. Here the momentum imparted was twice as great. Why did not one pop out with twice the speed? The answer lies in the fact that, fast as these transfers of momentum are, they are not infinitely fast. Of the

**PLATE 9
ELASTIC IMPACT
EQUAL MASSES**

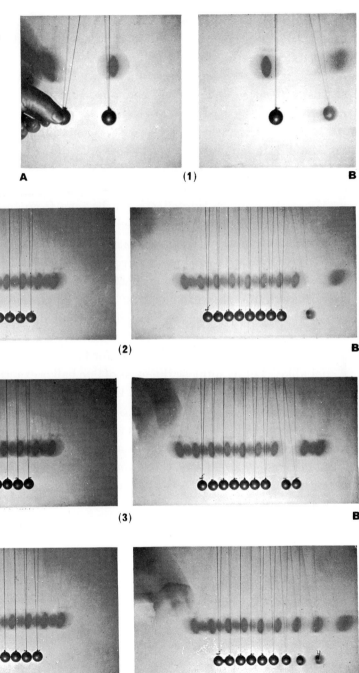

two balls dropped down, the one nearer to the row had to be stopped before it could stop the second one, behind and in contact with it. In stopping, it handed its momentum down the line, and, now at rest itself, proceeded to stop its neighbor, whose momentum was next passed down the line. The last ball at the other end is given the momentum of the first one stopped and straightway starts out with its velocity (equal masses, remember). It *has already left* the row when the impulse* from the stopping of the second arrives, and this impulse is given to the ball now at the end but which still appears to be the second from the end, since the eye has not perceived that the end ball has already left.

Many other interesting effects can be seen from this same apparatus. Its action can be completely disturbed by sticking any two of the balls together with a bit of wax. The motions now (Plate 9 [4]) are entirely different and very complex. As can be seen from the illustration, after the impact nearly all the balls in the row are affected and are moving out, those nearer the end having the greater velocities. The comparative simplicity of these cases depends on the masses of the balls being all alike.

THERE is one other important case for two balls where their masses are not alike. Let us suppose the mass of one ball is exactly three times that of the other. Plate 10 (A) illustrates the apparatus.† The mass of ball 2 is three times that of ball 1. A support, S, is arranged to hold the balls at equal distances from the center. When this is dropped by pressing the key, K, the balls collide, this time with equal but opposite velocities.‡ The heavier ball, 2, has, of course, three times the momentum of ball 1. That of ball 2 is to he left and that of 1 to the right. Their sum is

$$mv + (-3mv) = -2mv$$
$$\rightarrow + \quad \leftarrow \quad = \quad \leftarrow$$
$$\text{right} \quad \text{left} \quad \quad \text{left}$$

This total momentum must be found in the system after impact. Furthermore, since the balls are elastic, they must rebound with the same speed with which they came together. Traveling in opposite directions before

---

* Defined on p. 58.

† The equation from which the young man on p. 19 is fleeing represents the behavior of this case *in toto*.

‡ In accordance with the same argument we use on a pendulum later, pp. 68–69.

A

Stereophotograph of impact apparatus

B **C**

Snapshots before and after impact

PLATE 10   ELASTIC IMPACT—MASSES—UNEQUAL (1:3)

impact, they approached at twice the speed they would have had had either been at rest. Suppose, just for sake of argument, that the large ball should come to rest after the impact. The elastic condition would then demand that the small one rebound with twice the velocity with which it fell, i.e., it would bounce back to about twice the distance from which it was dropped. Now, if it did this, it would at the same time be carrying a momentum $-m \times 2v = -2mv$ to the left, which satisfies the principle of conservation of momentum, as well. Thus we might predict the result of this case in advance. Trying the experiment, and making snapshots reproduced in Plate 10 (B) before impact and (C) afterward, we see that the foregoing argument is justified by facts as well as by the logic.

OF COURSE here, as in all other cases of exact experimental science, experimental facts of some sort or another *must* be ascertained *first, before the controlling laws can be found.* Once these laws are found, then logical arguments based upon them can be made, and only then can the precise experimental facts in new and untried situations be predicted. Subsequent experimental verification of the predictions is what gives the scientific student a very large part of the enjoyment he finds in his work. The method in its entirety has been extremely effective in enabling mankind to acquire that control over the forces of nature that has been the outstanding achievement of human history during the past century and a half.

CHAPTER 9

WORK

FROM earliest times human beings have had to work. The idea is supposed to be repugnant to most people, but the fact of the matter is most people like to work, at least at things that interest them. Whether we like it or not, most of us have to do it anyway. This section is not intended to discuss the social or economic aspects of work, however, but only the technical sense of the word as used by physicists. The technical meaning of the term is comparatively modern—early nineteenth century, in fact.

ALTHOUGH we have made much of Galileo's clarity of insight into the fundamental property of inertia, and of Newton's definition of force as the measure of the effort made when inertia is overcome and a material object is given an acceleration, we have studiously avoided using the nomenclature of these early workers. In referring to the momentum (mv) of a moving body, Galileo used the terms "impulse" and "energy" indiscriminately. Descartes and Newton employed the term "quantity of motion" for momentum, in a sense that we still retain; but it was not until 1847 that Belan-

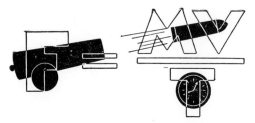

ger proposed that the word "impulse" should mean the product of a force by the time during which it was active. Now, force, defined in the Second Law of Motion, is measured by the change in momentum it produces, *divided by* the time taken by the force to produce this change, i.e.,

$$f = \frac{\text{total momentum change, } (mv)}{\text{time taken } (t)}$$

or

$$f \times t = mv .*$$

THUS we see that impulse (in Belanger's sense, $f \times t$) must equal the total momentum change produced by the force, and this meaning is not far removed from that of the word as used by Galileo. The case is altogether different, however, with the word "energy," which today means something entirely different from impulse. Quite a number of pages, alas, will be required to make this clear!

In many mechanical situations we are much more interested in the product of the force by the distance, $f \times s$, than in the impulse, $f \times t$. If we have to drag a heavy trunk across the floor, or move a piece of furniture from one place to another, it is clearly the distance we have to go, rather than the time it will take us, that is our chief concern. To this expression Coriolis (1792–1843) first applied the name of "work."

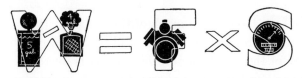

In sharp contrast to our loose usage of this word with reference to social and economic activities is this precise and technical meaning, given it first

---

* We formerly wrote $f = ma$ for the Second Law. Since acceleration, a, is by definition $\frac{v}{t}$ the foregoing form of expression of this law follows obviously.

WORK

by Coriolis, and now in universal use throughout the physical sciences: Work is the product of a force by the distance through which the force acts. $W = F \times s$; WORK equals FORCE times DISTANCE.*

A mosquito performs one erg of work

There are several different types of situations in which the idea of work is predominant. We drag a heavy object along a level surface against the frictional resistance, R. The latter can be measured by a spring balance in the rope on which we pull. It will be found that it takes a little more force to start the object sliding along than it does to keep it going but that, after it is in motion with a constant speed, the pull required is fairly steady. This average force R, times the distance the object is pulled, is a measure of the work that has been done. It is quite obvious that there is no way of getting back this energy (work) that has been expended. Our only satisfaction must be whatever we can get from having removed the object from its former location and having put it where we prefer to have it rest. In measuring the work we have done, we must, furthermore, be sure that we have kept the rope on which we tugged exactly in line with the motion of the object. Only in this case does the full value of

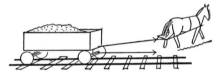

An ineffective way to work—easily corrected

our pull go into overcoming resistance along the path of motion. If circumstances prevent the pull being in all cases in the line of motion, then the force must be resolved into two vector components, which together are equal to it in the sense of our preceding studies (cf. p. 42). One of these components will be along the desired line of motion of the block, and the other at right angles to it. The work done is then the product of the former times the distance traveled.

Another case, somewhat different, would be involved if we dragged our heavy block uphill, not only against the friction of the surface irregularities, but also to some degree upward against the force of gravity. We would at once, in case of a job like this, search for a pair of wheels or some device to minimize the friction,

An ineffective way to work—not easy to correct completely

---

* The unit of work is called the "erg." One erg of work is done if a force of 1 dyne (p. 11 n.) is required to move an object, and does move it, exactly 1 cm.—no more, no less. The weight of a mosquito is about 1 dyne. To lift this insect 1 cm., thus requires about 1 erg of work.

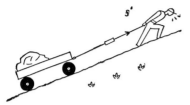

Against friction and uphill!

recognizing instinctively that the uphill part of the undertaking itself is quite enough to struggle with. In analyzing this case let us likewise take the easier mental path and neglect the frictional resistance on the uphill grade. Experience with these experiments, to which we always finally appeal to verify our deductions, has taught us that this simplification of our thinking is physically (in the sense of natural science) sound. In the figure we have sketched the hill, a straight slope that rises a distance h in covering a horizontal distance l. The pull of the earth is represented by a vector mg at the center of the body, vertically pointing downward. Not all of this force, which is the body's weight, is directed along the inclined plane, but only that part, or component of it, which we have sketched in and labeled f. The rest of the weight n can be taken as perpendicular to the surface of the plane. In accordance with our former conventions about vector sums it is noted that $f+n$, the sum, equals mg, the total weight.* Hav-

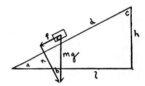

Force vectors, frictionless case of wagon on a slope

ing thus resolved the weight, we see that the work done against the force f, without considering friction, if the object is dragged from the bottom to the top of the plane, is $f \times d$, the length of the slope being d. If we were to be very precise we would draw a vector f' as in the figure at the top of the page, equal and opposite to f to represent *our* pull as we drag the object to the top. The work *we* do is really this vector f' times the distance d. The work $f \times d$ is really the work done *by the object in sliding down the incline a distance d*. Since $f = -f'$† (equality of action and reaction, Newton's Third Law, pp. 44–45) the amount of work we do in pulling the object up is equal to that the object does in sliding down.

In the figure to the right above there are two right triangles—one the inclined plane of sides h, l, d; the other a vector triangle of sides mg, f, n. In the two triangles the sides mg and l, n and d are mutually perpendicular. Thus, according to Euclid and all geometers since his day, the angles a and b

---

* In general, $f = ma$; but when the force is the weight of the body, the acceleration produced is g; consequently weight $(w) = mg$.

† Work itself is not a vector quantity, so the sign is unimportant.

WORK

are equal, and corresponding sides are proportional in the two triangles, i.e., the ratio $\dfrac{\text{force along plane}}{\text{weight}}$ in one of them equals $\dfrac{\text{side } h}{\text{side } d}$ in the other, or $\dfrac{f}{mg} = \dfrac{h}{d}$ ‡

or $f = \dfrac{mgh}{d}$.

Now the WORK done in lifting the weight up against gravity along the plane, being the product of force along the plane by distance along the plane, is $W = f \times d = \dfrac{mg\,h}{d} \times d = mg \times h.$

But this expression is the same as that we would have obtained had we asked for the work expended on the object if we had lifted it VERTICALLY up a distance h against the *undiluted* pull of gravity.

LOOKING at the entire matter in a slightly different way, we might have expected that in doing work against the force of the earth's attraction the only component of the distance that would count in the product of force by distance would be that portion of the distance in a vertical direction, i.e., the distance h. In quite the same way in which the force triangle was treated, so could the inclined plane triangle be managed. The distance d along the plane is made of two components: l horizontal, h vertical. Along the horizontal direction, l, the force of attraction has no component, since it is vertical; but along the vertical h, the full value of mg is effective; thus

‡ So frequently are the ratios of the sides of right triangles used in practical problems that they have been given special names and symbols. In the triangle whose sides are a, b, c and angles are A, B, C, C being a right angle, the various combinations of ratios of sides are named as follows:

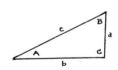

$\dfrac{\text{Side opposite } A}{\text{Hypotenuse}} = \dfrac{a}{c} = $ sine of angle $A = \sin A$.

$\dfrac{\text{Side adjacent } A}{\text{Hypotenuse}} = \dfrac{b}{c} = $ cosine of angle $A = \cos A$.

$\dfrac{\text{Side opposite } A}{\text{Side adjacent } A} = \dfrac{a}{b} = $ tangent of angle $A = \tan A$.

In this language (of trigonometry) the relations on the inclined plane become (see figure on p. 60)

$$\dfrac{f}{mg} = \sin b = \dfrac{h}{d} = \sin a,$$

(a and b being equal), or, f, the force along the plane, equals mg, the weight, times the sine of angle a, which is the angle of the plane with the horizontal. Using the other angle of the triangle, $f = mg \cos c$. This latter rule is very useful, giving the general result that the component of a vector in any direction other than its own is always equal to the product of the vector by the cosine of the angle this other direction makes with the direction of the vector.

work = mgh. No work can be done against gravity in moving horizontally (or, taking a larger view, in moving along the curved surface of the earth, or along any surface parallel to it).

A MARKED and very *important distinction* between the case of *work done against friction* (discussed on a preceding page [p. 59]), and *that done against gravity* is that in this latter case all of this work can be recovered. The body can slide back down the inclined plane and, 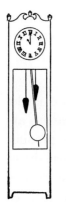 by so doing, it can, by means of a rope passing over a pulley at the top, pull another body, almost its equal in weight, to the top. Thus it will give back almost the same amount of work that was originally expended upon it. Friction, no matter how we try, can never be completely eliminated, either in the wheels or bearings of whatever mechanism we devise or, in this case, in the pulley or in the rope that must of necessity be bent and unbent as it passes over the pulley. Another example is to be found in the "weights" we wind up once a week in an old-fashioned clock. By their slow descent the weights provide additional energy to restore that lost by friction in the swinging pendulum. The motion of the latter is thus maintained for a week instead of a few minutes. A third example is furnished by a free pendulum. When pulled to one side some work is done on the bob since it is lifted slightly. As it swings back this work is returned and is responsible for the swing of the bob a nearly equal distance on the other side.

THERE is yet another way in which work may be done on an object. Although the object may rest on a surface both level and frictionless, nevertheless, if it be set in motion or if in any way its velocity be altered, *work is done against its* INERTIA. If friction and an inclined slope are also there, work is done against all three items. Since, by definition, a force is necessary to produce a change in motion, and since this force must act for an appreciable finite time to accomplish this change, the moving body must during that interval cover a finite distance while the force is active. Just as the total momentum change is given by the product of force by time, the work done is the product of force by distance. We suppose, for simplicity, a constant force, not a variable one.

WORK

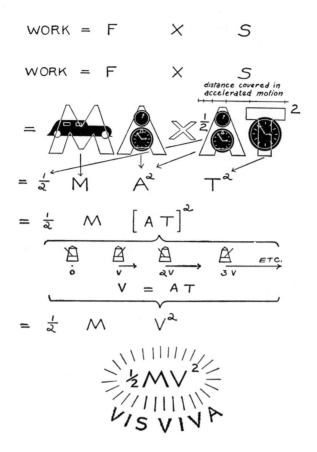

$W = f \times s$.

The force, however, is measured by the acceleration imparted to the mass, $f = ma$.

$W = ma \times s$.

The distance covered in accelerated motion increases as the square of the total time elapsed, and is proportional also to the amount of acceleration, $s = \frac{1}{2} at^2$.

(Cf. p. 18.)

$W = ma \times \frac{1}{2}at^2$.

Rearranging the factors in a different order,

$W = \frac{1}{2} m\, a^2 t^2$ or $\frac{1}{2} m(at)^2$.

The velocity acquired by an accelerating body is, by the definition of the term "acceleration," equal to the acceleration times the time. The parenthesis (at) above is just the velocity, v, that has been imparted by the force under discussion, in time, t, and over a distance, s.

Consequently,

$W = \frac{1}{2} m\, v^2$.

In the early nomenclature the term "living force" or "*vis viva*" was applied to the expression mv^2 by Leibnitz. Coriolis, however, applied the name to half of this quantity, i.e., to the right-hand member of the foregoing equation. Still other names, not especially interesting to us, were also used for this expression that measures the work that has been done in setting a body into motion. It is rather a pity that this ancient name has fallen into disuse. It is much more dramatic than the modern name, which is KINETIC ENERGY.

One final comment remains. Work done against inertia by setting a body into motion is recoverable in part when the body is brought to rest. Its recoverability is usually intermediate in degree between that of work put into a lifted object which is largely again available and that expended against friction which is entirely lost to future usefulness.

To a further study of this very important concept of energy—not only in its kinetic aspect—but in another equally important form we shall devote the next chapter.

CHAPTER 10

ENERGY, POTENTIAL AND KINETIC

AN ALTERNATIVE and synonymous term for work is ENERGY. We expend energy in lifting an object. This energy is measured by the work that was done upon the object while it was being lifted. In a very real sense, then, energy is stored up in an object that has been lifted; and, since this work is recoverable, a potential capacity for doing work has been put into this object. We use the term "potential" technically in this sense and speak of the potential energy of an object. Any chunk of matter, then, has potential energy if, for any reason by virtue of its position with respect to other material objects, forces from them act upon it, and if, furthermore, the chunk is held by some constraint from moving (accelerating) because of these forces. A lifted clock weight is a case in point. Here the attraction of the earth is the force. When supported at any distance above the surface of the earth, the weight can, by descending, expend its store of potential energy and do work on something else.

Barring losses by friction that always nibbles away at any store of energy as we attempt to utilize it, the energy the body will expend is just exactly that which was necessary to put it into the situation whence its potential energy is derived.

No potential energy

Magnetic

Electrical

Electrochemical

Chemical

Natural Resources

Food

A descending object when it has reached the surface of the earth can normally descend no further. Of course, it still could fall down a well, down the shaft of a mine, or down any other hole in the ground. Such situations are artificial. We thus consider that when an object is resting on the surface of the earth it no longer has any potential energy, owing to the earth's attraction for it.*

Now, our object might have been a piece of iron. If there had been a magnet in its immediate vicinity, then, whether it were at sea-level or elsewhere, it would have been pulled by, and would have fallen toward, the magnet. This would involve, however, the possession of potential energy of another kind—magnetic potential energy.

There are many kinds of potential energy besides that due to the force of gravity. Any kind of force that can cause changes in the motions of objects can be worked against. By so doing, reserves of potential energy may be accumulated to be expended later as the occasion requires. Electrical potential energy in large amounts is stored up through atmospheric processes in the clouds accompanying thunderstorms. We have never succeeded, however, in harnessing this force of nature. Electrochemical potential energy is stored up in batteries for cranking automobile engines. Some of it is restored to the battery as the car is driven, but everyone knows that the life of a battery is not indefinitely long, even with the greatest care. Losses that we are never able entirely to prevent ultimately necessitate the purchase of another battery.

This leads to another practical point. Energy, being of value in doing work, always costs money. Nature, it is true, is superabundant in her lavish generosity with it. To the sun we owe our manifold supplies. The sun's heat daily lifts countless tons of water by evaporation into the atmosphere. There it condenses and falls as rain or snow, much of it on elevated portions of the continents

---

* Strictly speaking, the "surface of the earth" should be at sea-level.

where its potential energy can be stored up by dams across streams, and thence directed to water turbines and transformed into other types of energy for useful purposes. In the winds, and in the tides also, are vast supplies of energy as yet but little utilized. The action of sunlight on growing things stores up chemical potential energy in starches and sugars—food for animals and men to keep their mechanisms heated and in operation. By subtle transformations through the growing plants  and animals of bygone ages vast stores of chemical potential energy have been buried and preserved for millions of years in the world's great resources of coal and oil. To utilize these stores, however, large amounts of capital and labor usually have to be invested. The mere problem of transportation of the energy requires co-operative effort on the part of human beings. These are the factors that contribute in large part to the cost to the consumer for what nature has freely supplied for man's common use.

WE NOW return to the consideration of another type of mechanical energy—another sort of capacity for doing work that is possessed by bodies *when in motion*. This is what we referred to on page 64 in the quaint parlance of a former generation as the *vis viva*, or living force in moving bodies, as contrasted to *vis mortua*, the potential aspects for doing work possessed by static material. Today, precise in language about such things, we do not call *vis viva* a force at all. It is a different thing from force, this capacity possessed by moving bodies for doing work against forces. We call it KINETIC ENERGY, in contrast to potential energy.

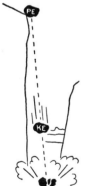

A rock rolling off a cliff, and falling more and more swiftly as it descends, possesses a tremendous capacity for doing work, even at the bottom of its fall on the instant before it strikes. Under the impact with which it strikes, its momentum is swiftly changed, and enormous forces are imparted to the region sustaining the shock. These forces, if controlled, can be made to do work. If entirely utilized, they can be made to do as much work as it took originally to lift the rock to the top of the cliff.

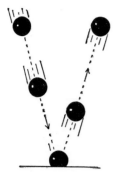

We have differentiated in an earlier chapter between two different kinds of impact—elastic and inelastic—recognizing that the former kind is never perfectly exemplified in any natural phenomenon. If a steel ball were perfectly elastic, on rebound it would reascend exactly to the height from which it fell. Its original potential energy was all lost as such the instant before it struck the equally elastic plate. In rebounding, the kinetic energy of motion is redirected by the impact without any loss (since perfect elasticity is postulated). After striking, its motion, now opposed in direction to the force of gravity, lifts the body swiftly at first but at a uniformly declining rate until the original height has been again reached. Kinetic energy has been transformed back again into the same amount of potential energy from which it formerly originated.

Of course, a complete rebound like this never happens. But always, if there is any rebound at all, a falling object reconverts some of its energy of motion back into the potential form. What is not thus reconverted is lost to usefulness. It is lost in just the same sense that the energy we spend in dragging a heavy object across a level floor is lost: it has become energy of another type, heat energy. Heat energy, as we shall see later in another place, is kinetic energy also, but kinetic energy associated with the minute particles, the atoms and molecules, of which all matter is composed. As such, we cannot conveniently lay hold of, or harness, it for useful work.

W E CAN, by means of apparatus so simple that anyone can construct it for himself, verify by measurement the truth of the statements we have been making about these things. The original experiments that first disclosed these truths were made by Galileo and by Newton. All of the experiments that have been made subsequently have given similar results; hence our confidence in these matters.

Suppose we take a heavy ball, suspend it on a thread, and swing it like a pendulum. Pulling it to one side, we let it fall. Constrained by the thread, it swings downward in a circular arc. By analogy with the motion down an inclined plane, it would seem that here there is obviously very little friction, also that the ball acquires a velocity in falling, considerably less than it would have had had it fallen

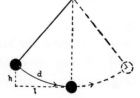

ENERGY, POTENTIAL AND KINETIC

vertically through the distance d measured along the arc to the bottom of the path. Since this path, while not exactly that of the former path down an inclined plane, can nevertheless be made to approximate, as closely as we wish, a path down a series of inclined planes, one after another, we might reason that the results will not be appreciably different from the former study. The velocity at the bottom of the swing we would then expect to be exactly the same that would have been acquired if it had fallen *vertically* a distance h (equal to the difference in levels from the top to the bottom of the arc). Such a velocity in the absence of any frictional loss would, if properly directed, carry the ball back to a height equal to that from which it originally fell. The directing string is still attached to the ball. If left alone, then the ball will swing an equal distance on the other side of its lowest point and so will rise to the same height as that from which it was dropped. Friction is small here, but it is never absent. The ball swings just a trifle short of the symmetrical position, and, if left to itself, after many scores of oscillations back and forth will finally come to rest, its original store of energy having been dissipated. Photographs of this experiment are given in Plate 11.

We see here one of the simplest illustrations of a shifting change back and forth, from potential energy of a lifted ball, to kinetic energy of a moving ball. The energy of the ball is all of the former type at the ends of the swing, all of the latter type at the lowest point of the arc. In intermediate positions both kinds are present. Since one type is always gained at the expense or loss of the other kind, the total energy must always remain the same, except for the slow losses in overcoming friction.

Even if the upward path of the ball is shortened by the string's swinging into contact with a peg (about which it winds on the upward swing), the same elevation of height is maintained. If the peg is so low that the ball cannot swing back up as high as the level from which it fell, the ball will still possess kinetic energy at the top of its now limited path and will go on swinging round and round.*

> * An interesting corollary to the motion will often be observed if the diameter of the restraining peg be not too small. The ball will then continue swinging round and round the peg until the string is wound up entirely. Certain ingenious types of anniversary clocks, needing winding only once a year, often use this device or a modification of it to retard greatly the motion of their pendulums. These often take the form of a weight twisting back and forth on a vertical wire. If the reader recalls the fact that angular momentum in a rotating system is always conserved save for frictional loss, and if he can understand that angular, or rotational, inertia depends on the distance of the mass from the center of rotation as well as on the mass itself, he can see that, as the string winds around the peg, the mass is brought

Before release on one side

After swinging to other side

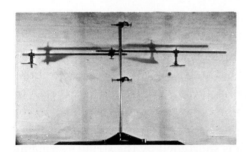 

Cord holding ball here is interfered with at mid-swing; interference occurs at a lower level in photograph below

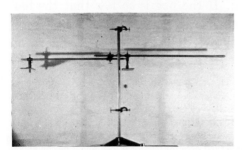 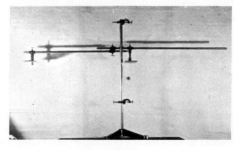

PLATE 11 FLASH STEREOPHOTOGRAPHS OF A PENDULUM

ENERGY, POTENTIAL AND KINETIC 71

As we have already shown on page 63 for any accelerating object, the kinetic energy represented by the work done by the earth's attraction in overcoming the inertia of a body and speeding it up to a velocity v, as it falls, in a time t, and over a distance s, is equal to $\frac{1}{2}mv^2$. We suggest that you turn back now and read this paragraph again; you will find you understand it better.

THIS kinetic energy can be made to do other sorts of work than that exemplified in lifting the object possessing it nearly to the level from which it fell. Suppose that a ball, rolling down a hill, or swinging down an arc, as in a pendulum, is caught in a basket attached to any kind of resisting mechanism. Obviously, to overcome the momentum of the ball the basket will subject it to a resisting force R, and this will bring the ball to rest in a distance s. Work will be done equal to $R \times s$, but now in no case will this work represent necessarily all, or even very much, of the energy of motion of the ball. Transformations into useless heat energy are often very large in this connection. Whenever a mass in motion is brought into contact with another system that slows it down and does not allow it to rebound to any appreciable extent, i.e., in *in*elastic (or largely inelastic) impacts, loss of kinetic energy is large.

TAKE the case of a bullet fired into a block hung as a pendulum so that it is free to swing—a method by which the velocity of small arms bullets is most frequently determined. The apparatus is shown in Plate 12. Application of the Law of Conservation of Momentum, to which considerations of friction are not pertinent, determines the bullet's velocity. If the block is at rest before the bullet hits it, the momentum of the system (block and bullet before impact) is that of the bullet alone, mV. The bullet lodges in the block, let us hope—especially if there be innocent bystanders around. The momentum of the system (block and bullet, after impact) is the product of the masses (m, of bullet, $+M$ of the much more massive block) times their common velocity, v, with which the block starts swinging the instant after impact.*

---

closer to the center all the time; the angular inertia thus gets less and less, so that the angular velocity must get greater and greater to maintain the momentum constant. The ball thus winds up *with increasing speed* around the peg. The experiments above described were performed first by Galileo.

* This common velocity, v, hard to observe directly, may be calculated from the horizontal distance the pendulum swings after the impact by simple geometrical relations that the reader, a little used

PLATE 12 BALLISTIC PENDULUM. For determining the velocities of small arms bullets by application of Law of Conservation of Momentum. Bullet is fired from the left, into the wooden block. The latter, hanging from cords from the ceiling, swings freely from the impact. A light cardboard marker is pushed to the right by the pendulum and records length of swing. From this and knowledge of masses involved the velocity of bullet is calculated. (See text also.) (Stereophotograph)

Conservation of momentum requires that

$$mV = (m+M)v$$

and, solving for the unknown velocity of the bullet, V, in terms of the known masses m and M of bullet and block, and the observed swing of the block when hit,

$$V = \frac{m+M}{m}v .$$

The energy relations, however, are altogether different. We will not go into the details of the analysis; but it can be shown that of the very considerable amount of kinetic energy, $\frac{1}{2}mV^2$, of the flying bullet before it hits the block, only a negligible amount is to be found remaining afterward in the swinging target $\frac{1}{2}(m+M)v^2$. Qualitatively, it can be seen that, since V is large (several hundreds of feet or many thousands of centimeters per second), its square, even if multiplied by the relative small mass of the bullet, will make the energy product very great. It will be much greater than the energy

---

to the practices of geometry, and clearly understanding the pages already read, can work out for himself. If the radius of the arc on which the pendulum swings is R, the horizontal displacement produced by the impact is D, and g is the acceleration of gravity,

$$v = D\sqrt{\frac{g}{R}} \text{ (approximately)} .$$

ENERGY, POTENTIAL AND KINETIC

product after impact, which, although containing the much larger mass of the block, has been multiplied by the square of a velocity of only a few centimeters per second. Accurate attention to details of the analysis shows the loss of kinetic energy to be the fraction $\dfrac{M}{(m+M)}$ of the original energy of the bullet. In such experiments the bullet's mass is usually a few grams and the block's a few thousand, so that in actual cases the swinging pendulum usually contains a few tenths of 1 per cent, or even less, of the energy originally present in the bullet. What has become of the rest of this original energy? It has been transformed into heat. Indeed, it is not unusual to find evidence of partial melting of the lead bullet on extracting it from the block, especially if the latter be of rather hard and resistant wood.

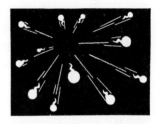

APROPOS of bullets, the earth is being continually bombarded by a veritable rain of projectiles, as we have formerly mentioned—the meteors that can be seen flashing across the sky on almost any clear moonless night. Estimates of their total number during every 24 hr., based on the number one sees in a small area in a few minutes, place the daily total at millions. Minute as are the vast majority of them, their total momentum and energy are not inconsiderable. What becomes of this momentum and energy? The momentum is added without loss to the earth. Except in the rare cases where the meteor is large enough to survive passage through the atmosphere and plunge into the earth itself, this momentum is simply imparted to the air and its directive quality largely lost. The energy in all cases is absorbed by friction during impact and converted into heat. The vast bulk of the earth is too huge to be appreciably affected by the momentum changes, or to be warmed to any sensible extent by the heat energy imparted.

IN THE modern world there are many important processes involving impacts. These usually arise when we have very great forces of resistance to be overcome. In spite of large amounts of energy transformed incidentally into useless heat, these processes are still very useful, as is well illustrated in the case of the pile-driver. To force a pile having a diameter of from 8 to 12 in. into a wet sand or clay soil by direct pressure alone is obviously quite

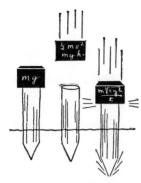

beyond the power of ordinary engineering machinery. The hammer of a pile-driver, resting on top of the pile, by its weight alone is hopelessly inadequate for the purpose. But by lifting the hammer some 15 or 20 ft. above the head of the pile and dropping it, the additional potential energy gained, mgh, is converted into kinetic, $\frac{1}{2}mv^2$, and the momentum, mv, corresponding, is all overcome in a few thousandths of a second upon impact. The momentary force thus applied to the head of the pile is enormous—a few tons times a score or more feet per second, divided by a few thousandths of a second, i.e., multiplied by thousands. At every blow this sudden force sinks the pile a few inches into the ground against the great resistance of the earth.

What does it matter if the useful work, represented by the earth's resistance times the few inches over which it has been conquered, accomplished by the pile-driver, is only a small proportion of the energy of the falling hammer? There has been achieved here something that would otherwise be quite beyond human powers. While an accurate accounting of the wasted energy in such an operation would be quite difficult to make, it is certain to be very great, especially if the relatively low efficiency of the donkey engine running the hammer hoist is included and the original chemical potential energy of the coal is compared to the small amount of energy extracted.

The case illustrates our characteristic indifference to energy economy in industrial processes, not always as well justified as here. This indifference is to be expected in an environment where natural resources of energy are still tremendously abundant. A more highly developed civilization on a planet scanty in resources, or where resources had been largely exhausted through ages of waste—a civilization faced perhaps with extinction unless its scanty supplies were conserved to the utmost—would be forced to develop many processes far more economical than those that we possess.

OUR own habitat, it must be remembered, has resources of energy, emanating from the sun, that are as yet still not utilized. The winds were formerly employed a little, especially in rural communities; but windmills have been largely supplanted by electricity or internal-combustion engines. The tides, lifted twice a day by the combined action of the attractive forces of the sun and moon, represent a colossal store of potential energy.

PLATE 13

ENERGY SOURCES

THAT WE OWE

TO THE SUN

The sun

Lifts the water, waterfalls, and tides,

Calls forth the winds,

Stores energy in plants and animals which, buried for millions of years, give coal and oil.

Photographs reproduced in Plates 13, 16, and 17 are from the Educational Sound Picture "Energy and its Transformations" by H. B. Lemon and H. I. Schlesinger, produced by Erpi Picture Consultants, Incorporated, and The University of Chicago Press (copyright University of Chicago).

The tides never have been harnessed and probably never will be until the more convenient resources of coal, oil, and water power prove insufficient. A source of energy still greater than the tides is the tremendous amount of heat supplied to the surface of our globe by the sun.

The energy of light and heat can be transformed into mechanical form by suitable devices through the mediation of machinery. Details are somewhat complicated in contrast to the automatic ease of the opposite process, wherein friction always is converting mechanical energy into heat. We shall discuss these energy transformations later: mechanical to thermal in the next chapter, thermal to mechanical after we have more carefully investigated the nature of heat.

FAR greater than all the above-mentioned sources of energy put together is one, the existence of which has been known for at least half a century, but which we thought (and hoped) might never be tapped by men. This is the energy locked up in the *nuclei of the atoms of matter itself*.

Within the past few years, stimulated by the urgency of a second World War, the concentrated efforts of physicists, chemists, engineers and production men, supported by about two billions of dollars in this country alone, have been successful in finding out how to utilize a very small part of the nuclear energy of a few heavy atoms. This energy, indeed, constitutes some of the very mass itself of the atoms involved, 1 gm. of mass being equivalent to 900 billion billion ergs of energy! Even the relatively tiny fractions of this source of energy so far released make all other sources seem trifling by comparison, as we shall see in a later chapter (p. 316).

The primordial character of nuclear energy—the fact that it is Nature's basic storehouse of energy—becomes clearer when we realize that it is nuclear energy that has supplied the sun with its vast outpouring of heat for the last two billion years and will continue to more than maintain it for perhaps ten billion more. Also it provides the light and heat of all the other bright stars.

At its present rate the sun is *losing mass* at the rate of 4,000,000 tons per second. Great as this loss seems to us, it means only that the sun in the next billion years will lose one fifteen-thousandth part of its entire mass. This, of course, does not worry us earth-dwellers in the least. What should be, *and is*, of the most intense concern today is the fact that in the recently developed so-called "atomic bombs" and in the "chain-reacting piles" used to make other artificial kinds of "fissionable" atoms for use therein, nuclear energy has now been made available for mankind's use. How has he so far applied this new source of energy at his disposal—a source which for destructive purposes beggars comparison with all former kinds of weapons? Is it going to be possible for men and nations in the future so to arrange their social, economic, political, and international affairs that future wars, more devastating to their participants than all the wars so far endured, may be avoided? Only if this can be done, will the peaceful and constructive uses and applications of these recent discoveries be of any benefit to mankind.

CHAPTER 11

WORK AND HEAT
CONSERVATION OF ENERGY

IN NEARLY all of the various mechanical processes that have been presented in the preceding pages, repeated reference has been made to the ever present annoyance of frictional effects.*

It is friction that causes falling bodies to cease their acceleration and to acquire a terminal velocity. Friction interferes with the ideal parabolic shape of trajectories and greatly limits the range of artillery. In every experiment we have mentioned we have found it necessary always to be apologizing for this omnipresent imp, that invariably prevents any apparatus from functioning ideally, as we would like. Impacts may be perfectly inelastic but never perfectly elastic. Momentum is conserved, but energy is invariably lost to usefulness; always, there is a little lost, usually a considerable amount, and often all. Furthermore, it is always true that in proportion as friction gets in its deadly work, something becomes heated.

* That friction has a good side also must not be forgotten—a side not germane to our story at this point. We content ourselves by an enumeration of a few of its virtues in the figure on page 78.

FROM GALILEO TO THE NUCLEAR AGE

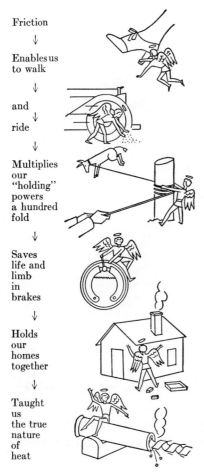

Friction
↓
Enables us to walk
↓
and ride
↓
Multiplies our "holding" powers a hundred fold
↓
Saves life and limb in brakes
↓
Holds our homes together
↓
Taught us the true nature of heat

His good points should also be presented

A clear conception with respect to the nature of heat is a comparatively modern achievement. As late as 1856 the *Encyclopædia Britannica* gave preference, among several theories of heat, to one that we now know to be entirely erroneous. Indeed, it was only at the beginning of the nineteenth century that there began to be performed those experiments destined ultimately to lead to a true understanding of the nature of heat and of its relation to mechanical processes. Here again, as we have mentioned several times before, experiments to acquaint us with interrelated groups of facts and to provide precise and accurate numerical descriptions came before any laws could be formulated or theories constructed. Prior to the experimental determination of the facts, heat was regarded either as some kind of a fluid, a "phlogiston" or "caloric," or as a manifestation of motion of minute particles of matter, or as a vibration within material substances, or as almost anything the human imagination could invent; but no means of deciding between the merits of different theories was available. It was anybody's guess.

A YANKEE, a native of Massachusetts, expatriated and in the lucrative service of the army of a foreign power, was the first to discover that apparently inexhaustible amounts of heat could be provided by the utilization of mechanical work expended continuously against frictional forces. Benjamin Thompson (1763–1814), later known as Count Rumford, said in connection with his observation of the heat developed in cannon-boring operations: "It appears difficult, if not quite impossible, for me to imagine heat to be anything else than that, which in this experiment was supplied continuously to the piece of metal, as the heat appeared, namely motion." The motion to which he referred was that of a rather blunt drill that was rotated

CONSERVATION OF ENERGY

continuously while being forcibly pressed against the block of metal. Rumford was doing work against frictional forces, and heat was the result. This phenomenon had been observed by others for generations on innumerable occasions. But Rumford measured the amounts of heat obtained in terms of water—a certain number of pounds  of water raised a certain number of degrees in temperature. He found the amount that could be obtained indefinitely increased as long as the horse that worked in the treadmill to turn the drill was kept employed.

It remained for a distinguished successor, James Prescott Joule (1818–89), a brewer's son from the vicinity of Manchester, England, to supply the other half of the contribution. Joule measured, not only the amount of heat supplied to the water, but also the amount of work expended in the process. He found* that the former quantity was always in direct proportion to the latter, i.e., that 1 calorie of heat (the amount involved in the raising of 1 gm. of water 1° C.)† always and invariably accompanied the expenditure of 41,800,000 units of work, i.e., of force times distance. Since unit force, called the dyne, about equal to the weight of a mosquito, is that which gives a gram of mass a change of velocity of 1 cm. per sec. each second that it acts, and since unit distance is the centimeter, the unit of work is sometimes called the dyne-centimeter. The more common term, however, for this unit of work is the ERG. Ten million (10,000,000) ergs of work are often called the JOULE.

UNITS OF WORK AND HEAT

One dyne (force) is about the WEIGHT of a mosquito.
One erg (work or energy) is the work of lifting a mosquito about $\frac{1}{2}$ in.
Ten million ergs (10^7) would be represented by a box packed solidly with "ergs," and having about 216 ergs along each edge, i.e.,
$$(216)^3 = 10^7$$
Ten million ergs equals one joule of energy.
4.18 joules will raise the temperature of 1 gm. of water from 19.5 to 20.5° C.
One calorie is defined as the heat necessary to raise the temperature of 1 gm. of water from 19.5 to 20.5° C.
Therefore, the MECHANICAL EQUIVALENT OF HEAT is 4.18×10^7 ergs per cal.

* We give here units current today rather than those used by Joule.

† For the reason for this peculiar definition of a unit of heat the reader must wait until chap. 17 is reached.

Thus the expression
$$1 \text{ cal.} = 4.18 \text{ Joules}$$
is a statement of the value of the "mechanical equivalent of heat."

A MODERN form of apparatus for the determination of this important constant is illustrated in Plate 14.

Two brass cups (c_1, c_2) are arranged to fit closely together. The inner cup is filled with water; the outer one is fastened to a shaft, s, that will revolve. A disk, d, is attached to the inner cup; and to this a weight, w, is fastened which prevents the inner cup, c_2, from turning as the outer one, c_1, rotates. The friction between the two cups generates heat which raises the temperature of the water in the inner cup. All the work done is here converted into heat, which is measured by the rise in temperature of the known mass of water.* The work is measured by the product of the weight (increased in the proportion of the diameter of the disk to which it is attached to that of the two cups) by the distance traveled by the surface between the two cups. The latter is given by the average circumference of the interface times the number of turns made by the shaft, which is measured by a revolution counter, r, at the bottom of the apparatus.

JOULE'S experiments in these matters occupied the greater part of forty years. Still more refined ones on the same subject were made by Professor H. A. Rowland, of Johns Hopkins University, about 1880; and the thermometers he used were checked and corrected almost twenty years later by Day. The results agreed with those of Joule by better than two parts in a thousand, although the later experiments were undoubtedly more precise. Still more accurate measures have subsequently been made, so that the final result today is known to be, in terms of water at 20° C.,
$$1 \text{ cal.} = 4.1813 \pm 0.006 \text{ joules.}$$

IT MUST not be thought that the experiments of Joule and his many successors in this important field of study were confined to the direct relationship between heat and mechanical work. Mechanical energy can be transformed into electrical energy, and thence into heat. A great variety of other intermediate steps may be involved. But the end result is always the same

---

* The heat absorbed in warming up the cups also must be included. (For method see chap. 17, p. 140.)

CONSERVATION OF ENERGY

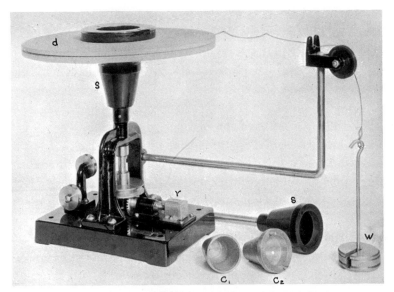

PLATE 14 A Modern Form of Joule's Apparatus

if every aspect of the processes is carefully watched and the contribution of each to the total heat checked up, and added to make the final sum. This fact led Joule to be one of the first to announce a general principle, in its scope one of the most far reaching that ever have been found to rule the world of nature, The Principle of Conservation of Energy. In the myriads of experiments since his time, in the thousand and one varied operations and processes of modern industry, in the utterly complex and ofttimes still obscure processes of a biological character, no single exception to this principle has ever been observed. Even where details are not yet clear, its accuracy and exactness is always assumed without question.

We can test the principle on a cosmic scale from observation of the motions of the members of the solar system. Here, also, it applies. That it also holds true far away in the island universe of stars to which our sun belongs we do not doubt, nor in those millions of other island universes from which our own is separated by distances inconceivable. We have never been able, nor shall we probably ever be able, to test the principle so widely. Therefore, our belief in its validity perhaps might be termed an act of faith—faith in the uniformity of nature and her laws, a faith to which each succeeding year of our collective study and enterprise brings additional conviction.

THE Principle of Conservation of Energy states that energy is uncreatable and indestructible; it states that in everything that occurs in the material world a transformation of energy is all that is taking place; it says that energy appears in many guises and disguises but that, no matter when or how it changes form, it is in sum total always of the same amount. If we isolate and keep track of any small amount of energy, and ring upon this all the changes ingenuity can devise, carefully watching every nook and cranny so that none of it slips unnoticed through our fingers, this same small amount with which we started will remain unvaried to the end. It is the only changeless thing perhaps in this ever changing world.

The vast scope of this idea dwarfs all other principles by comparison. Newton's laws of motion, conservation of momentum, and many other principles we shall yet come upon are but corollaries, special cases, restricted to certain situations, demanding idealized conditions for their perfect exemplification. It is not so with Conservation of Energy. The mathematical setting-down of its demands in what is often called the "energy equation" is the most quick, certain, and convenient way to attack the solution of any unknown problem. In realms where so little is yet known that speculation still is rife, no hypothesis would be entertained that did not take form in accordance with it. It implicitly contains the sum and substance of all that is known in natural science, ancient or modern, physical or biological.

CHAPTER 12

POWER

THERE is one idea that, more than any other, is most frequently confused with the idea of work—the one that goes by the technical name of POWER. Power is the rate at which work is done: it is the rate at which a transformation of energy takes place: it is the quotient of an amount of work performed divided by the time taken by the performance.

Suppose your weight is 150 lb. This weight is the force with which the myriad tons of Mother Earth attract you and hold you close. You can, by the expenditure of energy in biological processes in your body, energy provided by the food you eat, walk up a hill or a flight of stairs, climb up a ladder or a rope, and raise yourself against this force—thereby doing work against it. The work you do, the potential energy you acquire, is the product of the force of your weight times the distance you lift yourself. All this has been said before. In the strength and vigor of youth and splendid health, you may run up the stairs three steps at a time or go swiftly up the rope hand over hand. In middle life you will undoubtedly prefer the stairs to the rope, and be content to take the former two at a time or even one by one. In old age or in ill health, your progress to a higher elevation may be slow and painful; but by wisely "taking your time" you still can reach the upper story. In all of these cases you have done the same amount of work, but in youth and

Three ages of man and two kinds of work

strength you do it in a shorter time than when you are old and feeble. The "power" of youth is greater, that is all. Capacity for sustained effort and total work done, however, frequently becomes greater with increasing years. Taken in its entirety, the energy put forth by the man usually exceeds that of the youth: the race is not always to the swiftest runner if the distance is a long one. A wise restraint in rate of working and long-continued application on the job are likely to result in greater total work accomplished. A reckless expenditure of power may result in an untimely cessation of all activity. These illustrations, you understand, are intended to be suggestive of the meanings of the words "energy," "work," and "power," rather than to mean literally that we measure human efforts in this way.

OUR legitimate applications are, of course, purely mechanical. A common type of device to measure power is shown in Plate 15. The unit of power most common in daily language is the "watt." Many that use the word are ignorant of what a watt really is. It is a rate of doing work, or of utilizing energy, to the amount of a joule every second. A joule we remember (or cf. p. 79) is 10,000,000 ergs. A girl of 110 lbs. weight has a mass of about 50 kg. (50,000 gm.). Her weight in metric units is 980 times this (mg), i.e., 49,000,000 dynes. If she steps up 1 cm. (less than $\frac{1}{2}$ in.), she does an amount of work equal to 49,000,000 ergs, about 5 joules. If she takes 1 sec. of time to step up this $\frac{1}{2}$ in., she is "working" at a rate of about 5 joules per sec.; i.e., 5 watts measures the power of her effort. An ordinary 60-watt incandescent lamp consumes power at twelve times this rate, roughly corresponding to our young lady walking slowly, at the rate of one step a second, up a flight of steps, each about 6 in. high.

A horse-power (so called with little reference to horses) is an activity of 746 watts. A 150-lb. man, running upstairs at a rate that carries him upward a little less than 4 ft. (43 in.) per sec., is working at the rate of 1 h-p. This involves

5 joules of work

POWER

60 watts,
60 joules per second

taking three steps at a time about three times every second. Human endurance will not enable even the athlete to keep up this pace very long. In almost no other way can a man approach a horse-power of performance.

In British units a horse-power is equivalent to lifting 550 lb., or in dragging any object resisting with a force of 550 lb., *1 ft. every second*. This weight is about that of a bale of cotton, and a horse by means of a rope over a pulley can do such a task readily (see Plate 16). For a man to lift so large a weight, he would have to be given some mechanical advantage over it. He might use a block and tackle. If the latter had five strands of rope supporting the bale, the man pulling on the free end would have to exert a force of only 110 lb. This he could undoubtedly do, and at the same time walk away from the hoist as fast as the horse could with his single strand of rope. In the man's case the bale would rise with only one-fifth the speed, and thus his power would be but one-fifth that of the horse. Even this is a vigorous rate of working and could not be long sustained. In sustained physical effort a man's power is seldom as great as one-tenth of a horse-power.*

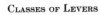

CLASSES OF LEVERS

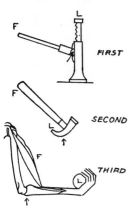

Note relative positions in each of force, *F;* load *L;* and the fulcrum, ↑

To give a mechanical advantage, there are a great variety of devices, some known from remote antiquity, to enable men to exert forces otherwise quite beyond their strength. The lever, of which there are several kinds, is the simplest type. The mechanical advantage of this simplest of all machines, as well as that of the most complex form, is readily obtained by consideration of the *force* that must be exerted at one end of the machine with respect to that which the machine puts forth at the other end. In an ideal machine, of course,

---

* Our examples of power have been chiefly in lifting objects and doing work against gravity. Equally important are examples of expenditure of power against resisting forces. An automobile plowing through a muddy or sandy road needs more power quite as much as when climbing hills.

PLATE 15

APPARATUS FOR MEASURING A PERSON'S HORSE-POWER

Over a pulley, behind the flywheel, is a broad leather belt, fastened under tension to two spring balances attached to the floor below. The slipping of this belt over the pulley as one turns the crank provides a steady frictional force to work against. This force is measured by the difference of reading on the two balances.

The distance through which the force is overcome is given by the circumference of the pulley times the number of revolutions that the crank is turned.

A clock measures the time taken to do this work, and (fortunately for the experimenter, who is afraid of multiplication and division) the result in horse-power has been figured out and set down in the table shown on the wall.

Selecting the nearest figure to the value of the force against which he worked in the left-hand column, and the number of revolutions per minute he accomplished across the top, opposite these two points in the body of the chart the experimenter finds the horse-power, or fraction thereof, that he produced. The normal performance for an average man and woman, an unusually strong man, and a horse, are indicated by the curves.

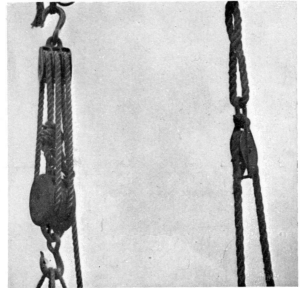

5 strands   1 strand

Mechanical advantage of a block and tackle

Equal loads. The horse lifts his five times as fast as does the man

PLATE 16 COMPARATIVE POWER OF MAN AND HORSE

PLATE 17 The Work Relations in the Use of a Lever

the ratio of these forces is exactly the inverse of the distances traversed by the points of their application. This is only another way of saying that the work *done on* the machine (e.g., small force acting a large distance, $f \times D$) equals the work *done by* the machine (e.g., large force acting through a small distance, $F \times d$).

The mechanical advantage is the ratio of the forces, F/f (equal to D/d only in an ideal machine). Frictional losses are always present, however, so that the work which comes out *is always less* than the work which goes in, i.e., $Fd < fD$. The EFFICIENCY of the operation defined as $\dfrac{\text{Work out}}{\text{Work in}}$, $\dfrac{Fd}{fD}$, is therefore never equal to 1, or, as we usually say, 100 per cent. It is nearly this amount in the simple lever that we chose for our illustration.

If you use a lever to pry up a stone, you choose the long arm and you arrange the short arm of the pry under the rock. The work you put in (your force times the distance you move your end) should practically, although not quite, be equaled by the work done on the stone, i.e., by its weight times the distance it is moved (Plate 17). If your lever arm is ten times that of the stone's, you will have to move your end ten times farther than the other end will move. However, the force available at the opposite end will be ten times the force that you exert. Your weight of 150 lb. thrown bearing down upon your end of the lever will exert ¾ ton (1,500 lb.) of lifting at the other end. Thus, by repeated operation, the rock can be moved a little each time by your moving a much greater distance each time.

The mechanical advantage of an inclined plane in rolling a heavy cask up into a wagon is measured by the reduced component of the weight of the

POWER

cask resolved along the surface of the incline—a problem we have considered before in another connection. The mechanical advantage of a double inclined plane (i.e., a wedge) inserted into a crack at the end of a log to split it is enormous under the effect of the impact of a blow from a maul. A moderately husky man with a 15-lb. maul and a wedge can exert  a splitting force of 10–15 tons with a well-chosen single blow. However, the considerations of the energy of impacts of this sort that we have already discussed indicate that the efficiency is very low, most of the energy being wasted in heat, although the result desired is nevertheless achieved. Pulleys, combined into block-and-tackle devices, capstans, winches, screws (that are simply inclined planes wrapped around cylinders), wheels and axles, and all sorts of geared devices are a few of the legion of methods that have been invented for the acquisition of a mechanical advantage, or disadvantage. Devices that convert a large force into a small one are chosen usually where one wants to give to some object ready to be accelerated a velocity otherwise inconveniently large. A large force applied by a man for a short distance can be transferred as a much smaller force to some object moved through a correspondingly larger distance. The throwing-stick used by native tribes of primitive peoples to hurl a spear is an example. The use of bats and racquets is somewhat related to this also.

In Plate 18 are shown working models of a few of the many kinds of mechanisms for transmitting force, and power, and effecting mechanical advantage and disadvantage.

Suppose that we compare the *power* at the two ends of any machine instead of the work:

$$\text{Power} = \frac{\text{work}}{\text{time}} = \frac{F \times d}{t} = F \times v,$$

since the distance covered divided by the time $\left(\frac{d}{t}\right)$ is the velocity of that part of the machine to which the force is applied. We find the power which we get out of any actual machine, $F \times v$, is less than the power we put into it, $f \times V$:

$$F \times v < f \times V.$$

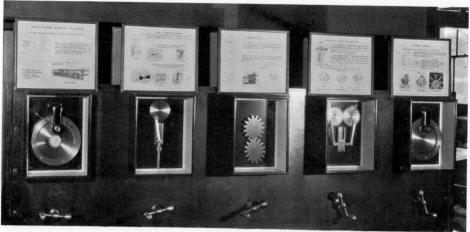

Courtesy of Museum of Science and Industry, Chicago

Mutilated-Gear Reversing Mechanism
Changes continuous rotary into reversing rotary motion

Slider-Crank Mechanism
The most extensively used mechanism. It transforms reciprocating (or back-and-forth) motion into rotary motion or vice-versa

Elliptical Gears
Used to transform a constant velocity in one shaft into a variable velocity in another.

Cartwright's Straight-Line Mechanism

Internal Gearing
A convenient form of speed-reducer or speed-amplifier

PLATE 18

The efficiency can also be expressed as a power ratio,

$$E = \frac{F \times v}{f \times V} < 100 \text{ per cent}.$$

Thus we have in summary a little triumvirate of products of force by various other simple things:

$F \times t$, force times time, called IMPULSE, measures momentum, mv.

$F \times S$, force times distance, called WORK, measures energy, potential or kinetic.

$F \times v$, force times velocity, called ACTIVITY, measures power.

CHAPTER 13

OUR TWO OCEANS—
ATMOSPHERE AND HYDROSPHERE

MUCH to our regret, we have to break the continuity of our story for a moment at this point—at least, it might as well be at this point as at any other. If by any chance you cannot restrain your impatience to be done with mechanics and to enter upon another theme for a change, you may skip this chapter for the present. You will find, however, that mechanics is not so easily to be avoided. It will still dog your footsteps as you study about heat. Indeed, the latter is introduced through the agency of friction that gives rise to it, and in the last analysis heat turns out to be a mechanical phenomenon of an ultra-microscopic sort.

With respect to electricity likewise, we recognize it by virtue of its mechanical effects, and the various interrelations of electricity and magnetism are largely mechanical. Indeed, in some way or another the language, the principles, and the laws of mechanics play a vital rôle in all but very few subjects in the entire realm of physical science. They are to be found implicit, if not explicit, in not a few places in biological sciences as well. The result of our better understanding of these laws and such control over natural forces as we have attained thereby has had a profound effect on the social

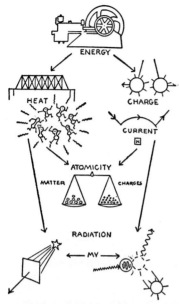

and economic structure of the world, so that purely historical, humanistic, and philosophic perspectives of our culture in the past can be fully appreciated only if we are acutely conscious of how mechanized our lives have become in contrast to the lives of former eras.

BY FAR the greatest part of the matter in our environment on the earth is not solid but fluid, either liquid or gas. Whether we live on land or at sea, we spend our entire existence at the bottom of a vast atmospheric ocean. Beneath this, and supporting it, over by far the greater part of our little planet, is a water sphere, the hydrosphere. We are much more limited, indeed, than fish, who in their fluid medium may travel up and down, as well as north, south, east, or west, i.e., in the three dimensions that make up volume; whereas we live in the two-dimensional flat land of area. Most of us never leave the bottom of our ocean's bed for long or far. The fact that most of us, like fish, have never been very conscious of our abode in a fluid medium does not alter the fact at all.

THE ocean of air, as well as the ocean of water, is material, as we understand and use the word. It is subject to the earth's gravitational attraction, and it thus has weight. This weight, though we are quite unconscious of it, is not inconsiderable. A box that is no larger than 3 feet on a side, inclosing 27 cu. ft., contains more than 2 lb. of air. An ordinary-sized room contains about 100 lb. of it, and a large lecture room more than a ton.

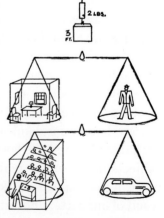

In considering the weight of a *free* fluid like air or water we are faced with a slightly different situation than any we have previously met.* If the liquid can be put into a bucket, it is not free and there is nothing different in the business of

---

* The weighing of a finely divided solid like loose sand would present quite a similar problem.

ATMOSPHERE AND HYDROSPHERE

weighing it from what we do when we weigh cabbages or rocks or any other solid material. We weigh the bucket with the liquid in it and then without it, and the difference is the weight of the liquid. In a sense the water in a swimming-pool or in Lake Michigan or in all the oceans is not free either—the container is larger, that is all. With respect to the total weight of large bodies of water we have little interest. Knowing that a cubic foot of fresh water weighs 62.4 lb.,* we could calculate the cubic feet in any large body and, by multiplying by this amount, find its total weight if we should be curious about it. But what shall we do to find out about the weight of this ocean in which we are perpetually submerged?

When we are submerged in water we at once become conscious of its weight—the more so the more deeply we are submerged—we feel the PRESSURE. What do we really mean by this word "pressure," and what has it to do with the weight of the fluid medium that is exerting it?

A fluid, be it either a liquid or a gas, differs from a solid in its complete lack of any *rigidity*. No portion of it is constrained to occupy any particular position with respect to any other portion of it. Within itself, in its own body, within the confines of its container of whatever size, it is perfectly free to flow and move about. This is what the word "fluid" means.

Thus, whenever a body of liquid is at rest, every portion of it must be in equilibrium, with any and with all the forces that may be acting on it completely balanced. If it were not, being fluid, it would flow until equilibrium was established; and not until then would it be at rest, as we assumed it was.

If we picture an imaginary cylindrical space of 1 sq. cm. in cross-section extending for a distance h cm. downward into a liquid which is at rest, the entire weight of the liquid in this column, a downward force equal to its mass, m (gm.), times g (cm./sec.^{2}), must be supported by an equal upward force

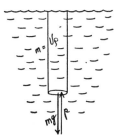

exerted by the liquid outside the column on the bottom surface of the column. This supporting force must also equal mg (dynes). Now the mass of the liquid is its volume times its density.† The volume of the column is the product of its base, 1 sq. cm. and its height, h; and if we call the density by the letter ρ, the force due to the weight of the column is

$$f = \rho g h,$$

---

* Salt water is a little more dense: 1 cu. ft. of sea water weighs 64 lb.

† DENSITY is defined as the mass per unit volume: $\rho = \dfrac{m}{V}$ (gm./cc.).

which is also the value of the supporting force of the liquid on the column. This force is distributed evenly over the 1 sq. cm. and this is the pressure of the liquid at this point by definition.*

Thus we have for the pressure at a distance h, beneath the surface of a liquid of density ρ,

$$p = \rho g h.\dagger$$

EXPERIMENT can easily establish the correctness of the statement we have derived. We will take a funnel which is provided with a flexible rubber cover over its larger end and connect the smaller end to one arm of a U-tube containing water which at first stands at the same level in both arms. The other arm of the U-tube is left open to the air. On insertion of this funnel beneath the surface of water in a tank, the rubber cover will be pushed in by the pressure of the water outside—more and more, the more deeply it is submerged. The pressure communicated across to the U-tube will cause a difference of the water-level in its arms that increases in proportion as we submerge the funnel.

Now, of course, we have been using some of the very principles we have been explaining when we transmit this pressure from the funnel to the U-tube; and in noting how a longer and longer column of liquid in the U-tube is supported by the pressure we obtained upon the funnel, we will find that this height of column in the former is just equal to the depth of submergence of the latter, since water was involved in both cases.‡

Furthermore, we will notice that the pressure, as measured in the U-tube, depends not at all on the orientation of the funnel but only on its

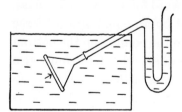

average depth. Of course we have already anticipated this. If the pressure at any point were not the same in every direction, the liquid would flow as a result of this lack of balance and would not then be at rest, as we have taken pains to insist it was, by assumption be-

---

* PRESSURE is defined as force (perpendicular component) per unit area: $p = \dfrac{f}{A}$ (dynes/sq.cm.). Here $A = 1$, thus, $p = f$.

† If the liquid has air above it, the total pressure is what we have just calculated plus the pressure of the air that rests upon the surface of the liquid. This point will be discussed at greater length on p. 97.

‡ Because of the quality of the flexibility of the rubber membrane over a wide range of stretching, this actual experiment is limited to small changes in pressure, if our statement is to be taken precisely.

fore we started the argument and by experimental arrangement before we inserted the funnel into the water.

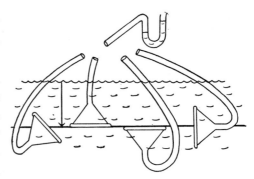

The hydraulic press, used in many ways, is an excellent example of the relation between pressure and force, and illustrates well how pressure applied at one location to a fluid is transmitted elsewhere to all points if the fluid be confined as it was in our funnel and U-tube apparatus.

In the hydraulic press there are two cylinders, each with a tight-fitting piston—one cylinder large, the other small. The cylinders are filled with liquid and connected by a pipe. A force applied to the small piston creates a considerable additional pressure on the fluid. Pressure equals force/area. Here the area is small; hence, for a given force, the pressure may be quite large. This pressure is present everywhere throughout the liquid, not only in the cylinder where it originates, but also in the connecting pipes and in the other large cylinder. In the latter the pressure acting on the bottom of the large piston produces on every square centimeter of it the same force that it received from each square centimeter of the small piston where it was created. On the large piston there are many square centimeters instead of few. On each of these the force is applied, and consequently the total force on the large piston is correspondingly many times greater than that which was applied to the small one.

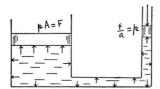

Principle of the hydraulic press

If the diameter of the larger one is ten times that of the smaller, its area is a hundred-fold larger; and a force of 20 lb. on the small piston becomes a force of 2,000 lb. (a ton) on the large one.

Objects submerged only to relatively small distances in water are subjected to forces that total surprisingly large amounts. At a depth of only $1\frac{1}{4}$ mi. or 2 km. (2,000 m.) we calculate the pressure to be in sea water

$$\underset{\rho}{1.1} \times \underset{g}{980} \times \underset{h}{200,000} \text{ cm.} = 216,000,000 \text{ dynes/cm.}^2,$$

which is about $1\frac{1}{2}$ tons per sq. in. Nothing but a heavy, solid, steel ball could withstand this crushing force.

Beebe's Bathysphere

A submarine boat, built as ruggedly as possible, can descend only a few fathoms (1 fathom equals about 6 ft.) deep. If anything goes wrong and it sinks too far, the hull crushes and, not only does the water rush in to drown, but with the water comes its enormous pressure to crush and kill as well. And yet the ocean depths far down we know are populated with finny creatures—fish that live in the darkness of an eternal night and that have developed lights instead of having lost their eyesight.* How can such delicate creatures withstand such frightful pressures? The answer is to be found in the fact that fish are not "water-tight." Water permeates freely within, as well as without; and whither it goes, the pressure goes everywhere and in all directions the same. Thus these organisms are subjected to just as much pressure from within as from without; and by a kind of nonresistance, there becomes nothing to resist—the forces equally great in all directions neutralize and balance. If the fish be removed too quickly from a great depth and if it has some space, like a bladder, inclosed with only a slowly permeable membrane, it might well explode, or at least swell up greatly without bursting. But, so adaptable is living tissue that fish have been brought up from these great depths for purposes of study and kept alive, for a short time at least.

WE OURSELVES live at the bottom of our ocean. It is not as dense as water; but, as we have noted, its weight is not inconsiderable and the pressure it exerts amounts to nearly 15 lb. per every square inch on the surface of our bodies. This is a crushing force on an air-tight tin can, as can easily be demonstrated; but we are not air-tight. Through the interstices of our porous tissues, as well as through mouth, nose, throat, and lungs, the air can penetrate. We are saturated with our fluid medium, as it were; it is in equilibrium, or very nearly so, at all times; and thus its multidirectional forces all balance off to zero.

* Cf. William Beebe, *National Geographic Magazine.* LIX (1931), 653–78.

ATMOSPHERE AND HYDROSPHERE

Sudden changes of pressure which do not give time for all parts of our organisms to adapt themselves to the change are accompanied by most serious results, if not by death. But men can, by slow stages, acclimate themselves to live, and to live vigorously, at pressures under which they otherwise would quickly succumb. Well-known instances are the climbing parties that frequently have attempted to conquer Mount Everest. In deep-sea diving also, this practice is followed.

We have said the pressure of our atmosphere amounts to nearly 15 lb. per sq. in. How, you may ask, do we know this? Surely when one lives at the bottom of the ocean, he cannot conveniently duplicate the practice we went through as we bent over a small ocean of water-like gods, as it were, and inserted in it a funnel with a rubber cover to act as a pressure gauge. No! What little conceit we might have as to godlike proclivities would have to do with our slowly acquired powers of reasoning and a certain ingenuity of invention —not with physical prowess. We can determine the pressure of our atmosphere without ever leaving the bottom of our ocean, by a simple application of the laws of fluids in equilibrium that we have already learned.

Our ocean of air is supported by whatever lies beneath it—land or water, solid or liquid. Suppose we take an open dish of a very heavy liquid, mercury. Why we choose this will develop in the sequel. This mercury bears the weight not only of itself but of the vast column of the atmosphere above it, as indicated previously by footnote on page 94. At a point beneath the mercury surface the pressure is that due to the weight of the mercury above this point plus that of the atmosphere that lies above the mercury. Suppose we now take a glass tube about 3 ft. long, open at both ends, and support it vertically above the mercury with its lower end submerged an inch or two. The mercury within the tube rises through it as we submerge it, and stands at the same level inside as out.* The pressure of the atmosphere is communicated to the surface of the mercury quite as freely through the open tube as it is outside of the tube. This circumstance we will now proceed to change by attaching a good vacuum pump at the top end of our glass tube and pumping out the air therein; then we will seal off this upper end when we

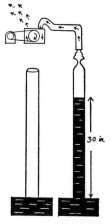

* A slight depression of the mercury level within the tube owing to capillary phenomena becomes quite negligible if the bore of the tube is not too small, not less than ½ inch.

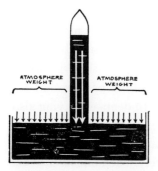

remove the pump. The atmospheric pressure is thereby removed from the surface of the mercury within the tube. The weight of air above this surface is no longer supported by the mercury but now must be carried by the tube and its support. What happens inside the tube as we have done this you undoubtedly have guessed. You might say that the mercury was "sucked up" into the tube when we pumped the air out. Your description is picturesque rather than precise. It is like that of the men who, before the time of Galileo, said that nature "abhorred a vacuum." Galileo's reply was, in effect, that nature undoubtedly did, but not more than 30 in. of it. The pump can continue to suck for all its busy, noisy life, but all evidence of sucking ceases after the mercury has risen 30 in. high.

The pump removes the air within the tube. The atmospheric pressure on the mercury surface outside is no longer balanced by a corresponding pressure inside the tube; the fluid mercury yields and flows until again all forces within its body equalize. It is driven up the tube by the outside pressure until the pressure due to its weight therein exactly equals the outside atmospheric pressure (Plate 19, The Mercury Barometer).

Normally, on the average, it rises about 30 in. or 76 cm. It is 13.6 times as dense as water. Thus the pressure it indicates as equal to atmospheric is $\rho \times g \times h$ again, in this case:

$$13.6 \times 980 \times 76 = 1.018 \text{ million dynes per sq. cm.}$$

As a weight this is a little more than a kilogram for each square centimeter. If we had used water instead of mercury, the experiment would have been awkward to perform, since it takes 34 ft. of water to balance atmospheric pressure.

THE height of this barometer is never constant. As we travel around over our atmospheric ocean's floor and ascend to higher levels of it, the pressure, of course, gets less. Indeed, for every 12 m. ascent near sea-level the mercury column drops by about 1 mm. It is quite easy to notice the change of reading as one carries a barometer up in the elevator of a fifteen- to twenty-story building.* Even in a fixed location the barometer is usually rising or falling. Seldom is it steady.

* Mercury barometers are quite unsuited for transportation—the unstable, heavy liquid, the rather fragile, long glass tube, are handicaps. Thin, metallic capsules, air-tight and with sides suitably connected

PLATE 19
BAROMETERS AND BALANCES

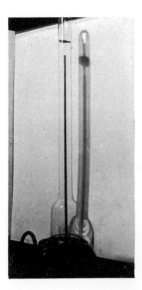

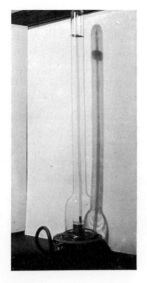

THE MERCURY BAROMETER. Here the barometer, inner glass tube filled with mercury and closed off at the top with its lower end submerged in an open dish of mercury, is inclosed in another larger jar from which the air may be exhausted.

On the left the outer jar is at the outside pressure, which supports nearly 76 cm. of mercury in the inner tube.

On the right evacuation of the outer jar has reduced the pressure so that it will support only about 1 cm. of mercury.

ANEROID BAROMETER. Under a lens the capsule, with corrugated top which, tightly sealed, yields slightly to changes in atmospheric pressure, can be seen. The rest of the mechanism is to magnify this motion and transmit it to a needle.

A PROBLEM OF WEIGHING. Is the air-tight brass bulb really heavier than the brass weight?

The bell-jar containing the balance also contains air.

The air has been removed from within the bell-jar.

PLATE 20 SENSITIVE RECORDING ANEROID BAROMETER AND THERMOMETER. (Note series of capsules.) Used by Commander Settle and Major Fordney in balloon ascension into the stratosphere in 1933.

OUR atmospheric ocean is a turbulent one. The rotation of the earth, the tremendously unequal heating of different portions at the same and at different seasons, the great variations of humidity in different places, the rather large inequalities of the earth's wrinkled surface—all contribute to make the atmosphere disturbed and tumultuous. A fluid at rest it never is. But the properties of moving fluids and the mechanical laws that govern them are likewise known and understood. To follow this thought would take us too far afield, however. The same is true if we should attempt a discussion of how a systematic study of the variations of atmospheric pressure over a continent like North America enables us to some extent to predict the weather for a few days ahead. Such predictions are by no means accurate, as our frequent jests at the weather-man's expense would indicate. The subject is important, nevertheless, and is worthy of a treatment by itself.

---

by levers that magnify their bulging out or bowing in, can be used to measure changes of air pressure. They are known as "aneroid barometers" and are the common type one sees around. All are graduated and checked against the ones using mercury. (See Plates 19, top right, and 20.)

ATMOSPHERE AND HYDROSPHERE

THE chief and fundamental reason, however, for the turbulent nature of our atmosphere is the fact of its being gaseous. Gases are less stable even than liquids. The outstanding mark of difference between these two kinds of fluids is to be found in their respective compressibilities. Liquids are so difficult to squeeze into any smaller volume that we may say for our purposes they are entirely incompressible. Anyone who ever had to pump up a tire knows what an enormous lot of air at atmospheric pressure can be squeezed within that inner tube protected by its sturdy casing and still not make the tire so very hard.

The most obvious result of this difference in the compressibility of these two types of fluid media is witnessed in their densities. Water, except for minor variations due to temperature, is always about the same in density. At sea-level or a mile below it there is little variation. Not so the air.

Being readily compressible, the weight of the upper regions of the atmosphere bears down and compresses that below it. This, in turn being compressed, is more dense, and heavier, and compresses still more greatly the layers underlying it. Thus, we get a rapid increase of density as we reach the bottom of the atmosphere. Under temperature differences, as we shall see later, gases expand and contract much more than solids and liquids.

Variation of density, air and water

Our barometer registers the pressure due to the total weight per square centimeter of surface of the air in its vicinity. This depends on height of the atmosphere, of course, but this height is a very indistinct affair—the atmosphere gets thinner and thinner and has no sharp outer boundary. Its weight depends on its density as well; but this, in turn, depends on the temperature at every point, as well as on the pressure of the overlying layers.

We can give you, thus, only a rather vague picture of this variable and turbulent atmospheric ocean from top to bottom after all. Let us confine our attention now for a few moments to the region where air and water meet.

At the interface between our two great fluid oceans, atmosphere and hydrosphere, centers practically all of mankind's activities upon the latter. The seven seas are covered with the fleets of human enterprise and industry. Boats used to be made of wood and other light materials. Now they are largely made of steel and other substances much denser than the water they rest upon. They may be made even of concrete. Most of us understand, more or less obscurely, how this is possible; but let us spend just a page or two on the matter to point out the simple origin and applications of a principle, discovered over two thousand years ago by Archimedes (287–212 B.C.).

WHEN we weigh a chunk of iron when it is submerged in water, we find it weighs less than it does in air. Archimedes found how much less. The apparent loss of weight is equal to the weight of the fluid it has displaced or crowded out of the space that was all fluid-filled before the iron was sub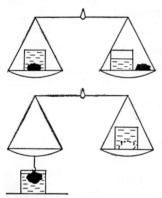merged. Notice our use of the word "apparent" relating to loss of weight. There is no real loss of inertia (mass) of the iron, and no real diminution of the earth's attraction for it gram by gram; hence there can be no real loss of weight. If a beaker of water is on the scales of a balance and we submerge a block of iron therein, the scales will show the sum of the weights of the water and of the iron as previously weighed in air. Only when the iron alone dangles from the balance and a bucket of water is held up from beneath does the iron seem to become lighter. If you perform this experiment with a sufficiently large block of iron hanging from a balance, you will notice that as you submerge it by lifting up a bucket of water around it the bucket in your hands gets heavier. Of course, the bucket must not be so full that it overflows as the iron crowds up the water in it. If the bucket had been initially brim full, then when the iron is submerged, the water will overflow in volume equal to the volume of the iron. The bucket will, under these circumstances, seem no heavier than before, since now the apparent loss of weight of iron, equal to that of the displaced water, has not been added to your arms but has been spilled over, perhaps into another bucket, and is now being supported by Mother Earth directly.

Let us follow the implications of this into our other ocean, the ocean of air, for a moment. The iron was submerged in this when we first weighed it. Did we obtain its real weight then? The answer depends on how

ATMOSPHERE AND HYDROSPHERE

we weighed it. If we weighed it on a spring balance, we did not. If, on the other hand, we weighed it on a beam balance and balanced it by putting on the other end some weights that were the same in volume as was the iron, then, and then only, was its true weight obtained. This assumes, of course, that the balancing weights are marked with their true weights, which we might be apt always to take for granted. In very accurate weighing we take nothing for granted, and all weighings must be corrected for the buoyancy of the weights and of the object weighed (Plate 19, bottom).

The buoyance in air does not amount to much more than a few grains in ordinary weighings, often less than the accuracy of the rather crude balances employed; so we will not trouble you longer about it.

Possibly our use of the word "buoyance" of a metal block in air surprised you. You think of buoyancy in connection with something that floats. Metal, such as brass or iron, certainly does not float in water; and as to floating in air, you probably never thought of anything doing this except balloons. Buoyancy, nevertheless, is always present when any solid object is immersed in a fluid. The buoyant force is just that apparent loss of weight discovered by Archimedes and which on a block of iron in air is indeed quite negligible. It is quite perceptible, however, on iron in water; and if iron is submerged in mercury, the buoyant force is greater 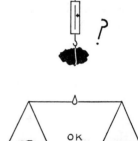 than its weight: it loses apparently more weight than it possessed; it goes up instead of down. Iron floats on mercury, just as wood floats on water. If the weight of the displaced fluid is less than the weight of the solid, the solid sinks; if, however, the weight of the displaced fluid is greater than the weight of the solid placed in it, the solid will sink, *but only until a certain fraction of its volume is submerged.* This fraction will be such that the weight of the liquid which has been displaced exactly equals, and so balances, the weight of the solid. The solid then comes to rest, i.e., floats always partly submerged, sinking a little deeper if more weight is added to it, floating higher in the water if some weight is removed.

It all reduces thus, you see, to a question of the relative densities or specific gravities of the solid and the fluid it is in. Solids sink in liquids less dense than themselves, and float in all others. How far they sink into the liquid on whose surface they float can be used to determine the density of

the liquid, or its specific gravity,* on a calibrated stem as in the common HYDROMETER.

In the case of a liquid, usually almost incompressible, the density, as we have seen (p. 101) is approximately constant from the top surface to the bottom. A given volume of it, displaced by any solid object that sinks completely in it, represents a constant mass and, therefore, a constant buoyant force, although, of course, one insufficient to equalize the weight, since the object sinks.

The object sinks always to the bottom if it sinks entirely below the surface. The only way in which it could sink part way would be for the buoyant force to increase as the body descended, i.e., for it to displace a greater weight of water as it went down until a point was reached where it displaced its full weight of water. Gravity being constant within very small limits, the mass of the displaced water would have to increase, which could happen only in one of two ways: the body might swell up and so displace more volume, or the fluid around it might get more dense and so have more mass in a given volume. With ordinary solids and a homogeneous liquid of course neither of these things can happen.

If a solid hangs suspended, like Mahomet's coffin, in the interior of a liquid, you may be sure either that there is some concealed constraint upon it or that the liquid is not homogeneous, and that the object is really floating on the surface of a heavier liquid that lies beneath a lighter one.

The case of the atmosphere is altogether different. Being gas and readily compressible, its density increases rapidly as the lower levels are reached. A balloon filled

---

* SPECIFIC GRAVITY is defined as the density of a substance compared to that of water.

On the metric system, since the gram is defined as the mass of 1 cc. of water, density and specific gravity have the same numerical measure.

$$\text{Specific gravity} = \frac{\text{density}}{\text{density of water}} = \text{density} \div 1 \ .$$

On the British system the density of water is not 1 but 62.5 lb. per cu.ft.

ATMOSPHERE AND HYDROSPHERE 105

with hydrogen or helium or any gas lighter than air thus displaces much more than its own weight of air at ground-level. But as it rises, even although the gas within it may expand its envelope so that it swells up, the density of the surrounding air decreases ultimately to a point where even the now much larger volume of the balloon can displace only its own weight or less. Balloons thus have a definite "ceiling" higher than which they cannot float.

WE WISH finally to leave one trick question in your minds—one you can answer, we feel sure, but one by which you may puzzle the uninitiated.

Suppose one has a cork floating on the surface of water in a jar from which the air can be exhausted. When the air pressure on the cork is removed by evacuating the space above the water, will the cork rise or sink?

PART II
HEAT

CHAPTER 14

TEMPERATURE AND EXPANSION

THE discussion of mechanical phenomena, in view of the ever present annoyance of friction, leads us inevitably to the consideration of heat. Heat always appears in connection with the invariably incomplete transformations of energy from one mechanical type to another, viz., from the potential to the kinetic form. In chapter 9 we discussed at some length the fact that there is a constant relationship between the number of ergs of energy or work that are lost to usefulness in any mechanical operation and the number of calories of heat energy that appear instead. The calorie was defined in terms of heat required to raise the temperature of an amount of water, without any explanation being given of why we should have adopted so peculiar a definition. The only comment was a footnote appealing to the reader's patience.

WHY all this mystery about so obvious a thing as heat, anyway? Doesn't anyone know what is meant by heat and cold? Surely we can tell by touch something about temperature, and our general bodily sensations seem sufficient to inform us when to bundle up our ears. Of course,

* In one of the earliest hot-air balloon ascensions a cock, a duck, and a sheep were sent up to discover how they stood the trip.

The dependence of our bodily sensations upon amount of moisture in the air. *Above:* In a moist room one can be comfortable at 65° F. *Below:* Feeling chilly at 80° in a dry steam-heated atmosphere, and yet feeling too warm at 80° in humid summer weather

our grandparents shed or put on the famous red flannels by the calendar rather than by trusting either the weather man or their bodily impressions. This is all very well as far as it goes; but have you not outside your window a thermometer to which you appeal for a little more precise information about temperature at times? Do you know that how hot or cool your living-room feels depends quite as much upon how moist the air is as upon its temperature as indicated by a good thermometer? Most living-rooms in winter are drier than the Sahara Desert; and for all of those that are, thermometers usually read near the eighties before anyone thinks of feeling too warm. Yet, how we grumble at the humid eighties in the summer time! Even granting we must have something to complain about, there is more behind this behavior than can be laid at the door of the cussedness of human nature.

The sense of touch or general bodily sensation is an utterly unreliable means of measuring temperature for everyday needs as well as for scientific purposes. In direct contact with solid objects the sense of touch registers not the temperature but the rate at which heat is withdrawn from or conducted to the hand. Our bodies, usually much warmer than the surrounding air and clad in garments getting ever lighter and more porous (except in the case of the conservative male attire), are always cooled largely by evaporation,* swift and chilling in desert drought, almost non-existent with high atmospheric humidity.

THE only other sense we have that can form any impression of temperature at all is that one most marvelous of all, the sense of sight. We can not tell the temperature out-of-doors by looking at the window. At low temperatures our eyes are blind to temperature. At quite high temperatures this is no longer true. A stove may be so hot that we can feel its radiant

---

* Explanation as to why this is true is given later, p. 165.

TEMPERATURE AND EXPANSION

warmth across a room, and yet we may not see it in the dark. Let it get a little hotter, however, and it begins to glow a dark deep red. But this you thought was light? It is, and it is heat as well, because the color of the light depends upon the temperature. Furthermore, where visible light and invisible heat blend and mingle as here, the matter of defining which is which becomes a rather arbitrary business.

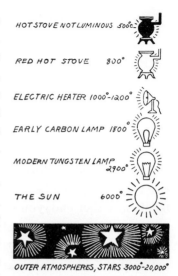

Everyone knows that a stove as it gets hotter gets cherry red in color and then takes on a yellower tinge. We take drastic steps to cool it off usually before this stage is reached. Let us turn our attention instead to the electric heater coil in front of a copper mirror, so popular on chilly days before the furnace is started. It acquires a pronounced yellowish color. Our incandescent lamps are hotter still. The obsolete carbon filament type, Edison's famous "light within a bottle," is very yellow compared with modern tungsten lamps. By the same token these latter must be more of a blue white. At temperatures higher than those of modern lamps most substances melt or evaporate. At higher temperatures there are the stars to give us testimony. Perhaps you know Vega, overhead early after dark in midsummer in the northern portion of the United States. If you do, you know it is a blue star. It is blue in spite of the fact that our atmosphere turns yellowish the color of everything that shines through it. Vega is very hot.

We shall see later (pp. 416–18) that we can, by instrumental means, tell temperature better with our eyes than with our fingers; but we can already understand that this applies only to a higher range of temperature than that involved in our immediate environment. For this lower range touch is inadequate and eyesight avails nought. Indeed, we have no way as yet to measure ordinary temperatures directly, nor is such a device likely to be invented very soon.

Our method, therefore, must be indirect, and all the common methods that we use depend upon the fact that inanimate objects, almost without exception, expand and contract with changes in temperature.

Next time you motor across a long suspension bridge regard the roadway carefully. You will certainly find one or more places where the ordinary

surfacing is replaced by a sheet of metal. Stop and investigate and you may be astonished to find there is no bridge girder underneath, but only empty space and that the structure is actually discontinuous at that spot. Examine the discontinuity on a hot summer day and later on a cold winter afternoon. You will find the gap has greatly changed in size. All bridges must be built to allow for expansion and contraction, or else they would not be open to traffic long. If the structure rests upon abutments at either end, you will find that on one of these it is not fastened but rests on a roller or some other yielding device to allow it to alter its length as the seasonal, and even the daily, temperature rises and falls. This is true not only of bridges but of any large engineering structure built of rigid metal. Building materials of other sorts usually more elastic do not need the same allowance in their construction for expansion and contraction, but nevertheless they expand and contract.

IF WE are to make expansion and contraction, then, a means to measure temperature, two fixed and definite temperatures must be selected between which the expansion can be measured so that by means of this property other temperatures may be determined. Early workers selected the lowest temperature during a certain very cold winter as one extreme and the body temperature of cows and deer as the other. This had two objections—not every winter was subsequently as cold, nor was the upper temperature perfectly constant in all animals. Later the temperature of an ice-salt mixture and that of the human body were selected. The former was quite easy to reproduce approximately, but with respect to the latter somewhat fevered subjects must have been used, since 98.6 and not 100 is the human normal temperature. This was the origin, nevertheless, of the Fahrenheit scale still in common use.

Various fixed points that have been chosen for the measurement of temperature

TEMPERATURE AND EXPANSION

Two other very suitable temperatures, however, are always generally available, those of melting ice and of boiling water—the former, constant under all conditions, the latter unfortunately depending on the atmospheric pressure (see later, p. 167). The connection between the boiling-point and atmospheric pressure, however, can be found; and its variation for pressures

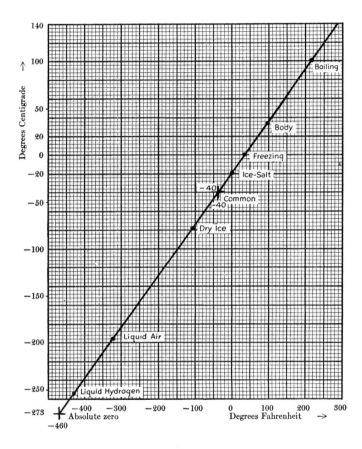

different from the standard pressure (76 cm. height of the barometer) is readily tabulated for reference. Melting ice and boiling water, then, give us the fixed points of the Centigrade scale, which, like the Fahrenheit, is also divided into a hundred degrees between the two. The range between its two chosen temperatures is larger on the Centigrade scale; its degrees are nine-fifths the size of the others. The zero points differ by 32° F. or 18° C. To change over from one scale to the other is a nuisance, but the accompanying graph will make the task easy for you. You will note that when it is 40° below zero you need not specify which kind of thermometer you mean; −40° is the same temperature on each scale.

WITH the fixed points determined and named 0° and 100°, we next need but to select a bar of any common substance, preferably of metal, measure its length in ice water and in boiling water, and divide the difference into a hundred parts. To find the value of any other temperature we again measure the bar at that temperature. Of the hundred intermediate lengths of the bar between the 0° length and the 100° length we can now pick out one that most nearly corresponds to our last measured length. We now take it for granted that the number that goes with this length (suppose it is the seventy-third of the hundred divisions) is the number (i.e., 73) that represents the temperature we wish to find (i.e., 73° C.). If we have a microscope and other apparatus at hand, it ought not to take us more than five minutes or so to do this, after we have spent a day or so finding the lengths corresponding to the two fixed points. The equipment needed is unsightly on the parlor table and inconvenient to work outside the window, observing from within.

A bridge-tender, collecting tolls from passing cars, could use his bridge in all seasons as a thermometer much more conveniently if only he could have once checked and established its length at two known temperatures which, if some other scale existed, need not be 0° and 100°. We cannot all have bridges or complex contraptions sitting around for such purposes. Even in a scientific laboratory, one would not find a thermometer like the one described above; it's too inaccurate. Solid objects do not expand uniformly at all temperatures. Of course, if there were nothing better, a laboratory could determine the degree of lack of uniformity and then, making its observations and calculations accordingly, get an accurate result.*

THE inconvenient lack of uniformity of expansion rate of most solids and liquids, while rendering them rather unsuitable for use as precise thermometers, nevertheless does not prevent their finding the greatest favor in temperature-controlling devices. The modern gas or oil heating system, that keeps your home at an even daytime temperature and at night

---

* An iron bar 1 m. long in expanding from 0° to 100° changes in length about 1 mm. The lack of uniformity over this range is represented by about 5 per cent of this.

TEMPERATURE AND EXPANSION

automatically cuts down the heat, functions by means of a little device called a "thermostat." Although the varieties of thermostats are almost as many as the number of inventors, the same principle is used in all.

Thermostat

No two substances expand alike. Brass expands nearly twice as rapidly as iron with rising temperature. Two strips, one of each metal, riveted or soldered together, if flat and straight at any one temperature, will become bent and curled up at any other—concave (hollowed out) on the iron side of the strip at higher temperatures, and on the brass side of the strip at lower temperatures.

Many thermostats are quite as simple as this. The bending of the strip one way or another with the temperature is made to make and break electric contacts and thereby electrically operate drafts and dampers and valves in heating plants. Other types of thermostats consist of little metal capsules, thin-walled, elastic, and filled with liquid. These can be made to swell up or shrink in size by the expansion of the liquid or its vapor as the temperature is altered. With them sufficient force can be developed so that they can do work and actually perform small tasks such as opening and closing shutters on automobile radiators.

The thermostatic mechanisms in the human body keep its temperature within a degree of normal in spite of activities and outside temperatures

In all nature the most marvelous thermostat that we know is the mechanism in our bodies that, winter or summer, indoors or out, waking or sleeping, maintains our body temperature within a fraction of a degree at 98.°6 F., 37° C. A most complicated mechanism it is, not only making use of the simpler physical properties that we shall discuss in later chapters but involving numerous and complex biological mechanisms as well.

CHAPTER 15

THERMOMETERS

ONCE again, we must pay tribute to the genius of Galileo. Historical studies of only comparatively recent times have made quite certain that it was he who made the first thermometer. For a long time it has been known that he chose the one type of substance, above all others, that had the ideal constant rate of expansion—a source of difficulty discussed in the preceding chapter. Galileo's was a *gas* thermometer of the utmost simplicity. He inverted a bulb filled with air and supplied it with a long stem that dipped beneath the surface of a liquid. First he warmed the bulb and drove out a little air. As the bulb cooled, some liquid ascended into the stem; and thereafter, as the temperature of the room rose or fell, the height of the liquid column fell or rose accordingly.

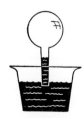

Galileo's thermometer

There is from our vantage point of subsequent centuries of experience but one objection. The air pressure on the liquid (in the vessel into which the stem dips) is not constant (cf. chap. 13, p. 98), and the liquid column responds to this pressure as well as to changes in temperature of the air within the bulb. Galileo invented a thermometer and a barometer in one, but of the latter virtue he never suspected his apparatus. Indeed, it can hardly be called a virtue, for, without some additional instru-

THERMOMETERS

ment to detect the two effects combined in a different degree, it would be impossible to separate them.

In the hands of succeeding workers only the form of his instrument remained. It is the form of our common thermometer today. The one great virtue of Galileo's experiment—his use of air—was, however, unrecognized; his device was next inverted, its stem pointing upward, and bulb and stem were filled with water—a liquid, not a gas. The Florentines used alcohol later and sealed the tube. Although their choice of fixed points was rather quaint (cf. drawing on p. 112), they observed that the melting-point of ice was constant, 13.5° upon their scale. So rugged were their instruments that one of them, accidentally uncovered after burial for over two centuries, still recorded melting ice as 13.5°.

TO FAHRENHEIT, who discovered a method of cleansing mercury from impurities, we owe our present type of instrument, as well as its very inconvenient scale, abandoned long ago in scientific usage. Alcohol still is used in modern thermometers, since its expansion is greater than that of mercury; but it has other properties that are disadvantageous. But irrespective of what substances are used, any combination of a liquid in a solid suffers from the same drawback that we mentioned in our hypothetical expanding-bar thermometer. The solid and the liquid both expand and contract in volume with temperature change. The liquid shows this property to a greater degree than the solid substance of the bulb. The height of liquid column in the stem, rising with the temperature, indicates how much more the liquid is expanding than the solid. But just as the change of length of a solid bar or bulb is not constant in rate, neither is the change in volume of a liquid a constant affair. Furthermore, liquids in general, much more than solids, show a variable expansion rate. When the solid substance is glass, we have made a very bad choice of solid, for the expansive properties of glass are most annoying. As Galileo used it, inclosing highly expansive air, this made far less difference. But when the glass contains a liquid differing but little from it in expansibility, the bulbs must be large and tube bores very fine to give an open and readable scale; and the inconstancy of rate of expansion both of solid and liquid make such thermometers utterly unreliable for accurate work. Furthermore, glass, after being heated, is slow to return to its subsequent dimensions after cooling. Sometimes it never does this. We have, in a former connection, mentioned that Joule studied the variability of one of his thermometers for thirty-eight years. We have just explained the reasons.

Modern science therefore has returned to the technique of Galileo. The standard thermometers against which all others are checked and calibrated are gas thermometers. Within the bulbs, that need not be made of glass, the gas is kept accurately at a constant volume. The variation of its pressure as the temperature is changed can be measured with great precision. The photograph (Plates 21a and 21b) shows a simple

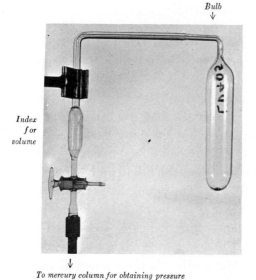

PLATE 21a Detail of Gas Thermometer

PLATE 21b A Standard Gas Thermometer

laboratory type that does not resemble any thermometer you ever saw; yet, like the master-clock at the observatory which those outside the staff seldom see, this is the type of instrument against which every other kind of a thermometer is, or should be, calibrated.

Not only do the common gases show constant and uniform expansion over great ranges of temperature (hydrogen and helium are a little to be preferred to air in this respect), but, if kept at low pressures, they do not liquefy or freeze even at the lowest temperatures we can easily obtain. For extremely low-temperature work (see later, p. 131) they are the only substances that can be used.

THERMOMETERS

WE MUST mention two other very important types of thermometers, widely used in large-scale industrial processes, convenient for both low and high temperatures. Both are electrical in character. They utilize electrical properties of metals which, like the length of a metal rod, alter with the

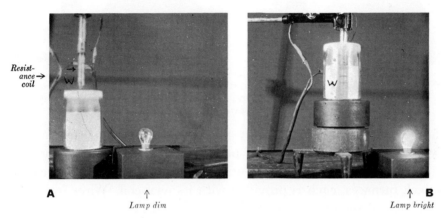

PLATE 22 An Electrical Indication of Temperature. Experiment to show that *electrical resistance* alters with temperature. In (A) the current through the small lamp is reduced so that the lamp glows but dimly owing to the resistance of a coil of wire (w). When cooled in liquid air (B) to $-320°$ F., the resistance of this coil is so reduced that the lamp glows brightly.

temperature. The "resistance thermometer" is one of these, and its method perhaps can be understood from the two accompanying photographs taken in the laboratory (Plate 22). A small flash-light bulb is connected to its battery, but not directly. Between the bulb and battery is a coil of fine wire that offers so much resistance to the flow of current* through the bulb that the latter, instead of glowing brightly, is so dim that it can hardly be seen in the left-hand picture. In the other illustration we have photographed the same apparatus when the resistance coil has been immersed in liquid air and cooled to about 320° F. below zero. It has so much less resistance at this lower temperature that it now is unable to impede the current through the lamp, which glows almost as brightly as if the coil were not present in the circuit.

Coils of wire therefore can be used to measure temperature if arranged in an apparatus that records or indicates their electrical resistance. It happens that this electrical property is one of the easiest to measure, and, furthermore, such measures can be made quickly and with very great precision.

*The electrical resistance of metallic conductors to flow of current is discussed in Part IV, chap. 28.

PLATE 23 ELECTRICAL RESISTANCE THERMOMETER AND EQUIPMENT. *T*, Thermometer coil inside porcelain tube; *B*, batteries; *G*, galvanometer to indicate current; *C*, box of coils and connections to measure resistance of coil in *T*; marked in degrees C, $-200°$ to $+1,000°$.

For this reason the devices are very practical. Resistance thermometers, usually in the form of little coils of platinum wire contained in porcelain tubes (Plate 23), are often found in furnaces. Where there are a large number of such furnaces, each is provided with its own coil. Wires from each lead to a central room where one person, by means of a single apparatus which he can connect to the coil from any furnace, can measure the temperature of each in turn.* Such thermometers are just as suitable and accurate at temperatures like that of liquid air, far below zero, as they are at white heat inside a furnace.

THE other type of thermometer is also a sort of long-range affair, but in a somewhat different sense. Resistance thermometers can be read electrically from afar; but they, like any other thermometer with which we are familiar, must be in contact with the object whose temperature is to be measured. Now we are to tell you of one where this is no longer necessary. Its use and development belongs to our own age and generation. The simple truth about its achievements is one of the fairy tales of modern science that are almost impossible for the layman to believe. You may believe what follows or not; nevertheless, it is true.

Another electrical property of metals must first be described. Imagine an iron wire, marked *Fe* (abbreviation of the Latin *ferrum*, meaning iron) in the figure. To its ends is attached a similar copper wire, marked *Cu* (ab-

---

*The measuring device *C*, in Plate 23, is an adaptation of the Wheatstone bridge for measuring resistance. A handle moving an index around a dial enables the observer quickly to find a "balance" where no current from the battery, *B*, flows through the current-indicating galvanometer, *G*. One reads the temperature of the coil in *T*, then directly from the dial, calibrated previously from $-200°$ C. to $+1.000°$ C.

THERMOMETERS

breviation for *cuprum*, Latin for copper). Inserted in the circuit thus created is an instrument, called a galvanometer, diagramed as a circle with an arrow that will indicate the passage of electric current through it and the two wires. But, you say, there is no battery or dynamo or anything else there to generate a current. That is where you are wrong. A current almost always will be found flowing in this circuit when A and B (the two junctions) are at different temperatures. We

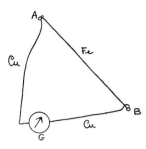

Circuit for a thermojunction

will dip A permanently into a bath of melting ice to keep its temperature fixed at 0° C. As we heat B with a flame (Plate 24), more and more a current will flow, at first in one direction as B warms up. This current will reach a maximum when B is 288° C. and then get less and less. When B reaches 576° C., the current disappears; but as B gets above this temperature, the current now flows the other way, and increases indefinitely in this direction until the heating reaches the melting-point of the wires, when the apparatus falls to pieces. We will not try to explain why all this happens. All that men know about it can be found in treatises on electricity and in encyclopedias under the names of "thermoelectricity," "thermoelectric effect," or "Seebeck effect," the latter name in our approved manner of memorializing a phenomenon's dis-

Galvanometer

PLATE 24 THERMOJUNCTION IN USE. A, cold junction; B, hot junction.

coverer. Without fully informing yourself upon these details, can you not see at once that here we have a means of measuring temperature? We doubt if you can appreciate how tremendously sensitive a means this is, however. To realize that, you should know that electric currents are very easy things for men of this modern age to recognize and to detect. They are "right up the alley" of the present generation. If they are a little too feeble to work with directly, as in the radio, almost any modern youth who has followed this fad will be able to tell you how to amplify them a couple of hundred thousand, or a half million, times. The faintest breath of electricity stirring anywhere around can be made to set off shrieking sirens, open doors, start elevators, dump a few thousand tons of coal, or perform any other little task you have around.

THE currents in this little circuit, then, can be next to nothing; but if they exist we can find and measure them. Now, when we used a flame to heat the wire, we were just crude and rough. We set up comparatively prodigious currents, hundreds of thousands—yes, millions—of times larger than were necessary. All we needed to do was to let our thermojunction, as we call it, simply "see" the radiation from the flame, and its tiny currents would have been set flowing by the sight—a thermometer that functions at long range it literally is. Ten or fifteen years ago these devices were made sensitive enough to count the number of candles in a birthday cake several miles away, i.e., to record the fact, for example, that you extinguished one of ten candles at that distance. The one we regularly use for lectures (Plate 25), a very coarse, crude affair, responds even to a cake of ice,* a thing generally conceded to have little warmth. In the hands of one or two wizards in their use, thermojunctions, or batteries of them, called "thermopiles," have performed marvels. In Plate 24 is a fair-sized thermopile made by Coblentz, one of the wizards. In the enlargement (Plate 26) the row of tiny junctions may be seen above the arrow. Single junctions often made of wire finer than the finest hair can be planted in the bodies of the smallest insects, functioning there as tiny thermometers in the normal sense, taking up instantly the temperature of their environment.

When talked to, these thermojunctions can measure the heating and cooling produced in the air by the sound waves from the voice, and then

---

* The thermojunctions lose heat to the ice of course, are cooled thereby, and generate the corresponding degree of current.

THERMOMETERS

PLATE 25 Thermopile (inside of Case T) "Responding" to a Cake of Ice. Note deflection due to electric current in G.

electrically draw faithful pictures of those features that make one voice different from another. In their long-range aspect astrophysicists use them in telescopes to explore and search out the heat of stars and planets. They can be made sensitive to the heat of stars of the sixth magnitude—stars under ideal conditions just visible to the unaided eye. With them also we can go out into space and measure temperatures in other worlds than ours.

Applied to the moon, they show that at high noon there after fourteen days of sunlight, uninterrupted by any cloud, unsoftened in intensity by any atmosphere, the surface rises to the blistering heat of 240° F! During the long lunar night, of course, the temperature falls to several hundred degrees below zero.

The planet Mars also has been studied. Provided with an atmosphere of sorts—rather scanty com-

PLATE 26 High Sensitivity Thermopile by Coblentz. Enlargement of apparatus shown at c in Plate 24. Arrows point to row of junctions.

pared to ours—its surface temperatures are not quite so extreme. At sunrise the temperature is about that of dry ice (solid CO_2), $-85°$ C.; but by noon it warms up to about 10° C. above zero (50° F.). The polar cap, of snow and cloud, almost never disappears at Mars' north pole. The south pole enjoys a greater diversity of temperature, and ranged on one occasion from $-100°$ C., on August 14, 1924, to $-15°$ C. on the following October 22.

Jupiter, Saturn, and Uranus also have been studied. Of these great worlds we never see the surface, only the upper regions of their perpetually cloudy skies. The upper air at home, on Earth, we know is cold; but these larger outer planets are much farther from the sun than we are; consequently, it's not surprising to learn that the upper levels of their cloud surfaces are $-130°$, $-150°$, $-170°$ C., respectively, although it is rather startling to know that man's ingenuity has enabled him to get there, so to speak, and measure these temperatures.

IN THE form of thermojunctions and thermopiles, thermometers find their highest expression; they become exalted from bits of metal into the playthings of gods. It seems a far cry from Galileo's inverted bulb of air to these tiny affairs often visible only under a lens and yet potent enough to reach far out into space and explore temperature there. It is a far cry, and yet it is but the result of patient careful steps taken one after another, of phenomena diligently examined, recorded, thought about, and analyzed, of a quality of attentive curiosity that pokes around in odd corners and stirs things up, watching everything that takes place with hawklike scrutiny.

CHAPTER 16

TEMPERATURES OF VARIOUS PLACES

IN THE preceding chapter we have shown how it is possible to learn something about the temperatures of other members of our solar family of planets and satellites. Even the stars, from the nearest of which our sun is tremendously far away, yield to the telescope and thermocouple some information as to the temperature of their outer atmospheres. Hotter than Vega some of them are, with incandescent gaseous outer envelopes as high as 25,000° C. From this value downward to about 3,500° we find representative classes of stars at all intermediate temperatures. Our sun is, after all, only moderately hot, as stars go. Its 6,000° C. are fully adequate to warm us at a favored distance from it. The nearer planets are much too close to the sun for comfort; whereas, upon the outermost ones, his rays shed but little light and heat.

With respect to the interior temperatures of stars, or of the sun, or even of the earth beneath our feet, we have no direct knowledge. As we go down

a mine the temperature rises about 1° F. for every 100 ft. descent. At a very short distance down, the daily changes or seasonal changes at the surface have no influence, so that the temperature year in and year out is constant. As we descend into the deepest mines, we find that they become about as hot as the most torrid surface temperatures. These facts lead us to believe that far down in its interior the earth is very hot. A rate of increase such as we observe near the surface, if continued to the center, would lead to a value of over a hundred thousand degrees F. for the central temperature of the earth. That it is as high as this seems quite beyond belief. What the facts are, however, we have no means of knowing. Lavas and other material ejected from volcanoes, which come from regions not far below the surface of the earth, are already heated to incandescence and lend support to the hypothesis of a moderately high central temperature. The most recent studies and theories suggest 4,000°–5,000°, and add the additional item of interest that during the last two billion years only a layer of 200 km. thick on the outside has appreciably cooled off.*

ONE thing is certain. The interior of the earth, however, is not molten, as was formerly supposed. Hot as it is, the pressures of the overlying billions upon billions of tons of weight is so great that melting is not possible unless this pressure is relieved. Of this solidity we have experimental evidence. The earth, as it spins upon its axis, wobbles slightly, as does a spinning top. The period of the wobble is very long, two hundred and fifty centuries are required to see one wobble completed. This fact in itself betrays that the earth spins as a rigid unyielding body, in its entirety quite as rigid as if it were made of steel. But in addition to this evidence of the manner of the earth's rotation, gained from observations on the stars, we have also the results of experiments upon Mother Earth deliberately planned by her human off-

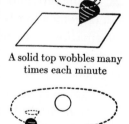

A solid top wobbles many times each minute

The solid earth but once in 25,000 years

---

* For a most interesting account of how physicists, armed with thermometers and given a mine to explore, can read past thermal history of the earth's surface for the last twenty to thirty thousand years, cf. Hotchkiss and Ingersoll, *Journal of Geology*, XLII (1934), 113.

TEMPERATURES OF VARIOUS PLACES

spring for the purpose of measuring her solidity. The tides raised by sun and moon upon surface waters find a counterpart in tides raised similarly within the solid ground itself.* If the whole earth were water, we could live only in boats. In boats we would never be aware of tides at all, for we would be rising and falling gently with them all the time.

If, on the other hand, the earth were composed of some absolutely unyielding substance, in the rise and fall of the ocean tides we would see the full effect of the sun's and moon's attraction. Since we have knowledge of the laws of gravity, and of the masses and the distances of sun and moon, we can calculate how great the tides should be. These calculations should check with our observations if the body of the earth were totally unyielding. They do not. The water tides appear to rise and fall only about seven-tenths of their calculated value. This can mean nothing less than that the solid ground beneath our feet, with reference to which we measure tides, yields just as does the water, but to a less degree. Thus, by subtracting the observed tides in water from the calculated values, we are able to find the value of the "earth tides" even in spite of the fact that our homes and laboratories, like boats, are floating, as it were, upon the heaving bosom of the continents.

The earth and water tides

In just the same way that we can test the quality of a spring by putting a measured force upon it and observing how much it yields, so we can know the rigid quality of the earth from the knowledge of how much it yields (the earth tide) to the known forces of sun and moon pulling upon it.

You doubtlessly have been thinking that experiments and observations necessary for this information must have been taken along the ocean shore, probably at some place like the Bay of Fundy, where the tides are great. The facts are quite otherwise and give us again an example that the truth is often stranger than fiction and difficult for the layman to believe.

* These matters are more fully treated in chaps. iv, v, of Stetson, *Earth, Radio and the Stars* (McGraw-Hill, 1934).

THE water tides in question, from observations on which the earth tides can be found, are most modest affairs. Imagine a 12-inch pipe 500 ft. long buried horizontally in the ground, and with a pit at each end so that one can descend and look through the pipe. Close the ends, projecting into the pits (Plate 27), with plates of glass and fill the pipe half-full of water. In the little lake thus created, the ebb and flood of the tides can be observed! The amount of rise and fall is about $\frac{1}{1,000}$ in., but nothing more elaborate is needed than an ordinary—indeed, a rather low-power—microscope. It was with apparatus such as this that Michelson* in 1914 measured the rigidity of the earth and found that, in spite of what is probably a very high interior temperature, it is not molten but is as solid as steel.†

If, then, the earth, so cool outside, is so hot within, even though still solid, what about interior temperatures of the sun and of the still hotter stars? This is asking us too much. We admit that science is pretty good on the whole, and that scientific procedures can work wonders, but we have not yet advanced as far as this. If you would take the *guess* of a few individuals who, more than anyone else, for years have thought about these matters, you will be asked to entertain the idea that the interior heat of the stars is to be reckoned in scores of millions of degrees. We reiterate that we have no direct means of knowing this or even of calculating it. Such estimates as have been made involve so many assumptions that they are little better than guesses.

OF ONE thing we may be sure. Very much of the matter in the physical universe exists in temperatures to be reckoned in millions of degrees. At least 50 per cent of all the matter we know anything about is in the stars' interiors. That in such an environment matter is of far less complexity and variety, of much simpler structure and form than that with which we are on familiar terms, is very likely. We exist in a relatively frigid world, where matter, so to speak, has "frozen" into a vast variety of forms. The high temperatures within the stars tend to break down to a very considerable extent the complex structures that go to make up the atoms and molecules

---

* *Astrophysical Journal*, XXXIX (1914), 105–38.

† Further evidence of the great rigidity of the earth is furnished by the waves that originate from earthquakes. The transmission of earthquake waves goes to show that at least seven-eighths of the earth today is solid and very rigid. The central one-eighth, or core, may not be so, but very little certain evidence is, as yet, available about this portion.

TEMPERATURES OF VARIOUS PLACES

of which terrestrial matter is composed. It is in the details of these structures that the general reader of works on modern physics is most interested. We intend to discuss them later (Part IV). The only point we wish to emphasize at the moment is that our position in one of the colder portions of the universe—a position that seems quite comfortable and natural to us who have evolved upward in it—is, nevertheless, a very exceptional position from what little we know of the rest of the vast pageant that Nature has staged.

End of pipe projecting into pit with microscope looking at water surface

An analogy may help to make our position clear. Water is one of the most common things in our environment. As solid, liquid, and gas we are familiar with it. Years of patient study of it and of many other substances have brought us to the knowledge that it is a complex substance composed of two quite different things—two gases, hydrogen and oxygen—which we have independently recognized elsewhere in nature

Close-up of water half-filling the pipe and needle. Distance between needle and its reflection in under surface of water measures changing height of the water tide

PLATE 27 TIDES OBSERVED IN A WATER PIPE

as existing both separately and in combination with other substances. We are at a sufficiently warm temperature in our environment to be familiar with the fact that water exists in these three very different forms—ice, liquid, and steam. This fact in itself has been of very great use to us in getting at the underlying composition of the substance.

But suppose we had lived at lower temperatures and had been familiar with water only in the crystalline form it takes as snowflakes. No two of these are ever exactly alike, it is said; certainly there are a myriad of different shapes all built, it is true, upon the same hexagonal pattern. With only snowflakes available, the analyses that would have led ultimately to the discovery that ice was composed of two gases, oxygen and hydrogen, would undoubtedly have been far more difficult to make, so that we would have been a much longer time in acquiring this knowledge.

The case is somewhat similar with respect to the structure of all the different forms of matter with which we are surrounded. These forms are, in large part, due to the unusual circumstance that we live on the frigid exterior of a tiny planet, attached to the train of an insignificant star lost in the depths of space. Had we grown up in the interior of some one of the countless million other suns, or even in the interior of our own, matter would there have taken on a somewhat more simple guise. Thus is knowledge of temperature a very important thing for us, not only of the temperatures which are characteristic of much of the matter in space which are so transcendentally far above our own narrow range, but also those temperatures which are even lower than those to which we are accustomed.

Some common means of maintaining low temperatures

If we mix ice and salt, as we have seen, we get a temperature which is that of good snug winter weather, 0° F. (cf. also figure p. 112). Dry ice, rather new as a refrigerant for ice-cream chocolate bars and other things, has been known for years as a laboratory product—carbon dioxide in the solid state. It is the gaseous form of this compound (carbonic acid gas, so called) that bubbles out of fizzy drinks, both natural and synthetic, alcoholic and otherwise. Solid carbon dioxide, or carbon dioxide snow, as it is often called, is much colder than cold winter weather. One hundred degrees below zero is not encountered often, even by polar expeditions. Yet this is the temperature of this chilly white solid that sears and destroys, just as surely as does fire, human flesh left more than an instant in contact with it. At temperatures as low as this the mercury within the bulb of a thermometer will freeze as solid as a rock. One can drive a tack, using mercury at this temperature as a hammer.

THE very common air we breathe, by processes we now thoroughly understand and practice, can be condensed into a clear liquid, faintly blue in tinge, that floats on water, that is a non-conductor of electricity, and that runs into and saturates wool clothing easily, as if it were an oil. Its density is only a little less than that of water, but its boiling-point is 532° lower on the Fahrenheit scale, i.e., 320° below zero. For this reason it is about as easy to keep air liquid at ordinary temperatures, which are as furnace heat to it, as it would be to keep a quart of water indefinitely within a red-hot oven. Nevertheless, it is not an uncommon substance around schools and laboratories. A kettle of it boils furiously as it sits upon a cake of ice, for relative to it ice is very hot indeed (Plate 28A). Rubber, ordinarily so soft and elastic, when cooled in liquid air becomes as brittle, hard, and sharp as glass (Plate 28B). Alcohol chilled to these temperatures becomes first a syrup thicker than the proverbial January molasses (Plate 28C) and then solidifies. Lead wire becomes elastic like steel, and bells cast of this dull metal become as musical as if made of bronze. Meat and flesh become as bone; and living things, needless to say, become quite dead, with some notable exceptions, however, in certain bacteria and spores.

Air consists chiefly of two gases, nitrogen and oxygen. Liquid air likewise is a mixture of these two substances in the liquid form. Nitrogen forms four-fifths the bulk, but boils at $-195.8°$ C.; whereas oxygen boils at $-183°$ C. Liquid oxygen is a little heavier than water, but liquid nitrogen is lighter. This leads to a beautiful result when liquid air is poured on water. It floats at first (Plate 29A), being mainly composed of nitrogen; and it boils furiously all the time. The nitrogen, although larger in quantity, nevertheless is the first component to boil off. As the remainder gets progressively heavier from the increasing percentage of oxygen remaining in it, it finally sinks down in great balloon-like globules (Plate 29B). Then one sees the spectacle of one liquid boiling and giving off its vapor while sinking down within another.

The oxygen component of the air is what supports our life. It also supports combustion. When men go down to sea in submarines, or up into the outer reaches of our atmosphere in stratosphere balloons, sealed up hermetically, they soon would perish were it not for the life-giving bottles of liquid oxygen that bubble gently away and renew the failing atmosphere. A cigar that has been well chilled for half its length in liquid oxygen burns with the vigor of a railroad flare, or a Roman candle, when lit by the un-

A KETTLE OF LIQUID AIR. Boiling on a cake of ice

Unsilvered Dewar vacuum bottle half-full of liquid air (here and below)

B FROZEN RUBBER. Note jagged end and shattered fragments broken by a blow from a hammer

 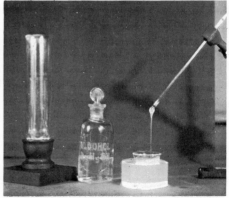

C FROZEN ALCOHOL. Frozen in beaker. A thawing lump has been picked up by the rod and is running off like a thick soup

PLATE 28 THREE STEREOPHOTOGRAPHS OF LIQUID AIR

A Fresh liquid air, largely liquid nitrogen, floating on water

B Old liquid air, largely liquid oxygen, sinking in water

C Strange behavior of a cigar, well chilled in liquid air, upon being lighted

D A quick smoke—amount of cigar consumed in 6 sec.

PLATE 29 LIQUID AIR A MIXTURE

suspecting victim (Plate 29C). His "smoke" is all consumed, almost before he has had time for a single puff, although, of course, he is too startled to puff (Plate 29D). We urgently warn the reader not to try practical jokes with this substance upon his friends. If they should be so fortunate as to escape the undertaker, a doctor's services would probably be needed and a friendship would certainly be marred.

Most important of all the scientific uses of liquid air is its function as a stepping-stone for the attainment of even lower temperatures. By its means the liquefaction of hydrogen (boiling point, $-423°$ F., $-253°$ C.) is possible; and the latter liquid in turn has rendered service in the liquefaction of helium, the most difficult of all substances to condense. Entirely new phenomena very significant as to the nature of matter have been discovered at this temperature of liquid helium, $-269°$ C., $-453°$ F.! This is within a very few degrees of the absolute zero of temperature, a point whose significance will be discussed in the next chapter.

THE advances of modern science, therefore, in the production of a wide range of experimental temperatures are thus seen to be not inconsiderable. From almost the lowest possible temperature—a temperature close to that of interstellar space—up to temperatures of about $3,000°$ C. in electric arcs and furnaces—temperatures that are higher than those of the surfaces of the cooler stars—we are able to perform controlled experiments. Within the stars we are able to observe processes taking place at temperatures as low as the highest furnace temperatures in our laboratories, and up to as high as $25,000°$ C.

The modern sons of that mythical Prometheus who first stole the fire from the gods have already traveled far along the road of knowledge about it and its uses.

CHAPTER 17

QUANTITY OF HEAT AND TEMPERATURE

ONE of the most significant things about the most humble thermal experiments, such as are daily witnessed in every kitchen, is the great diversity among different substances in the rate at which they get hot when put over a fire. Who is there who has not scorched his mouth on a potato when other food, even the proverbial "hot dog'" taken out of boiling water at the same time, is quite cool enough for comfort. This is another aspect of the same situation. In general, it is observed that those substances which require a long time to heat seem to retain their heat much longer than do others that may be heated more rapidly.

The time taken by objects to heat up or cool down, within a given range of temperature, depends on several factors, among which are the characteristics of their surfaces, the difference in temperature between them and their surroundings,* and especially upon the mass and also upon certain intrinsic properties of the substances themselves.

THE following experiments will serve to make our meaning clear. Suppose we have, as illustrated in Plate 30, a block of iron of a definite mass, and for comparison a bucket of thin sheet iron which, when filled with water, will have exactly the same mass. Next we will take two identical gas burners and suspend the iron block in the flame of one and the bucket of water in the

---

* Newton's law of cooling: The quantity of heat passing out per unit time from a hot object to its surroundings is proportional to the temperature difference between it and its surroundings.

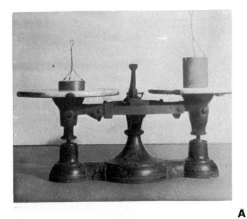
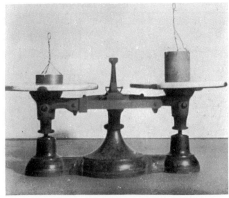

A Equal weights (masses) of iron and water

B

IRON AND WATER HAVE DIFFERENT THERMAL PROPERTIES
PLATE 30

When equal masses of iron and water are heated in similar flames of equal size and intensity, the iron rises in temperature much more quickly than the water

QUANTITY OF HEAT AND TEMPERATURE

flame of the other. Obviously, we are supplying heat to both objects at the same rate; but we will observe that the water rises in temperature very much more slowly than does the iron block. Indeed, the latter will have become sizzling to the touch while the water is still only tepid and quite cool enough for the hand to bear. A much more precise study can be made if thermometers are placed in contact with each object and the temperatures read at frequent, regular intervals while the flames are being applied. We observe the temperature as recorded by the thermometer in the iron rises rapidly, and in the water very much more slowly.*

THE same idea could be unearthed by a quite different method of procedure which is far more convenient, considerably more accurate, and therefore, as the saying goes in scientific circles, far more "elegant." We can compare the heat contained in quite a variety of substances in the following way: Let us suppose, as shown in Plate 31, we have a number of substances all of the same size (volume) and heated to the same temperature, 100° C., by being placed in a vessel of boiling water. Now we have previously made ready and have in waiting, one for each object in the boiling water (Plate 31a), a number of small vessels containing the same amounts of cold water at the same temperature, 23° C. We next proceed to remove our scalding-hot objects, one by one, and to plunge them each into its own separate vessel of cold water. In each of the latter we have placed a thermometer, and now we observe how much these hot objects, all of the same size and all at the same initial temperature, will heat the cold water in which they are submerged. The results vary widely; indeed, if care be used, it will be found that no two of these objects produce exactly the same degree of heating of the cold water into which they were plunged. The actual heating of the water in each case illustrated in the figure, for blocks of equal volume, is recorded in the first line below the illustration. They have obviously each transported a different total quantity of heat from the hotter vessel to the colder one. The same statements would be true if the various blocks were so chosen that they all had the same masses, differing, therefore, in size. The rise in temperature per gram of mass in our experiment could be found by dividing each figure in the first row by the one under it in the second row. These answers likewise would all be different.

* An even more accurate method of determining these intrinsic differences is by comparison of rates of cooling.

FROM GALILEO TO THE NUCLEAR AGE

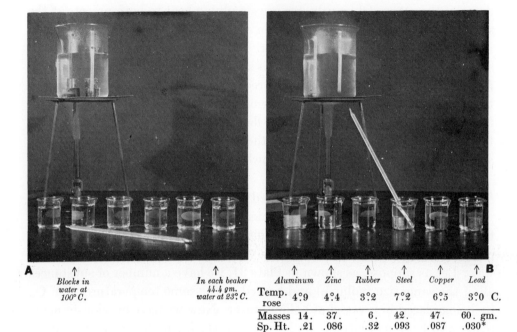

	Aluminum	Zinc	Rubber	Steel	Copper	Lead	
Temp. rose	4°.9	4°.4	3°.2	7°.2	6°.5	3°.0	C.
Masses	14.	37.	6.	42.	47.	60.	gm.
Sp. Ht.	.21	.086	.32	.093	.087	.030*	

A ↑ Blocks in water at 100° C.   ↑ In each beaker 44.4 gm. water at 23°. C.   B

PLATE 31a SPECIFIC HEAT, SOLIDS. (A) Blocks of various substances being heated to same temperature, 100° C., in boiling water. Below: separate beakers, each with same mass of cool water, all at 23° C., in readiness. (B) Each block has now been transferred quickly to its own beaker. The thermometer has measured the heating of each amount of water by its corresponding block.

Equal weights of water and iron; equal rise of temperature desired. Water requires ten times the input of heat that iron demands

We could equally well have had small vials of different liquids in our hot-water bath, which, of course, need not have been raised to the boiling-point of water, except as a matter of convenience. If equal masses of these, all at one and the same temperature, be dumped into cold water, each into its own little portion of the latter (Plate 31b), the thermometers in the cooler vessels will not read the same immediately after the completion of the experiment. Furthermore, if one takes the

* Because of the simplicity of operation in the foregoing experiment, some heat was lost from the blocks that was not recorded in the rise of temperature. Therefore this set of values is about 2–8 per cent too small.

QUANTITY OF HEAT AND TEMPERATURE

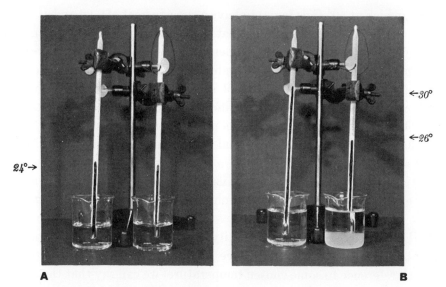

PLATE 31b SPECIFIC HEAT, LIQUIDS. Each beaker contains (in A) 50 gm. of water at 24° C. Into the left beaker (in B) there has been poured 25 gm. of water; and into the right one, 25 gm. of carbon disulphide. The temperature of the liquids added was 40° C., and that of the resultant mixtures 30° and 26°, respectively.

same mass of some of the hot water itself, at the same temperature as the vials of other liquids, or of the blocks, and adds it to one of the cold cups of water, a greater rise in temperature will result *than in any of the other* experiments described.

Water was shown in our first experiment (Plate 30) to be able to soak up heat at a greater rate than iron, since, under a given input of heat, it experienced the least rise in temperature. Similarly, the second and third experiments show that, at the same temperature, and pound for pound, water contains more heat than any other common substance, since it has more to give up when it encounters a cooler environment.

DISREGARDING surface differences, this different capacity of various materials for soaking up, containing, holding, or absorbing heat we call by the name SPECIFIC HEAT. In measuring it and describing it, we always refer to the corresponding property of water as the point of reference. Thus we find that iron has a specific heat of only 0.11 that of water, which we take to be 1.000. Only one metal of the entire list of known elements has a specific heat that approaches that of water. This is the very active alkali metal lithium, which at 50° C. is found to have a specific heat of 0.96 and at 190° C. a value of 1.37, greater indeed than that of water. Highly specialized methods must be used for making such experiments on this tremendously active chemical element, which combines with water instantly if given the chance. The techniques outlined above would not avail at all, but of course there are many modifications of them which differ not the least in fundamental principle. Of all liquids, only liquid ammonia has a specific heat slightly higher than that of water. This is a material sometimes used in refrigeration machines and not to be confused with a watery solution by the same name bought at grocery stores for laundry use. This latter, strong and irritant as it is to eyes and nose, is only a weak solution of the vapor of ammonia.

THUS we see that the total quantity of heat transferred in any operation is to be described only in relative terms. With reference to water, and also with reference to some chosen temperature, we can say that the amount of heat depends upon the mass, the specific heat, and the temperature to which the given material in question has been raised or lowered from some chosen temperature. By definition, therefore

Quantity of heat = mass × specific heat × temperature change.

Our heat unit, the calorie, was selected (cf. p. 79) with these ideas in mind. A calorie is the amount of heat that will raise 1 gm. of water 1° C.

Thus, to raise a mass of iron of 25 gm. from 20° C. to 85° C. will require

$$25 \times 0.11 \times 65 = 179 \text{ calories.}*$$

To raise from the freezing-point to the boiling-point 1 liter (1,000 cc.) of water (about a quart) is, since the density of water likewise is taken as unity,

$$1{,}000. \text{ gm.} \times 1.000 \text{ (sp. ht.)} \times 100° \text{ C.} = 100{,}000 \text{ calories,}$$

which would be written as 10^5 cal.

* If you should have had the temerity to check our figures on this, you would have obtained 178.75. Since each of the factors on the left is known only to two figures, we recognize the foolishness of calculating the answer to more than three; hence we rounded it off.

QUANTITY OF HEAT AND TEMPERATURE

Solids				Liquids		
Aluminum	0.208	}		Alcohol	0.548	@ 0° C.
Copper	.091			Benzol	.340	@ 40°
Gold	.030	} @	0° C.	Ether	.529	@ 0°
Lead	.029			Turpentine	.411	@ 0°
Iron	.104	}		Petroleum	0.511	@ 40°
Lithium	.96	@	50°	Ammonia	1.126	@ 20°
Silver	0.055	@	0°			
Water	1.000	@	15°			
Ice	0.487				Gases	
Sea water	.94	@	17°	Water vapor	0.44	
Rubber	.487			Hydrogen	3.4	
Glass thermometer	.199			Oxygen	0.22	
Wood	.327			Nitrogen	.24	
Paraffin	0.694			Carbon dioxide	0.20	
Asbestos	0.195					
Quartz	.174					
Rock	0.2					

TABLE OF SPECIFIC HEATS*

NOW, there is another and quite a different way in which substances may soak up heat and, paradoxical as it may appear, *suffer no change in temperature at all.* Perhaps you never noticed that a little piece of *ice* will cool off a cup of coffee very much more than a considerably larger amount of "ice-cold" *water.*

It is possible, but we think not likely, that you may be one of those individuals who still labor under the delusion that you can make potatoes or vegetables in boiling water cook more quickly by turning up the fire. The water then boils much more vigorously and, in some people's estimation, is correspondingly "hotter." If you think so, the best way to convince you of your error would be to leave the matter to the dispassionate testimony of a good thermometer. Indeed all of *our* knowledge about such matters has been acquired by this and similar instruments, and you can have no valid

* From Smithsonian Tables, 1934.

ground for clinging to a misconception if we accord you the same privileges that we had when we cleared our minds of any such erroneous ideas.

Suspend the thermometer in the midst of the boiling water, so that it is not resting on the slightly superheated bottom of the kettle. Read it when the water is gently bubble-boiling over the lowest flame that will maintain this condition, and then again over the hottest fire you can produce. You will find no difference in its reading. We are so sure of this we would not only swear to it, we would even bet on it. Nevertheless, if you have doubts, try the experiment yourself, and, furthermore, do not read any further until you have done this.*

NOT only does water, while it is boiling, absorb heat continually and get not one small fraction of a degree any hotter, but the same is true of ice as it melts. One will find by experiment that a large bucket filled with ice cubes and water can be placed over a fire and, if kept well stirred up and agitated, will read the temperature of the freezing point, 0° C., no matter how hot the fire, as long as any ice remains unmelted.

The fact that heat could be so absorbed without there being any temperature evidence of its presence was a fact that centuries ago astonished its discoverers. Such heat that seemed to disappear was said later to have become LATENT, since it was found it could be recovered. When steam condenses back into water, it gives up the same amount of heat that was put into it by fire when it was converted from water to steam. This condensation in your steam radiators is the chief cause of their maintaining their heat so well in spite of the coldness of the room, and also why so small a radiator can heat so large a space.

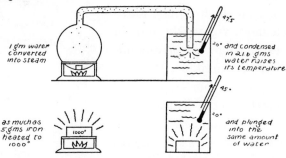

* The cook with sufficient knowledge of physics, to make the boiling water *hotter*, and thus to cook more quickly with it, acquires a pressure boiler with tight cover and safety valve. Under a steam pressure greater than atmospheric, the boiling temperature can be raised many degrees above 100° C.

QUANTITY OF HEAT AND TEMPERATURE 143

TO EVAPORATE 1 gm. of water at the boiling-point requires that 540*
cal. of heat must be put into it. This is enough heat to raise 21.6 gm. of
water from 20° C. to 45° C. We have asserted, also (cf. fig. on p. 142) that if
1 gm. weight of steam (at 100° C.) be passed through a tube into 21.6 gm. of
water at 20° C. and be condensed thereby, the water will be heated very
effectively, as this small amount of steam cools down to the final water temperature, i.e., to 47°.43 C.† Five grams of iron, introduced into the same
amount of water, would need to be at an incandescent yellow temperature of
nearly 1,000° C. to warm the water as much as did 1 gm. of steam.

IT IS the same with ice, which in melting absorbs 79.63 cal. per gm.‡ Each
gram absorbs enough heat to raise it, were it in the form of water, from
room temperature (about 20° C.) to the
boiling-point. Thus the secret of the potent
cooling quality of ice introduced into a hot
solution finds its explanation.

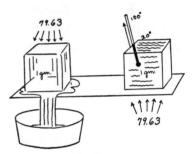

The quantity of heat required to melt
1 gm. of ice will raise 1 gm. of water
from room temperature to the boiling-
point

Most astonishing of all the tricks that
Nature's magic may be made to perform in
these connections is the exhibition of something which is ice cold getting almost too hot
to touch when it freezes! We cannot perform
the experiment with water because it is difficult to cool water below its freezing-point and
not have it freeze.

Jules Verne has a story about some wanderers upon another world who
discovered a lake unfrozen, although the temperature was below the freezing-point. (To imagine this as possible, requires a degree of license that we
cannot allow ourselves, much as we would like to lead you on to see a little
of the fascination held by Mother Nature in some of her simplest aspects,
and much as we would like to show you that these aspects really are simple,
too.) Jules Verne's travelers, noting this unfrozen lake—unfrozen in spite
of the low temperature because no breath of wind or any other disturbance
ruffled its mirror surface—tossed in a pebble. Instantly, with a quiver and a
huge sigh, the lake froze solid!

* 538.7, to be exact, a quantity called the HEAT OF VAPORIZATION.
† Can the reader guess whence came the extra 2.43° C.?
‡ This is called the HEAT OF FUSION.

Now this phenomenon has some basis of truth in it, and, again, the truth was so strange that even Verne, one of the most imaginative of writers, never even suspected it. Verne did not know that the lake got *hot* in the process of freezing, and that unless every gram of ice in freezing gave up the 80 cal. of heat that it absorbed in becoming water it could never have become ice again. *If* water *could* be supercooled to something over 230° below zero, i.e., to a temperature less than that of liquid air, *without* freezing, then, upon freezing, all at once it would warm itself up nearly to the normal freezing-point. The specific heat of ice, you see, is not nearly as large as that of water, being, at $-80°$ C. only 0.350, and at 0° only 0.487 that of water.

Of course, in what we have been saying there are one too many "ifs," "coulds," and "withouts." We cannot demonstrate our point with water, but we can with other substances, such as, for example, photographic "hypo" used for fixing plates and films.

HYPO* comes in crystals, and these crystals contain much water. The crystals are water and hypo frozen up together, as it were. If we heat the crystals over a slow fire, we can dissolve the hypo in its own water of crystallization and then we get a thick heavy solution. This solution, if filtered through hot filters and kept very clean and very quiet, may be slowly cooled down even to room temperature and below, to that of an ice box, without recrystallizing or "freezing" again. In this condition the slightest jar, or more certainly, the introduction of a tiny crystal of hypo, will cause the liquid to solidify instantly. Ice cold it may have been as a liquid, but now in becoming solid it gives back some part of the heat that went into melting it originally, and it becomes almost too hot to handle (Plate 32).

In the next chapter we shall possibly be able to take you further within the borders of this realm of heat and show you what a beautiful interpretation has been made of all of these and many other surprising and curious facts of nature. It will there be seen that, instead of just having to present to you an array of things to be learned and remembered, we can adopt a point of view from which all of the phenomena we have been discussing become inevitable, predictable—the natural, logical way in which material objects should behave.

* Technically known as sodium thiosulphate, $Na_2S_2O_3$.

QUANTITY OF HEAT AND TEMPERATURE

A B

PLATE 32 HEATING OF HYPO BY "FREEZING." (A) A supersaturated liquid solution of hypo at room temperature, 20° C. (B) Introduction of a hypo crystal has caused the solution to solidify into a solid crystalline mass. The temperature has *risen* to 42° C.!

BEFORE leaving the subject of water in its various states of solid, liquid, and gas, we wish to re-emphasize its fundamental importance to us as living creatures. In a universe whose temperatures range from hundreds of degrees below zero to scores of millions above, we live, as we have already intimated, in what seems to us a very favored spot. And yet in our immediate vicinity are other worlds in which life, as we know it, would have the greatest difficulty of survival, even granted it might at one time have originated. Possibly only on the planet Venus are conditions at all like ours, and even there the continually cloudy skies that always veil the surface of the planet prevent our knowing much about its atmosphere or temperature and surface character. That Venus has an abundant atmosphere appears quite certain. Unless there is abundant water likewise, life, evolved into forms at all like our own, would seem to be impossible.

The equable climate that we enjoy we owe entirely to this strange substance—water. Strange it is, indeed, and eccentric in nearly all of its

Southern Hemisphere Northern Hemisphere
Abundance of water

properties. It covers four-fifths of the surface of the globe. Only blistering lithium, eccentric hydrogen and suffocating ammonia approach it in specific heat.* The vast amount of water on the earth, together with its great specific heat, makes it a huge reservoir for the storing-up of excess warmth and the giving of it back gradually to us as it is needed. In evaporation, likewise, its latent heat has a value unapproached by any other common substance, and here also is its value as a temperature-controlling thermostat greatly enhanced.

High specific heat of water. Radiator cooling diminished by adding alcohol to water, since the mixture has less specific heat than pure water

Latent heats of water

In cold weather, of course, it freezes. Even in this process it gives up heat in large amounts, which it reabsorbs on melting. Still more important to us, however, is another remarkable fact in connection with this freezing, a fact with respect to which water is most eccentric.

Most substances, on freezing, become denser than they are in liquid forms, and the frozen solid sinks to the bottom of the liquid. Not so with ice. Water is unique in having the most irregular of all expansion coefficients. As it is cooled, it contracts as do other substances, *until* it gets within *4° C. of its freezing-point.* Then its coefficient of expansion becomes negative; and as it gets still colder, it actually expands. As it freezes into ice, it expands still more. Thus the frozen solid *floats* on the surface of the unfrozen liquid. Were this not so, ice would start to form at the bottom

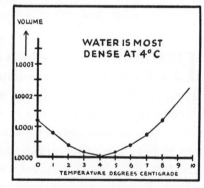

Expansion of water: Life as we know it on earth could not exist without water. The curious fact that water, unlike anything else, contracts in volume down to 4° C. and then expands when the temperature falls lower results in ice being less dense than warmer water. As a result, ice forms on the surface and *floats*

---

* Did you ever notice how much cooler the radiator in your car keeps with pure water in it than when this is diluted with even a small amount of alcohol or glycerine?

QUANTITY OF HEAT AND TEMPERATURE

of our lakes and ponds and seas. Since water is one of the poorest conductors of heat, the summer sun would thaw this submerged ice but little; and each successive winter would increase the store, so that in temperate and arctic climes, at least, all the water on the earth would shortly be locked up in the solid state. A little surface melting would take place each summer, but not enough to moisten appreciably the atmosphere, let alone to saturate it and provide rain to wet the soil.

If it were not for this unique property, ice formed in the winter would sink to the bottom, where thawing during subsequent warm weather would be greatly reduced. Soon a large portion of the water on this planet would be permanently congealed in the form of ice and conditions for life would be very difficult, if possible at all

As it is, beneath a surface of thick protecting ice that keeps out the cold or keeps in the heat, whichever you prefer to say, the underlying water seldom becomes very cold. In summer, likewise, because of the low conductivity of water, the deeper portions of oceans and great lakes seldom get very warm. The vast regions of the sea, where life seems first to have originated, enjoy an almost constant temperature. Even in winter, in ice-locked shallow waters, life, submerged in this perfect thermostatic fluid beneath, can still go on.

Low thermal conductivity of ice

The perpetual fluidity of water thus assured, vast circulating streams of it pour southward from the polar seas and cool the summers of adjoining coasts. Similarly, great streams of warm water circulate from the tropics ever northward and profoundly modify the rigors of the winter season in the higher latitudes.

Thus we see that with respect to its relative abundance, the values of its specific, and both of its latent heats, its expansion coefficient, its thermal conductivity and the fluidity that results at our earth's average temperature, one might almost think that water had been especially prepared to make this planet's surface ready for the evolution of organic life upon it. This is a statement, however, that we are not at all prepared to make. This would seem to give to

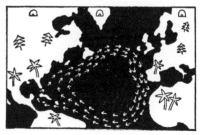

Fluidity of water on the earth provides for a great modification of temperatures

our especial forms of life a significance that is quite incompatible with the cosmic perspectives that the study of science brings. If one judges the importance and virtue of living matter by the collective behavior of one of its most exalted forms, one might be forgiven, perhaps, a sneaking doubt that there be such value in the result to date that water should have been specially created for its existence.

CHAPTER 18

THE KINETIC THEORY OF HEAT

WHAT interpretations can be made of the vast array of observational knowledge of heat and temperature that we briefly have outlined in the four preceding chapters? The study of mechanics led us inevitably into a consideration of thermal effects, and we might well suspect some connection. Many of the most able physicists of early times reacted precisely in this way. There are passages in the writings of Francis Bacon in the sixteenth century and in those of Hooke and Boyle in the seventeenth that are pregnant with the modern point of view. Of early speculations that heat was some kind of a manifestation of motion there were many.*

Francis Bacon writes in his *Novum organum:*

When I say of motion that it is the genus of which heat is a species, I would be understood to mean, not that heat generates motion, or that motion generates heat (though both are true in certain cases), but that heat itself, its essence and quiddity, is motion and nothing else. . . . . Heat is a motion of expansion, not uniformly of the whole body together, but in the smaller parts of it, and at the same time checked, repelled, and beaten back, so that the body acquires a motion alternative, perpetually quivering, striving, and struggling, and irritated by repercussion, whence springs the fury of fire and heat.

Hooke, in his *Micrographia*, says:

First, what is the cause of fluidness? And this I conceive to be nothing else but a certain pulse or shake of heat; for heat being nothing else but a very brisk and vehement agitation of the parts of a body (as I have elsewhere made probable) the parts of a body are thereby made to loose from one another that they may easily move any way and become fluid.

* Quotations from H. Buckley, *A Short History of Physics* (London: Methuen, Ltd.), pp. 126–27.

And later:

> Now that the parts of all bodies though never so solid do yet vibrate, I think we need go no further for proof than that all bodies have some degree of heat in them, and that there has not yet been found anything perfectly cold. Nor can I believe indeed that there is any such thing in nature as a body whose particles are at rest or lazy and inactive in the great Theatre of the World, it being quite contrary to the grand Economy of the Universe.

That heat could be generated by mechanical processes appears to have been clear to Boyle from the following remarkable passage:

> If a somewhat large nail be driven by a hammer into a plank, it will receive divers strokes on the head before it grows hot; but when it is driven to the head, so that it can go no farther, a few strokes will suffice to give it a considerable heat; for whilst at every blow of the hammer the nail enters farther and farther into the wood, the motion that is produced is chiefly progressive, and is of the whole nail tending one way; whereas, when that motion is stopped, then the impulse given by the stroke, being unable either to drive the nail further on, or destroy its entireness, must be spent in making a various, vehement, and intestine commotion of the parts themselves, and in such an one we formerly observed the nature of heat to consist.

IN THE absence of any precise experimental facts, of any way of exactly measuring temperature, the precise relationship between heat and work was unknown; consequently there were at that time no means of testing these hypotheses to prove or to disprove them. As the science of chemistry came into being on an experimental basis, the obvious temperature changes that accompanied so many chemical reactions led to the postulation of a quasi chemical theory of heat that regarded the latter as a material substance with certain hypothetical properties. Facts about specific and latent heats were learned, and the properties of this hypothetical "caloric," or heat fluid, were designed to be in accord. The caloric theory dominated the scientific world until nearly the middle of the nineteenth century, although its chief opponents by the end of the eighteenth had performed experiments that clearly substantiated the older point of view that heat was motion. These men also had shown by crucial tests that the caloric theory could not be made to square with the facts of nature.

Sir Humphrey Davy killed the "caloric" theory by melting two pieces of ice by friction at temperatures below the freezing-point

THE KINETIC THEORY OF HEAT

That matter was not continuous substance but consisted of excessively minute and discrete particles was likewise a speculation that runs far back into antiquity. Democritus, the Greek philosopher, as long ago as 460 B.C., held such ideas. More modern atomic speculations were made in the attempts to explain certain things in the behavior of gases, and of course these speculations found excellent substantiation in the early pioneer work in chemistry done by John Dalton and his followers.

RATHER than continue to trace the development of the modern conceptions of thermal theory in a chronological way, let us go at it, as it were, afresh. This fresh start has a distinct advantage, since in so doing we may be able to point a moral on the way. But lest you think by this that we intend to preach about something, we hasten to assure you that such is not the case. A "moral" was not the word to use; an adventure rather let it be. Pretending to know nothing about the subject except a few isolated and dry and dusty facts, let us first set up a scheme for their interpretation.* Next we will see whither this scheme, carried further in some of its implications, will lead us. If it leads us to discover unexpected verifications of its possibilities, good! If it leads us further to the discovery of some new facts hitherto unsuspected about nature, better still, for this is indeed adventure! If it does this repeatedly, we shall probably end up by believing in this theory ourselves. If we do, we shall have plenty of good company—that of all of the men of science in the world. For here, strange to relate, is one field of human enterprise established several generations back, where the storm and stress of modern discoveries have but reiterated and reemphasized the undoubted correctness of an earlier point of view—indeed, have provided the most improved types of direct sensory proof. The point is, not that these especially were needed, but that it is very gratifying to see with your eyes and to hear with your ears things that have been before your mind's eye clearly and for quite a long time.

INCIDENTALLY, if you will read thoughtfully between the lines in what is to follow, and meditate a little as you proceed, you will discover on nearly every page an illustration of the method by which science has accomplished her achievements. This method is vaunted not so much by

---

* A schematic diagram of the typical growth of a scientific theory is indicated in an accompanying figure (p. 152), which will repay following through in detail.

THE GENEALOGY OF A SCIENTIFIC THEORY

FACTS	HYPOTHESES AND THEORY
1, 2, 3, 4—Isolated and apparently unrelated facts are drawn together by the ingenious formulation of a ———	
	Provisional Hypothesis which "explains" them—1, 2, 3, 4—but which further implies that probably
(5), (6), (7), (8), are also true. Further experimentation attests the correctness of 5, 6, 7, but (8) proves recalcitrant and to be included requires the further formulation of a ———	
	Modified Provisional Hypothesis consisting of the original 1, 2, 3, 4, 5, 6, 7, (8) + the addition of 9, 10, 11, 12. It implies that
(9), (10), (11), (12), are also true. During the experimental verification of 9, 10, 11, 12, hitherto unsuspected facts 13, 14, 15, are discovered. 13 is seen to agree at once, but 14 seems unrelated. 15 appears to be contradictory until 16 is subsequently discovered. This later turns out to be another aspect of 15, and both shed light on 14. Now all agree with the ———	
	Modified Provisional Hypothesis.
Further experimentation with improved techniques has also been going on to show that the original 1, 2, 3, were not exactly right but that 1', 2', 3', more nearly represent the truth. Similarly 8, 12, 16 become 8', 12', 16', which have been concurrently found to be contained in the ———	In the meantime the Modified Hypothesis has been mathematically transformed into a more general
	———THEORY which contains 1', 2', 3', 4, 5, 6, 7, 8', 9, 10, 11, 12', 13, 14, 15, 16', and as consequences 17, 18, 19, 20, 21, 22.
	As a result of the theory, in a number of different directions, new types of investigations are suggested and
(17), (18), (19), (20), (21), (22), are predicted quantitatively in advance. Experimental verification confirms 17, 18, 19, 20, 21, 22, and	

GENERAL ACCEPTANCE OF THE THEORY COMPLETES THE STORY

THE KINETIC THEORY OF HEAT

men of science themselves to whom it is after all an everyday affair, as by others, not of this group, who believe it is a method by which all of the ills that human, social, and economic flesh are heirs to may be conquered. From this opinion we dissent; this conviction we do not share. Problems are of many kinds —a method suitable in the attack upon one kind may be of little value in dealing with one of another sort.

TO PROCEED now with a brief review of the interpretations of thermal phenomena, we will begin with gases. Invisible and tenuous, mysterious and many voiced when in motion, they are, nevertheless, unquestionably material. We can weigh them and measure their inertia. When they are violently aroused, there is nothing that can withstand their mighty forces. The chief attribute which they possess to distinguish them from any other form of matter is their ability swiftly to fill any space, no matter what its size or shape, that is made available for them. Besides this, their density is usually very low, i.e., the mass of a given volume is quite small compared with that of other states of matter. In a third way also, they are conspicuously different from solids or liquids: they are highly compressible. Thus they may be given much greater densities than normal by squeezing them into smaller compass. Very significant also is the fact that, whereas more than one solid or liquid substance cannot occupy the same space at the same time, except in very special cases of solutions, gases diffuse one into another with great ease and considerable rapidity. A given volume of space appears to be able to hold any number of different gases simultaneously.

If, as a hypothesis, we assume that matter consists of innumerable minute particles, which we will call "molecules," and if we further suppose that in the gaseous state these particles are separated by distances relatively great in comparison with their sizes, we could account readily for the great compressibility of gases.

The three different "states" of matter—solid, liquid, gas —occupy the space in a containing vessel quite differently

Also, this would explain why several different substances can occupy the same space at the same time, although alone this hypothesis would not

Bromine introduced into a vessel of air at 8:48 has completely diffused through it by 9:15

By collision with air molecules (white)

the rate of diffusion of bromine molecules (black) is impeded

With a pump we now evacuate the vessel
(Continued on next page)

PLATE 33 OBSERVATIONS ON DIFFUSION OF GASES

Illustrations in this plate and also in plates 34, 38, 40, and 41 are reproduced from the Educational Sound Picture "The Molecular Theory of Matter" by H. I. Schlesinger and H. B. Lemon, produced by Erpi Picture Consultants, Incorporated, and The University of Chicago Press (copyright University of Chicago).

F

Now introducing bromine at 9:14:18

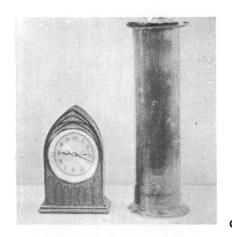

G

The bromine fills the vessel completely before the second hand is observed to have moved at all

H

Unimpeded by collisions with other molecules

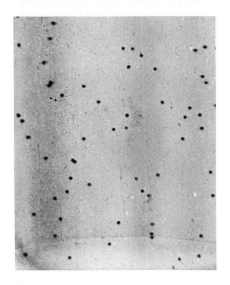

I

The bromine molecules are able to scatter throughout the vessel in a small fraction of a second

PLATE 33 (Continued) OBSERVATIONS ON DIFFUSION OF GASES

account for the automatic character of diffusion, whereby two gases, originally separate, move about, each into the interstices of the space occupied by the other. Nor will our hypothesis account for that obviously related property that individual gases have of expanding to fill uniformly any volume of space accessible to them. The assumption of molecules is not sufficient unless, in addition, these molecules are imagined to be *in constant motion*. To put matter in motion requires the exertion of force and the doing of work; and since diffusion and spreading out into all the space available is an automatic process, we must suppose that the molecules are endowed with a great deal of agitation all the time. In other words, a state of agitation is just as essential a part of the hypothesis as is the assumption that matter exists in this finely divided granular or molecular state.

AT ONCE we see, then, that in the business of filling an empty space, or of diffusing into a space already occupied by another gas, we have only one idea involved—that of particles in motion. Now it is possible to observe diffusion by use of the element bromine, which at ordinary temperatures is a highly colored volatile liquid that evaporates into a vapor very readily. If a bulb of it, introduced into a closed vessel of air, is broken, the colored gaseous vapor is visible (Plate 33A, B) and within a few minutes will be observed to have filled the entire vessel, already previously filled with air (Plate 33C, D). If, on the other hand, we should repeat the experiment in a vessel which had first been evacuated of all its air (Plate 33E), we should expect that the bromine would diffuse and fill the entire space much more quickly, since its flying molecules would not now encounter other obstructing flying molecules of air. Experiment fully bears out the expectation. The bromine diffuses and fills a previously evacuated vessel almost instantaneously (Plate 33F, G, H, I).

ANOTHER expectation would be that these flying molecules, upon colliding with the walls, would rebound elastically, since the gas stays within the interior of the vessel and does not condense, except in very special circumstances, upon the walls. Since the gas as a whole has mass, this mass must be made up of the sum of the masses of its individual particles. These, in rebounding from the walls, suffer a change of momentum which must exert a force on the walls—a tiny bump for every impact. If the impacts come very frequently, the result would seem to be a continuous force, just

THE KINETIC THEORY OF HEAT

PLATE 34 STEADY PRESSURE FROM A SERIES OF BLOWS. A machine-gunner fires a continuous spray of bullets at a distant target. The target is backed by a coil spring, and an indicator arranged to show movement of the target. Under the spray of bullets the target is pushed back against the spring continuously and steadily, as shown by the indicator. With more bullets per second and same powder charge, or the same number of bullets and a greater powder charge (greater velocity of bullets), the indicator would move up to the right, indicating a greater steady pressure. It would move to the left with fewer bullets, or less powder, or both.

as a stream of water from a jet, although consisting of individual drops, nevertheless exerts a steady push upon an object in its path. Another illustration of this is given in Plate 34.

We should now have sufficient familiarity with mechanics to calculate how great this force should be. Take a long deep breath now, settle down, and concentrate on what follows. Suppose in the container whose sides are a, b, d, there are myriads of molecules, N in number, flying around at random in all directions. On the average, one-third of them at any time will be moving parallel to any edge; and if, on the average, their velocity is c, any one of them will pass from the face bd to the opposite face and back again in a time that is $2\dfrac{a}{c}$.

(Velocity = distance ÷ time, or time = distance ÷ velocity)*
(c) = ($2a$) / (t) (t) = ($2a$) / (c)

The number of times per second each molecule hits the wall is the reciprocal of this, $1/t = c/2a$, times per second. Now each time the molecule hits the face bd it imparts a momentum $2mc$ to the wall. [Remember velocity and momentum are vectors; the molecule approaches the wall with momentum

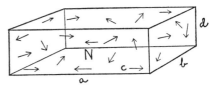

* We need not be concerned if, on the way, the molecule we were watching collided with others, since we have seen (p. 52) that when equal elastic objects collide they simply exchange velocities.

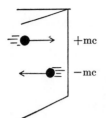

($+mc$) directed to the right and leaves with momentum ($-mc$) directed to the left. The change is $+mc - (-mc) = 2mc$.] Thus for the total momentum imparted to the wall each second we have,

Total momentum per sec. = mom. per hit × hits per sec. × total no. of molecules moving in this direction

$$\text{Mom. per sec.} = (2mc) \times \frac{c}{2a} \times \frac{N}{3} = \frac{1}{3} mN \frac{c^2}{a}.$$

By our definition of force (p. 58), it equals the rate of change of momentum, i.e., the total momentum imparted per second. We have then just calculated the force the gas exerts on the confining walls of its containing vessel! Usually, however, we speak of the gas "pressure" rather than of the force. By pressure we mean the force on each square centimeter of surface (cf. chap. 13, p. 94, n.). There are bd sq. cm. on the face of the container under consideration, so that the pressure p equals what we have just computed, divided by bd. There was already an "a" downstairs in our answer. When we now divide by bd, it joins the a. But abd is the volume of the box. So we obtain for the pressure,

$$p = \frac{f}{bd} = \frac{\text{Mom. per sec.}}{bd} = \frac{1}{3} m \frac{N}{abd} c^2 = \frac{1}{3} m \frac{N}{V} c^2. \tag{1}$$

You probably recall that to reside downstairs on the right of an equation is the equivalent of living upstairs on the left. Maybe you learned this in the form of a theorem that equals multiplied by equals gave results that were equals. In any event, let's move the V across,

$$pV = \frac{1}{3} mNc^2 = \frac{2}{3}\left(\frac{1}{2} mNc^2\right) = \frac{2}{3} \text{ of the total } \textit{molecular kinetic energy}. \tag{2}$$

On the basis that a gas consists of flying molecules, these symbols have led us to a *new idea* with respect to the behavior of gases: viz., that the product of the pressure a gas exerts by the volume it occupies should equal two-thirds of the kinetic energy of all its molecules.

LONG before the acceptance of so strange a hypothesis as the one we have made, two experimenters by the name of Boyle and Charles, in observing gas phenomena, discovered that the product of the pressure of a given mass of a gas by the volume of its container always remained the same *if*

THE KINETIC THEORY OF HEAT

the temperature was constant. If, on the other hand, the temperature declined, so did the product equally, provided that the temperature was measured from a zero point that was $-273°$ C. We have used already two temperature scales but now must introduce a third, the Absolute scale, where T° K. (Absolute) $= 273 + t°$ C.

Boyle's and Charles's Laws therefore say that experimentally,

$$p = R\frac{T}{V} \text{ or } pV = R*T .$$

Centigrade (C.)	Absolute (K.)
100°	373°
50°	323°
0°	273°
−50°	223°
−100°	173°
−150°	123°
−196°	77° liquid air
−200°	73°
−250°	23°
−273°	0°

CONVERSION OF CENTIGRADE TO ABSOLUTE

Our hypothesis has said:

$$pV = \frac{2}{3} \text{ total } KE$$

OF COURSE, by now you are feeling pretty dizzy and wondering what we are driving at anyway. It's all over. We have arrived at our destination. Our kinetic hypothesis finds support in experimental fact PROVIDED we accept the implication that *temperature*, on this new scale, and *molecular kinetic energy* of the flying molecules are ONE AND THE SAME THING. There was nothing fundamental about our choice of the Centigrade scale or the Fahrenheit, or about the scale of the early Florentines. Convenience alone dictated the matter.† But here seems to be a scale of temperature *with meaning*. Measure all temperatures from a zero point of $-273.18°$ C. and the temperature then will be an indication of the magnitude of the heat energy of agitation of the substance's molecules.‡

---

* The constant R is the same for all gases if the amounts of gas used are properly chosen, so that the same numbers of molecules are always involved.

† For a similar situation in electricity see Part III, p. 190.

‡ Transformation from Centigrade to this new "Absolute" scale is easy. One has but to add 273 to the centigrade temperature to get the absolute. For example: water freezes at (0+273) 273° K., and boils at (100+273) 373° K. Liquid air at $-196°$ C. is at (273−196) 77° K. See table at the top of this page.

NOW, please bear in mind that we did not *start out* to "explain" temperature. We started out with a hypothesis to explain how gases could diffuse, how they could quickly fill all space available to them, how several could occupy the same space at the same time, and why they were so easily compressible. Small particles flying about in a lot of otherwise empty space would do this. But they would do other things as well. They would bump

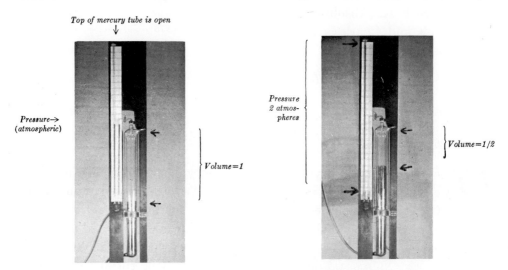

PLATE 35 Boyle's Law

around on the walls of the containing vessel and exert what would appear to be a continuous pressure there. We calculated this. We found:

First: An expression that had V, the volume of the vessel, downstairs on the right-hand side. This told us our picture implied that the smaller we made the volume, the greater would be the pressure of a given mass of gas upon its walls. Of course, theory or no theory, we all know that this is true provided the temperature is constant. This relation between pressure and volume was discovered first by Boyle and is known as his "Law."

Instead of saying, "The pressure increases in proportion as the volume diminishes," it is equally correct to say, "The product of pressure by volume is constant"; so we moved the V upstairs on the other side. This now says PV is constant, but we must not forget the proviso *of a constant temperature*.

Second: In this form

$$pV = \frac{1}{3} mNc^2$$

THE KINETIC THEORY OF HEAT

our calculation gives us, on the right-hand side, an old familiar friend from mechanics, one of the "energy twins" KE. Except for the fraction $\frac{1}{3}$ instead of $\frac{1}{2}$, mc^2 is one molecule's kinetic energy and N times this is the sum of the kinetic energy of all of them. Now the experimental facts for gases discovered by Charles, and independently by Gay-Lussac, show that if we do *not* keep the temperature constant, the product of pV is not constant but *declines with the temperature.*

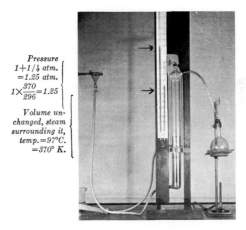

PLATE 36 CHARLES'S LAW

IF WE cool a gas 1° C. below zero, the value of pV falls off by 1/273 of its amount at 0° C.; 2° cooling produces 2/273 decrease; 10°, 10/273; and presumably, if the same rate holds, cooling to −273° C. would cause the pressure volume product to decline by 273/273 of its value, that is, to *nothing at all.*

Now we should hardly be rash enough to suppose that we had destroyed the gas by simply cooling it off. Our hypothesis would suggest that this paradoxical behavior simply meant that we had destroyed *all the motion* of the molecules, i.e., that they no

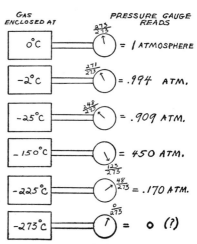

Rate of decline of the pressure of a fixed amount of gas as it is cooled.

longer possessed kinetic energy at this low temperature. You might well ask, then, "Have not all the molecules settled down to the floor under gravity? Then why not get a broom and dust pan, sweep them up, and throw them away, leaving a perfect vacuum in the container? For several rather obvious reasons the question, although logical, is not practical!

Thus, if we adopt a new zero of temperature, an "Absolute zero" identified by no kinetic energy of agitation, all other temperatures find a rational meaning in terms of molecular energy of motion. Heat we know from previous experience to be some form of energy, and this tells us what form it is.

Our hypothesis has thus shown itself to be one of unusual value. It sheds new light on matters previously obscure. The fact that we did not design it primarily to explain these obscurities makes it all the more effective. It appears to contain more than we put into it.

IN THE preceding chapters we have outlined many other aspects of thermal phenomena. Let us now put our hypothesis to further test and see if it has any light to shed in other dark corners of our understanding. What about the distinction between heat quantity and temperature?

If, instead of choosing N, the total number of molecules in our container, always to be the same, we take greater or smaller amounts of material and thus larger or smaller numbers of molecules, we see at once that the total amount of heat absorbed by an object will depend, at least in part, upon the total number of molecules whose kinetic energy has to be increased. Heat capacity depends first of all upon the mass of the body heated.

When we are going to compare equal masses of different substances with respect to their capacities for absorbing and soaking up thermal energy, is it not reasonable to suppose that these differences are reflected in their molecules also? If so, then differences in specific heats would seem to mean that, at least in the solid and liquid conditions that we chiefly studied, there are very different degrees of effectiveness of response to this thermal shaking-up in different substances.

In the gaseous condition we can better differentiate between the molecular characteristics of different substances.* The gases helium, argon, neon,

---

* The distinction between molecules and atoms must be borne in mind. All material substances are composed of small discrete particles between which, in the gaseous state, there is much space. These are its molecules. They may be simple entities or may contain many different parts.

Whatever their complexity, all molecules are but combinations of one or more fundamental entities—atoms of the chemical elements. These latter are but ninety two in number and are to be found

THE KINETIC THEORY OF HEAT 163

and certain vapors such as mercury have very simple molecules, one atom in each. The impression we get with respect to their general contours arising from these first introductions to them leads us to think of them as spherical. If equal numbers of molecules are taken, the relative specific heats of these monatomic gases are all *exactly the same*. All the energy imparted to them seems to appear as energy of motion.

Another group of gases, like hydrogen, air, carbon monoxide, chlorine, etc., are similar in that they all consist of molecules that contain two atoms each, and, as a result of an early bowing acquaintance with them, we think of them as being little dumb-bell like affairs.

These diatomic gases likewise show among themselves exactly the same value for their specific heats, characteristic of all diatomic gases, but these values are two-thirds as large again as those for the monatomic gases. Carbon dioxide, water vapor, and more complex molecular forms show still larger values for their specific heats. There are also systematic variations which we can likewise trace even down into the liquid and solid states of other substances.

It must be borne in mind that we have not said that heat motions of the molecules consist merely of a flying about. For simple monatomic molecules this is all we need to imagine. But molecules shaped like dumb-bells might whirl about an axis as well as shoot forward. They might also oscillate along their axis of figure. If molecules may be whirling or vibrating, as well as moving forward, all of these things represent heat energy that has been absorbed; but *only the motion of translation of the molecule as a whole will be registered as pressure on the walls* or recorded by a thermometer as temperature.

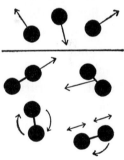

Monatomic and diatomic molecules

---

in the molecules of all substances as individual aggregates having a definite and unchanging mass or weight.

The molecule of common salt contains two atoms—one of the element sodium, one of the element chlorine—in close combination.

Air is a mixture of several substances; the elements nitrogen, oxygen, argon, helium, etc., and compounds such as carbon dioxide. Its individual particles are molecules of nitrogen and oxygen each containing two atoms of these elements, respectively. The "molecules" of argon and helium are simply single atoms of these elements. The molecule of carbon dioxide contains one atom of the element carbon and two atoms of the element oxygen. For reference and the evidence, see any text on general chemistry.

CONDUCTION of heat from one place to another is too familiar to all of us to require comment as to the fact. Your spoon is placed in a cup of hot coffee. The bowl of the spoon gets hot at once under the violent bombardment it receives from the water molecules. The handle also soon warms up as the agitation of the molecules in the bowl slowly spreads out through the body of the metal. In the cooler air surrounding the handle are more sluggishly moving molecules. These are set in motion by the more lively movements in the surface of the spoon handle that they touch. Heat spreads from the hot coffee, which is cooled thereby, into the handle, which passes it on to the air in turn.

LATENT heats, likewise in terms of our hypothesis, find beautiful interpretation. In a liquid the molecules are held close together and are effectively confined by each other's attractive forces. Solid and liquid states most conspicuously exhibit these cohesive tendencies. Not so with gases, where the molecules have been pulled apart against their mutual attractive forces—pulled apart we say—forces have been overcome, *work* has been *done;* energy that has been absorbed has gone into potential form, and is no longer represented in the kinetic aspect.

In crystalline solids molecules are more rigidly locked together than in the liquid state. Recall the supercooled hypo solution. Its molecules were, as yet, unlocked in the crystal cells and were moving sluggishly about, free to wander anywhere. Once set off to build up the solid crystals, down they tumble, each into its own appointed place, literally falling into it, accelerating as they go. Caught and bound in fixed locations by one another's forces, they are now in a state of more lively motion. In spite of their lack of freedom, they now vibrate vigorously around their chosen spot instead of wandering more slowly and aimlessly. Their velocity is greater than it was before their fall, their kinetic energy likewise; their temperature is therefore greater. The energy that was put into them to drag them out of their crystals into the more free liquid state has now been given back. Latent heat, potential energy absorbed, has now been returned as sensible heat—kinetic energy of vibration.*

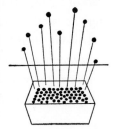

Evaporation: Molecules pulled apart into gaseous state from the liquid have had work done on them, represented by latent heat of evaporation

* The reader is cautioned not to confuse vibration of a molecule (or atom) in a crystal which is registered as temperature with vibration of parts of a molecule (atoms) within the molecule, which, while representing thermal energy, does not count as temperature.

THE KINETIC THEORY OF HEAT

Evaporation and the attendant cooling that it produces (cf. p. 110) is not unrelated to the idea of latent heat, and this likewise finds simple interpretation in terms of the kinetic hypothesis. Consider a molecule that has wandered into the surface layer of a liquid. It will be pulled strongly back into the body of the liquid

Freezing: Molecules that have "fallen" from the liquid state (left) into fixed positions in a crystal (right) lose potential energy thereby and give up heat (KE)

where there are many more attracting companions, compared with the small number roaming about in the gas or vapor above the liquid. The surface of a liquid is thus a seat of forces tending to pull molecules away from it into the interior of the liquid. Thus, liquid surfaces tend to become as small as possible, just as if they were elastic and under tension. Isolated small bodies of liquid assume for this reason the spherical form, since this has the smallest possible surface for a given volume.

For a molecule to *escape* from a liquid through this restraining layer—to climb out over this potential hill, so to speak—it must have a goodly supply of kinetic energy. Only the faster moving molecules thus can, and do, escape. Their loss lowers the average of the velocity of those remaining behind, and it is the average speed that we have associated with temperature.

If a volatile liquid then is fanned so that all escaping molecules are blown away and cannot wander back, the liquid will be found to be always cooler than its surroundings (Plate 37). If, on the other hand, the liquid is contained in a closed vessel, originally devoid of any of its molecules, they will soon accumulate in the space above the liquid. The pressure due to the vapor gradually increases. The more there are there, however, the greater becomes the number that, in their random journeys, encounter the liquid surface and are drawn again therein. Ultimately the number leaving the liquid in a given time is balanced by the number returning, and a condition of equilibrium results. When this is attained, the vapor is said to be *saturated*, and its pressure is called the vapor tension of the liquid (Plate 38).

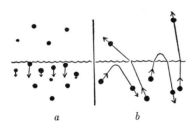

(a) Instantaneous picture, (b) motion picture, of the forces that act on molecules in or near the surface of a liquid (surface tension)

If, after these conditions have been attained, the temperature of the liquid should be raised, a larger supply of more swiftly

PLATE 37
COOLING BY EVAPORATION

A A closed bottle of ether has been standing in the room and acquired the temperature of the room, which is 26° C.

B On being poured out into an open dish the ether soon becomes cooler than the room. Temperature now 16° C.

C An electric fan is now blowing gently over the surface of ether in the dish. The liquid is cooled still further: thermometer reads −9° C.

THE KINETIC THEORY OF HEAT

moving molecules is provided and more can then escape into the space above. But as the number there increases, a new saturation point and a new vapor tension will soon be reached at this new temperature. Thus we see why it is

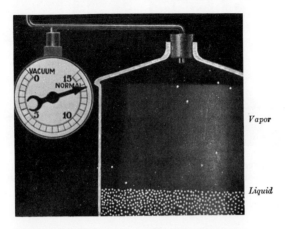

PLATE 38 SATURATED VAPOR EXERTING ITS VAPOR TENSION AS SOON AS THE NUMBER OF MOLECULES RETURNING TO THE LIQUID EQUALS THE NUMBER LEAVING IT

that the vapor tension of all liquids rises so rapidly with the temperature. An interesting illustration is the case of water. The table shows for various temperatures the pressure of water vapor in equivalent millimeters of height of a column of mercury. You will recall that the normal pressure of the atmosphere is equivalent to 760 mm. of mercury, i.e., that the pressure of the atmosphere will support just so high a column of this very heavy liquid.

You will note from the table that at 100° C., which we have *arbitrarily* chosen as one of our fixed points of temperature, the vapor pressure of water also is equal to 760 mm. Of course, this is the reason why water boils at 100° when the atmospheric pressure is normal at 760 mm. Since its own vapor tension is just equal to the pressure of the external air around and over it, the myriads of escaping molecules need no longer be confined to the top surface of the liquid. There are always other surfaces within the liquid, tiny particles of foreign matter, greasy spots, and

$t°$ C.	p, mm. of Hg
$-$ 10	1.96
$-$ 5	3.02
0	4.58
5	6.54
10	9.31
20	17.55
35	42.23
50	92.6
75	288.3
100	760.
150	3568.
200	11650.
300	64300.

CHANGE OF PRESSURE OF WATER VAPOR WITH TEMPERATURE

The small hydrometer has been so made that it displaces just a little less than its own weight of liquid and so sinks immediately upon immersion (cf. Archimedes' principle, p. 103).

The liquid is seltzer- (or soda-) water, i.e., water in which carbonic acid (carbon-dioxide) gas has been dissolved. This gas tends to escape from within the liquid and to accumulate wherever there is a *surface* available. The hydrometer in the left graduate has been scrupulously cleaned. To it the liquid closely adheres and there are no surfaces whence the gas may emerge. The hydrometer in the right-hand graduate is greasy—the liquid does not adhere to it. A free surface exists around this hydrometer from which gas escapes, forming bubbles that stick to the glass of the hydrometer. In a very few minutes these bubbles have grown to sufficient size to provide enough buoyancy to bring the hydrometer up to the surface. The application of this experiment to what is said about boiling at the bottom of pp. 167 ff. is obvious, of course.

PLATE 39 STEREOPHOTOGRAPH OF A CLEAN (LEFT) AND GREASY (RIGHT) HYDROMETER

THE KINETIC THEORY OF HEAT

corners on the surface of the containing vessel. Even the most scrupulous care cannot prevent them. At all of these surfaces, now, molecules rush out over the boundary of the liquid and accumulate their pressure, which is now equal to that of the atmosphere outside. Thus the vapor can expand as it is accumulated, and it swiftly comes out of the liquid in the form of bubbles (Plate 39).

WE HAVE come, it would seem, a long distance on this hypothesis made originally to account for the behavior of gases. In the realm of heat it has interpreted the simplest and the most complex processes with equal ease. The curious phenomena that accompany the change of a substance from one state to another yield to its simple explanations. In one field especially, it has pointed the way to new achievements and *predicted accurately in advance* under what conditions they could be accomplished. For the story of the liquefaction of air, of the more difficult hydrogen, of the final triumph of the conversion of helium into the liquid form, and of the attainment of a degree of cold thereby that is within a few degrees of Absolute zero, we must refer to more technical works. It has been in this field that some of the greatest triumphs of the kinetic molecular hypothesis have been realized.

WITH respect to processes going on in stellar atmospheres, the kinetic theory has become one of the cornerstones of the study of astrophysics. The earlier theories as to the age of the sun were based upon it. Their inadequacy was due to no inadequacy in power of the kinetic picture but to lack of a sufficient background of the facts in this most difficult field to which it might be applied. With the discovery of entities far less massive than molecules and atoms, entities that form the building-blocks of atoms, minute discrete charges of electricity, the kinetic theory entered upon the conquest of another world. In such matters as the evaporation of electricity from heated metals (cf. pp. 288–91), basic for understanding the development of the vacuum tubes of radio, the kinetic theory has continued to achieve success. It would seem destined to go down in the pages of history as one of the greatest achievements of human imagination in its groping attempts to understand the physical world.

We need no longer attach to it, therefore, any provisional and hypothetical character. The effects of thermal agitation can be *heard* in background noise in highly sensitive radio receivers and can be *seen* under the ultramicroscope. If precipitates of the utmost fineness of division be watched by means of indirect, dark-field illumination under the highest power oil-immersion lenses of the modern microscope, the effects of molecular motion will be seen in the continual incessant agitation of the tiny particles of the precipitate. Measurements of sizes and velocities reveal a relation between size and liveliness, highly significant from the standpoint of energy. The squares of the velocities vary inversely as the masses of the particles; i.e., the kinetic energy of all kinds of particles are always equal at any given temperature. By the same token these energies are equal to those of the forever invisible liquid molecules of the supporting solution.

If tobacco smoke, illuminated by a powerful transverse beam of light, is watched through a microscope of moderate power directed at right angles to the illuminating ray, these particles likewise will be found executing their kinetic dance in gases just as other particles do in liquids. The same laws are found quantitatively exact. Such motions are called the "Brownian movement," after their discoverer, a botanist named Brown, who saw them first. Fifty years elapsed before men rightly interpreted their nature. These motions can be photographed and shown to large groups of people through the agency of motion pictures (Plate 41b).

Convincing as such actual photography may be to the layman, such proofs of the reality of molecular motion are not needed by the scientific worker. By his long training and application to these matters he has been able to see these things with the eye of the mind, an organ of vision far less liable to error than the physical instrument of sight (Plate 40).

PLATE 40 A Dish of Water. *Left:* As seen by the layman. *Right:* As it appears to a physicist

THE KINETIC THEORY OF HEAT

SINCE it is beyond our scope to take you through the details of many different sorts of experiments and the train of reasoning through which the many interesting facts, relating to sizes, speeds of motion, distances apart, numbers of molecules for a given weight of substance, and the like, have been calculated, we can only append these items in still another table!

Number of air molecules in 1 cc. Obtained from experiments on electrically conducting solutions and knowledge of the elementary electric charge; or from observations on Brownian motions	$27{,}000{,}000{,}000{,}000{,}000{,}000. = 27$ trillion $= 2.7 \times 10^{19}$
Average velocity of molecules in a gas Obtained from simple kinetic theory, see Equation (1), p. 158	$c = \sqrt{\dfrac{3 \times \text{pressure}}{\text{density}}}$ is somewhat greater than the speed of sound in the gas. I.e., for air at 0° C., $p=76$ cm. 50,000 cm./sec., about 1,500 ft./sec.; for hydrogen at 0° C., $p=76$ cm. $c=185{,}000$ cm./sec., something over 1 mi. per sec.
Average distances traveled by molecules in a gas between collisions, L Obtained from experiments on viscosity in gases Or $\begin{cases} \text{heat conductivity} \\ \text{rate of diffusion} \end{cases}$	For air: 0° C., 76 cm. pressure $L = .0000085$ cm. $= 8$ millionths cm. $= 8.5 \times 10^{-6}$ cm. For hydrogen: $L = .000016$ cm. $= 16$ millionths cm. $= 1.6 \times 10^{-5}$ cm. *In interstellar space* (where there are a FEW molecules) a molecule travels about 50,000,000 mi. before it hits another
Numbers of collisions per second of molecules Obtained from same source as preceding	In air under ordinary conditions $5{,}300{,}000{,}000.$ per sec. $= 5$ thousand million per sec. $= 5.3 \times 10^{9}$ Between molecules *in interstellar space*, collisions occur about 0.00000003 times per sec. $= 3 \times 10^{-8}$ i.e., once a year
Diameters of molecules From combinations of preceding results	Nitrogen: 0.000000018 cm. $= 18$ thousand millionths $= 1.8 \times 10^{-8}$

SOME FIGURES RELATING TO MOLECULES

For the sake of some readers who may have some former background of knowledge in these connections, we are including, wherever possible, a brief statement of the kinds of phenomena that are found to be able to disclose such information.

It truly is a Lilliputian world, this world of molecules. What contrasts it shows to the world we know! Within the transparent air about us, quintillions of these tiny entities (Plate 41a), thousand millionths of a centimeter

PLATE 41a $10,000,000,000,000,000,000 = 10^{19}$, Molecules. If represented by grains of sand, the number of molecules in *one cubic centimeter* of air would make up a huge block of sand equal to a *cubic mile*

in size, travel distances many hundreds of times—even a thousand times—as great as their own diameters before colliding with each other. Most remarkable perhaps of all is the fact that these tiny things travel at prodigious speeds—speeds that approach a considerable fraction of a MILE each second. Of course, it is a rather crooked mile, however, with a turn at each collision—no less than 5,000,000,000 turns each second.

You may by now be in revolt. You don't believe a word of it! After all, no one will blame you much. One could not convince a fish of the existence of a giraffe. To go higher, one could not have convinced a man of early Stone Age culture of the existence of an airplane. Even if shown one, he could not really comprehend what it was he saw. Most of us will never go to the North Pole, to Antarctica, nor fly above Mount Everest. Pictures of these places we can see; we can converse with those who have been there, and our geographic knowledge began to be cultivated at so early an age that we can realize there are such places.

It is but lack of familiarity with these ideas about molecules that causes our skepticism. There are many men and women who daily spend their lives in realms even more re-

PLATE 41b Seeing Temperature. A snapshot of particles of a very finely divided resinous gum (gamboge) suspended in water and taken through an ultra-microscope (over one thousand times enlargement). When photographed by a motion-picture camera or when viewed directly, the perpetual agitation of these particles due to the agitation of the surrounding water molecules, which we call "temperature," is clearly visible

THE KINETIC THEORY OF HEAT

mote than these. The sum total of these persons is utterly negligible among the millions of human beings; and yet their influence, or rather the influence of the knowledge that they cull from the book of nature, when this knowledge is turned to practical uses, has within the past two hundred years completely altered the lives of a very large percentage of the rest of humanity.

This world of tiny things is invisible to and unseen by most of us but is just as real as we are. Since its exploration has been opened up, and we have found that it forms but the frontier of a world still more tiny, where much of mystery yet remains, inevitable effects are to be expected in the future on our social and economic structures. A great industrial revolution swept the world as a result of a clear understanding of the phenomena of heat and the engineering developments of thermal motors. What changes will result from the exploitation of the subatomic world of electrical phenomena which the pioneers of exploration are but now entering, no one can predict. It is to this region that we shall next devote our attention.

PART III
ELECTRICITY AND MAGNETISM

Courtesy of the Mayor and Town Clerk, Colchester, England. (From the painting by Hunt)

PLATE 42   GILBERT DEMONSTRATING BEFORE QUEEN ELIZABETH

CHAPTER 19

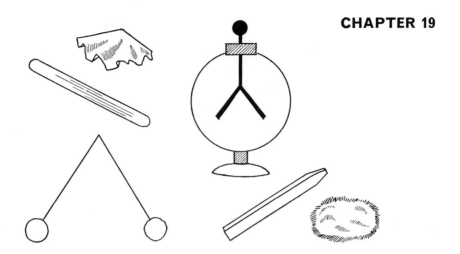

ELECTROSCOPES:
PITH BALLS AND GOLD LEAVES

THE development of experimental knowledge of thermal phenomena, followed by the interpretations of the kinetic theory of heat, together with the invention of the steam engine and its rapid improvement, ushered Western civilization into the AGE of STEAM, the nineteenth century. In an even more dramatic fashion the experimental efforts of a few men in the field of electricity during the nineteenth century, together with the penetrating theoretical advances of a few others, swung open toward the end of that century a new doorway through which our culture passed into the AGE of ELECTRICITY, the twentieth century.

It has been said metaphorically that the grains of pigment on a canvas cannot judge of the picture that embodies them, and it would seem to be equally true that we who call forth light by the unthinking flipping of a switch, who daily step in and out of electric vehicles, who talk across a city or a continent realizing nought but the necessity of speaking a number, who sit of an evening before an instrument and at the turn of a dial listen to a speech from Washington, a symphony from Philadelphia, or jazz from Hollywood, are all unconscious of the magic and the mystery within our lamps, more wonderful than was Aladdin's.

Electricity is not to be comprehended in a passing glance. The Greeks gave it its name. As far back as Thales (640–560 B.C.) they knew that

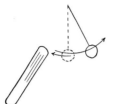
amber, when rubbed, acquired the power to attract to itself bits of straw and other light objects. In their language, ἤλεκτρον meant amber; so the objects were "amberized," or as we say, using their word, "electrified."

Court surgeon to Queen Elizabeth and supreme experimenter and critic of his day was Gilbert (1540–1603). (See Plate 42.) He knew that many other things than amber had this property. (We now know that all substances possess it.) Suppose we again perform some of his experiments. Hanging up a light ball of pith carved out of the heart of a humble dried potato, in the absence of better material, we bring near it a glass rod that has been rubbed with silk. The ball is first attracted, then vigorously repelled. We try the same trick with a stick of sealing wax* rubbed with fur, and get the same result.

If, however, we bring the glass rod near the ball after it has been touched with the wax, it is attracted. If contact with the rod is prevented and then the stick of wax brought up, the ball is repelled. Reversing the procedure, we first let the ball touch the glass and be repelled. Now we bring the wax near and witness attraction, but prevent contact. The return of the rod to the ball's vicinity still produces repulsion, as formerly.

ALL this seems pretty complicated for a beginning. It is more simple if we may be allowed to adopt as a hypothesis that there are two kinds of electrification: a "glassy" kind, vitreous; and a "waxy" kind, resinous. Or we can save type and words by adopting the usual convention—calling these plus (+) and minus (−), respectively.

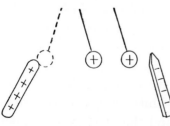
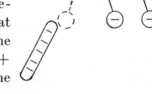

Now, when the ball touched the glass rod, some of the rod's + electrification was communicated to the ball and was then repelled by that on the ball. The − kind on the wax, however, attracted the + on the ball. If, however, the ball touched the

* Amber is often used but is much more expensive.

ELECTROSCOPES

wax first, it acquired $-$, and then again this $-$ was attracted by the $+$ charged rod. It would appear, if there are two kinds of electrification, that like kinds repel and unlike kinds attract each other, as is often said of people. This is all right as far as it goes; but, as we have seen, these hypotheses we invent have a bad habit, good for our purposes, however, of implying more things than we originally intended.

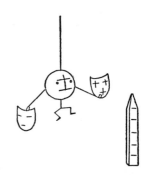

The very first thing that happened after preparing our potato ball and hanging it up was that it was attracted by either kind of electrification that was first brought near. To carry on with our hypothesis, it thus suggests that the ball took on a negative aspect when the $+$ glass was brought near; but a positive aspect when the $-$ wax approached it. This seems to give our lowly tuber a degree of penetration that we are reluctant to admit. Could it be possible that in some way the proximity of either kind called forth the other kind (which would be, of course, attracted) from within, or in some way upon, the ball?

If you yourself try these experiments with sufficient patience, you can verify everything we have described. It will take patience, however; less patience in winter than in summer. Not any old thread is equally good, either, to support the ball; it ought to be of real silk, from worms, not synthetic. There are simplifying tricks in all trades, furthermore. The experiment is greatly benefited, and the various happenings take place with greater certainty and uniformity, if the ball is covered with tinfoil, or painted with a metallic paint. Whenever contact is made with charged glass or wax, these implements are wiped off, as it were, upon the ball by letting it roll along them.

If you are clever in these matters, you perhaps can guess the answer to the question proposed above. If not, you must read further.

We will now abandon the ball apparatus, but not the question. We will make a more ambitious apparatus and, by the same token, one more sensitive. Alongside a flat *metal* plate we will hang a strip of *metallic* gold leaf. Instead of a silk thread, we will use an amber plug and surround the leaf with a box of metal and provide glass windows for us to look through. Thus we shut out air drafts that

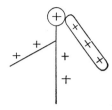

might wreck our gold leaf. A metallic knob is placed on top of the plate to which the leaf is attached, where it protrudes through the box. This adds quite a professional air, and so we will give our device its professional name. It is a "gold-leaf electroscope." Now we will take up some experiments that will help answer our question.

When the silk-rubbed glass rod is brought near the electroscope, the leaf diverges from its supporting plate, and upon making contact between rod and knob the leaf stands off at nearly right angles to the plate. Following our hypothesis, we have by the contact charged the electroscope with $+$, both leaf and plate; $+$ on leaf and $+$ on plate repel; hence the light leaf defies gravity and stands away from the plate. We then touch the knob with the hand, and the leaf collapses. Evidently the electrification has been removed. Similarly, when the fur-rubbed wax is brought near the knob the same sequence of events is again observed—divergence of leaf on approach, strong repulsion upon contact. The latter we understand, but not the former. In attempting to answer our question (p. 179) we seem to have raised another. Is it really another, or is it the same one in different guise?

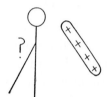

ALL we seem to have learned yet, in spite of all our efforts, is that like kinds of electrification repel and unlike kinds attract. But the *uncharged* ball was *attracted at first* by the approach of *either* glass or wax. The uncharged leaf, which is at one end of its plate, gives evidence of becoming charged when either glass or wax approach the knob at the other end. Maybe there was electrification OF BOTH KINDS and IN LIKE AMOUNT, both on the uncharged ball and on the uncharged leaf. The electrification was already there, only we did not perceive it, since both kinds, if they were there in equal amount mixed up together, neutralized each other and externally gave no sign of their existence.

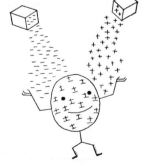

With this addition to our hypothesis we can now do some explaining:

ELECTROSCOPES

As a matter of fact, we can make several different explanations. Let us try one of them: First with respect to the ball: as the + rod approached, it attracted the − already in the ball to the side of the ball nearest it, and left the far side deficient in − and therefore positively charged. Similarly, in the 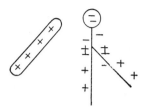 case of the electroscope, as the + rod approached this instrument, it pulled the − up into the knob away from the leaves, leaving the leaves +. On contact, either with ball or electroscope, much more − was pulled away into the glass rod and removed when the rod was removed, leaving a permanently and positively charged ball, or knob-plate-leaf system, behind. Well, this sounds ingenious, we must with due modesty admit. What if we use the − wax instead? That's easy also.

As the − wax approached the ball, it drove some of the − already there away to the far side of the ball, leaving the near side + and hence in a condition to be attracted toward the − wax. Quite similarly as the − wax 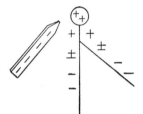 approached the knob of the electroscope, it drove the − away from the knob into the leaves, which diverged as it approached. On contact the wax left some of its negative charge on the ball or upon the electroscope, which thus acquired a permanent − charge. How is that?

IF NOT quite silenced now, hopelessly confused, or bored, you may feebly remonstrate and ask by what right we let only the negative charge be pushed and pulled about in this unseemly manner. Indeed, this was not necessary. It is one way of explaining, *but not the only one.* You yourself could have done just as well, or at least you could now, by letting the + be moved. Then you would have expressed the ideas of the second paragraph preceding this one in the terms of the one just preceding. To do this you would have interchanged all the + and − signs and words, and used the word "rod" instead of "wax." The idea in the last paragraph you could have taken care of in

a corresponding fashion by similar changes in the paragraph preceding it. Then you would have explained everything too, in terms of moving positive charges instead of moving negative ones, as we did. You, likewise, would have obtained an answer to the question first proposed. It is just like ours in assuming that both + and − in equal amounts are present in an electrically neutral object. It differs from ours only in a detail: as to which kind of charge, + or −, we are to consider mobile. Indeed, if you are ingenious, you could let both kinds of charges move about simultaneously in opposite directions, and your explanation would be logically just as good as either of those proposed.

THE nineteenth-century studies of electricity were in this very dilemma. There were "two-fluid" theories and "one-fluid" theories; and in the one-fluid theory that found favor the positive fluid moved and not the negative. The idea of fluid still remains in technical jargon. Electricians to this day refer to their commodity as "the juice." The idea of positive charges being mobile persists in certain of conventional representations that we shall refer to later (see footnote, p. 239). It remained for twentieth-century science to discover, first, that electricity was not a fluid at all but was composed of myriads of discrete particles, i.e., was of atomic character just as matter is; and, second, that there was only one kind of these atomlike charges that moved about in metallic conductors, and that this kind was the negative. Negative electricity has several other rather unique characteristics. In Part IV we will detail the evidence for this mobility of negative charges and many of its other unique qualities. We will ask you now to *accept our hypothesis*, rather than your alternative one, not on faith, but pending proof which later we will submit. We shall, from now on, speak of − electrification as if it were produced by supplying negative charges to an object, and of + electrification as if produced by the removal of some of the negative.

CHAPTER 20

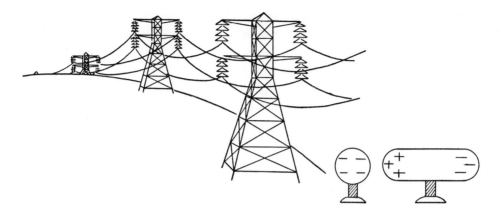

CONDUCTION, INDUCTION

WE HAVE mentioned conductors rather incidentally in the preceding chapter. For purposes of understanding electrical phenomena we can divide all substances into three classes. There are those that conduct electricity readily, and there are those that do it very badly, so badly that we say they do not do it. These latter we call "insulators." There are still other substances that are indifferent, neither very excellent nor very poor in this respect. With these we are not much concerned.

All metals are rather good conductors, although even among them there are conspicuous differences of degree. Silver, copper, gold, aluminum, in this order, are most excellent conductors. Zinc, nickel, iron, platinum, lead, and mercury are fair to middling; e.g., silver conducts nearly one hundred times as well as does mercury. Graphite and carbon are indifferent conductors. Boron, selenium, and sulphur are among the poorest, the latter being a hundred *billion* times worse than silver, which means that sulphur is about as good a NON*conductor* as can be found.

IT WAS for this reason we used it in our electroscope as insulation, since only insulated conductors can be charged effectively. If a body which we desire to charge electrically is not insulated (by sulphur, glass, rubber, silk, amber, or mica—all of which may be classed among the best insulators), negative charges placed upon it will leak off to the ground and be dissipated.

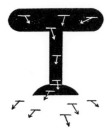 Or if the attempt be made to charge it positively, by removal of some of its negative charges, other negative charges will be attracted to it, and, finding their way up from the ground, will at once replace what has previously been removed.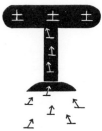

When negative charges are placed upon well-insulated conductors, the action of their mutual repulsions and the freedom of motion furnished by the metal results in these charges being spread out all over the surface. If we could see them on it, the surface would look as if it had been peppered with them. They distribute themselves uniformly, however, only if the surface is flat or *uniformly* curved. On a pear-shaped  conductor the charges tend to cluster at the more pointed end. On a sharpened point they crowd together very thickly in spite of their mutually repellent character. This is why the electric supply to and from apparatus like X-ray machines (apparatus that carries these charges in great quantities) is carried by good-sized metallic tubing rather than by wire. Otherwise the electricity may leak off, literally pushed off into the surrounding air by its own self-repellent character. This is also why lightning rods are pointed, in the hope that about one's home it *may* leak off quietly into the air above and so discharge the great amounts that sometimes accumulate in the vicinity of thunderstorms. (See later, p. 194.)

YOU must not gather from what we have said that only conductors contain, when neutral, equal amounts of + and − electricity. This is equally true of insulators; but on them the negative, as well as the positive, is associated closely with the atoms and molecules of the material, and *neither* kind is free to move about. By frictional processes, such as rubbing—or, indeed, by any other process that brings the surfaces of different substances into close contact (Plate 43), the negative may be transferred, not wiped off, but picked up, as it were, from one substance by another. Just as flies entangled in a thin syrup may be picked up and removed by the more sticky substance on fly paper, so can charges be passed from one material, either insulator or conductor, to another.

PLATE 43 ELECTRIFICATION BY CONTACT ONLY. Friction produces electrification solely as a result of the close contact between the surfaces of the materials rubbed, which the frictional process engenders. Here there has been no rubbing. The bucket on the electroscope contains water, in which, on the left, a paraffin block is submerged by an insulating handle. Upon the withdrawal of the paraffin, on the right, the water is left charged, as is shown by the electroscope. The paraffin is charged oppositely of course, as a supplementary experiment could show.

Thus, when glass is rubbed by silk, both insulators, negative is picked off the glass by the silk, leaving the former + and the latter −. You could prove that the silk had caught up the − in the following way. Suppose we bring it near an electroscope that had previously been charged, let us say positively, thereby causing its leaves*to diverge. Now as the − charged silk approaches, the leaves will be found to diverge *less and less*, showing that some of

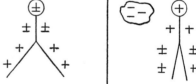

the negative remaining in the electroscope as a whole has been repelled down into the leaves, making them less + and therefore less mutually repellent.

Similarly for the case of wax and fur. If the fur (for insulation's sake, mounted on a glass or rubber support, and thus kept away from the conducting hand of the manipulator) be rubbed on the wax, the fur will show positive charges just as definitely as the wax shows negative.

* We now, for variety, speak of an electroscope that has at the lower end of the plate that previously held a single leaf two leaves, which when charged will diverge symmetrically away from the vertical line through their point of support.

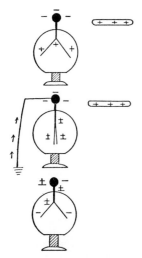

Now let us use what we have learned, and charge our electroscope by quite a different method, that of INDUCTION. Starting with it neutral, we bring our + glass rod near, calling forth − up into the knob of the electroscope. Then, maintaining the position of the glass rod nearby, we *ground* the electroscope with a wire to some water pipe, or by touching it with our finger. A path is now provided for some of the earth's vast store of negative to flow up as near as possible to the attracting + on the rod. Just enough of it does flow up to compensate for that previously pulled away from the gold leaves. Thus the latter regain what they lost, become neutral, and collapse.

Next we break the ground connection. In so doing, we have trapped some of the negative pulled up from the earth. As long as the rod is held near the knob, this trapped charge remains on the knob and the leaves are still neutral; but when we take away the rod, the excess negative, no longer held on the knob, and unable to escape back to earth, distributes itself over the entire electroscope system—knob, plate, and leaves. The latter, now having an excess of negative, diverge.

By bringing up a + charge in this manner we have charged our electroscope *negatively*.

WE IMAGINE that you can see no reason why we would not have obtained a similar result if we had used the − wax rod, and then, going through the same operations, obtained a *positively* charged electroscope. This is quite right. With a pencil and paper you can probably sketch the sequence of events yourself, the repulsion of − into the leaf as the − wax approaches, its flight still farther away to the earth when the ground is made, its inability to return after the breaking of the ground connection, when the repelling force has been removed, and the result, a negative-deficient knob which pulls some negative away from the leaf system and leaves a system positively charged in its entirety.

By these induction effects, using an electroscope that is originally charged, we can determine the sign and very roughly the amount of charge on some other object that is brought near. Details of this would make a

CONDUCTION, INDUCTION

boresome repetition of what we have been doing now for several pages. You can satisfy yourself that if an electroscope has a + charge upon its leaves originally, the leaves will diverge still farther on the approach of another positively charged object; whereas they will collapse and possibly

Operations of the Faraday ice-pail experiment

even rediverge on the approach of negative. You can also figure out that just the opposite takes place if you start with a negatively charged electroscope.

By attaching an insulated pail to our electroscope, we can show that whenever charge of one sign appears, an exactly *equal* amount of that of the opposite sign appears somewhere else. It makes no difference whether the electrification is produced by friction, by induction, or by any other of the manifold ways we have found for effecting its separation. Faraday (1791–1867) was the first to establish this fact. Since he employed a bucket that was regularly used to carry ice about the laboratory, you will find the experiment still called the "Ice-Pail Experiment" in the conventional textbooks on physics.

Incidentally, in these connections it has been found that electrical charges distribute themselves on the *surfaces* of conductors. Inside of a metal box there will never be found any free electrification unless one deliberately carries charges within upon one's clothes or other objects. When

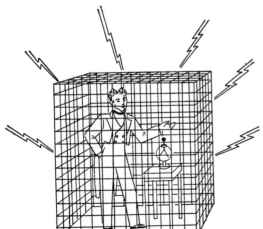

one does this even, the charge cannot be made to remain inside if contact with the interior of the box is established. The most violent electrical effects can be taking place outside the box, even over its exterior surface, without the most delicate electroscope detecting evidence of any electrification within. The box need not be solid metal, either, but can equally well be constructed of wire like a cage. Such Faraday "cages," as they are called, are used not to confine electricity within them but as a refuge and a secure haven whenever one wishes to escape from it. As we shall see later (p. 194), they are used as sanctuary if one wishes to protect himself or any of his apparatus from all electrical disturbances.

THE electroscope, which has been described in great detail because of the very many ways in which we shall need to employ it later to find out more about the nature of electrical phenomena, is, after all, only a qualitative instrument. Precise measurement is absolutely necessary if we are to be able to obtain verification of any deductions we may be able to make and so test the value of our hypotheses in these matters; and for precise measurement we need a more accurate apparatus—one that will not only detect charges of electricity but will also measure their amount by the forces which are exerted by these charged objects on other charged objects in their vicinity.

THE ELECTRO*meter*, as its name indicates, is a precise measuring instrument. It is a most fundamental piece of apparatus in every physical and chemical laboratory. It is taken aloft on expeditions into the stratosphere to investigate the nature of cosmic rays. Even biologists are coming to rely upon it more and more. In principle it is simply a refined electroscope.

Instead of gold leaves, a light needle is hung inside a protecting case from a fine wire supported by the usual insulated knob. On either side of the needle are arranged two pairs of insulated knobs or two pairs of plates, cut out as quadrants of a circle.

Connections are made as indicated in the diagram, so that opposite knobs or plates are connected together and can be charged independently from some available source of electricity. If the needle is uncharged, the induced charges called forth on it by the charged knobs or plates are all symmetrical as to position, and the forces due to them all balance, so that the needle is unaffected thereby.

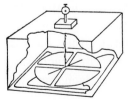

If, however, a small charge is placed on the needle, this symmetry is destroyed and the needle is very strongly pulled by one pair of knobs or plates and simultaneously repelled by the other, and so twisted around on its supporting wire. The stiffness of the

CONDUCTION, INDUCTION

latter can be previously determined, and the magnitude of the mechanical forces due to electrical charges thus measured with precision.

By apparatus such as this, it has been found that the *force between two charged bodies is exactly proportional to the charge in each one;* i.e., if the amount of charge on either one is doubled, the force is doubled. If the charge on both is doubled, the force between them is quadrupled. Furthermore, the force between them depends upon their distance apart. *As the distance between two charged bodies increases, the force diminishes in proportion to the square of the distance;** i.e., if the distance is increased threefold, the force diminishes ninefold.

These relations were discovered first by Coulomb (1736–1806), a French mathematician and physicist, about the time of the French Revolution. A concise statement of Coulomb's Law is given below:

$$f = \frac{Q_1 \times Q_2}{r^2}†.$$

Q_1 and Q_2 refer to the charges on the two bodies, and r is the distance between them in air.

Now if Q_1 and Q_2 are both equal to 1, and r is 1 cm., then $f = 1$. Thus by Coulomb's Law we have stated what we mean by *unit electric charge* in terms of its mechanical effect.

IT APPEARS that we have a unit charge if, when this charge is placed on an object which is 1 cm. distance from a similarly charged object, they repel each other with 1 dyne of force. This is the force, you will recall, that gives 1 gram of mass a change of velocity of 1 cm./sec. each second it is acting (Part I, p. 11, n.).‡ Of course, we do not have to resort to experiments on moving objects at this juncture to determine the force. We have all sorts and kinds of balances by means of which we can measure forces. Indeed, the needle hanging on the wire in the electrometer is in itself a little balance. By previous calibration its deflections can be recorded directly as forces.

---

* If $f_1 = \frac{QQ'}{r_1^2}$ and if $f_2 = \frac{QQ'}{r_2^2}$ (i.e., charges remain the same and only distance is altered), then, by division, $\frac{f_1}{f_2} = \left(\frac{r_2}{r_1}\right)^2$.

† The similarity in form of this expression to that which expresses the law of gravitational attraction, Part I, p. 29, is to be noted.

‡ Perhaps you recall it better as a force about equal to the weight of a mosquito.

That the mechanical forces between charges may be very considerable can be shown by placing a metal plate which is a good conductor in contact with another that is a rather poor conductor and charging the two plates oppositely by means of attaching them to the lighting circuit. When the switch is closed, the two plates stick very tightly together.

WE WISH to emphasize, however, that here, as in other connections previously (e.g., in the choice of a temperature scale, Part II, p. 159), we have chosen a unit of electric charge quite arbitrarily in terms of some convenient set of circumstances. We have said the electric charges were atomic in nature, without giving, as yet, any of the evidence. If this is true, there is probably a *natural* unit of electricity, the charge on *one* of these electrical atoms. What relation this might have to the unit charge chosen above is an interesting question which we will answer later. For the moment

we will only say that the above defined ELECTROSTATIC UNIT of charge contains twenty-one hundred million* of these tiny atoms of negative electricity (electrons) that have come to play so important a rôle in our modern understanding of electrical, chemical, physical, astronomical, and even geological, phenomena.

* One can hardly comprehend such a number of charges whose combined effect upon an equal number one cm. distant equals a mosquito's weight. Packed solidly throughout a cubical box, there would have to be 1,280 of them along each side. $(1,280)^3 = 2,100,000,000$, approximately.

CHAPTER 21

THUNDERSTORMS

NOW let us turn from the considerations of electric charges so small that thousands of millions of them combined will produce forces only equal to the weight of a mosquito, and consider a thunderstorm and especially its attendant appalling electrical phenomena. Little wonder is it that earlier cultures saw in such storms the activities of gods. Thor with his hammer, Jove armed with the thunderbolt, were supreme, even among gods, with such weapons. To us who are a little less defenseless than men of former ages before the forces of nature, the pageantry of storm holds more of beauty than of terror.

DETAILS of the study of thermal effects in mixtures of gases and vapors that we did not go into in earlier chapters inform us that the rate of decline of temperature as we ascend above the earth's surface is less within a cloud than it is outside. Clouds are really blankets—in a physical, as well as a poetical, sense. They are formed by the condensation of water vapor into fog or mist in the higher regions of the atmosphere, which are regions of lower temperature. Indeed, the level bases of potential storm clouds (cumulus clouds) attest to a uniformity of rate of fall of temperature at in-

creasing elevations.* Once within the fog blanket, the situation is altered from what it was outside. At any given level it is warmer within the cloud than it is outside. Pressure being everywhere the same at any given level, the expanded warmer air is less dense, lighter, and tends to ascend. There being nothing to prevent it, this takes place; and thus inside the cloud is established an updraft of air just as if it were a great chimney. Outside warm air from all directions below the cloud is sucked under it and then flows upward. Carried swiftly higher and higher, condensing its moisture into visible fog as it goes, it causes the cloud, at first tiny, to grow, piling visibly its substance of condensed water vapor higher and higher. The larger the cloud becomes, the greater becomes its chimney effect and the more rapidly does it grow higher into ever colder regions in the upper air. The more suddenly and the more completely the warm, moisture-laden air is carried up and chilled, the more is its water content wrung out of it, and the larger become the liquid drops of water.

NOW we have already seen in Part I, chapter 4, that the terminal velocity of an object falling through the air becomes greater the larger and heavier the object. We have observed in these connections that minute drops of fog settle back earthward very slowly. They cannot settle at all through this upward breeze; but, growing as they ascend, their terminal velocities are increasing, and soon they are large enough to make headway downward. Rain then is formed and falls. If the air below the cloud is very warm and fairly dry, this rain may all evaporate before it ever reaches earth. We all have seen dark curtains of rain being thus re-evaporated, especially over the high western plains in the summer time.

If there be a sufficient supply of moisture, however, these large drops fall swiftly through the air currents in the cloud and are blown out of their spherical shape (see p. 23) into momentarily distorted plate-like and filament-like forms. These

* More precisely, when such clouds are numerous, this shows that the temperature of condensation of moisture out of the air (the dew point) is reached at a constant elevation above the ground.

break apart and reform into smaller drops, and in the course of this process *they become electrically charged*. A majority of them in this manner lose some of their negative electricity to the surrounding air, and this lost negative charge is carried with the air and finer mist toward the top of the cloud; whereas millions of positively charged drops descend as rain. If massed electrical forces due to this separation  get too great, the charges leap off their carriers, air and mist, and reunite with those on the drops. Lightning occurs between different portions of the cloud or between one cloud and another.

Often the local winds are very great and the clouds are blown apart. Have you not sometimes observed at the top of a great fleecy cloud a horizontal plume stretching away from it as a high level wind blows the top of the cloud away as rapidly as it builds itself up into this region? In this way the negative charges may get separated from their parent-cloud and scattered far and wide. The cloud, however, continues to form, accumulating more and more positive charge. Some of it, of course, carried on the rain, reaches the earth and is dissipated there; much of it remains aloft, however, on smaller drops buffeted about in the violent air currents in such clouds—air currents that aviators have learned to dread and to avoid as far as possible.

OUR cloud, now with a large excess of positive charge, approaches. Under it, along the surface of the conducting earth, there is induced a corresponding negative charge. The cloud settles earthward, the electrical tension between its charge and that induced by it increases, and ultimately the insulating properties of the air between become insufficient. Processes  commence, the details of which are too complicated for us yet to go into. More swiftly than thought, these processes build up a conducting lane through the atmosphere, and through it the electricity pours; the separated charges in part reunite, creating a white heat in the air along their path—a lightning bolt. The air, suddenly heated and expanded along this narrow lane, is quickly cooled thereafter and, contracting,

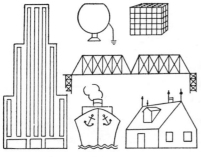

"Faraday cages"

crashes back. The terrified mortal, near one end of this path, which may be a mile or more in length, hears the sound from the nearer end first, then successively from regions more and more remote—a roll of thunder, which later re-echoes from cloud to cloud, perhaps for many miles.

We have previously referred on page 184 to the fact that free charges on the surfaces of conductors tend to cluster more thickly at pointed ends. This gives us one clue as to how we may protect ourselves and our homes from lightning. From what was said about Faraday cages on page 188, we can get another hint. If we were to inclose our homes in a network of closely laid cables, so that we virtually lived within a conducting surface (a Faraday cage), we could laugh at the storm. Indeed, such a structure could be struck by lightning repeatedly and we would never know it except for the commotion caused in the nearby air. Modern steel frame buildings with foundations sunk into ground saturated with water are virtually such cages. We know of no case ever reported of damage or injury from lightning to the interior of such buildings or to any person within them, although such structures, if very high, are quite apt to be struck, and indeed often are recorded as having been hit.

SUMMER cottages, isolated farmhouses and barns, are subjected to a very much greater hazard. These also may be afforded a large measure of protection by a properly installed lightning-rod system. We emphasize the importance of proper installation. A good ground connection—indeed, several such—are very important. They can be made by digging down to ground water, always to be found at greater or lesser depths, and burying in these holes non-corrosive metal plates to which the cables forming the rod system may be attached. The more completely these cables can be laid to form a network over the structure, the more perfect will be the protection. It is usually regarded as sufficient, however, to have them follow all the vertical corners of the building up to the roof, and then to follow all ridges, either those of the main roof or of projecting dormer windows. In addition this system of cables is provided with vertical extensions ending in sharp points.

THUNDERSTORMS

Points are placed at each end of every ridge, adjacent to and projecting above every structure that extends above the general roof level, i.e., above every dormer window and beside every chimney.

SUCH records and investigations as have been made by insurance companies in the United States and Canada of the history of structures thus protected in regions where thunderstorms frequently occur show two things: first, they are not very apt to be struck; and second, if they are struck, they are seldom, if ever, set on fire, or suffer any damage either to themselves or to their occupants at the time.

The reason for the first result is not far to seek. As the induced charge upon the ground surface under a heavily charged storm cloud piles up, it finds readily conducting paths to the lightning-rod system and, spreading over the latter, accumulates with great density upon the pointed rod tips. As we shall see later, with sufficiently large accumulations the electricity quietly leaks off into the surrounding air and, finding its way upward, tends to discharge both cloud and ground of the accumulated electric stress. If this action is going on to any extent, a faint bluish glow, called a "corona," will be seen at night on the tips of the rods. Sailors have always called by the name of "St. Elmo's fire" an appearance of blue light sometimes observed from the tips of masts and yards of vessels when violent electrical storms were impending. This is probably nothing but another form of this phenomenon of gaseous "ionization"* that occurs at the ends of pointed conductors when they are heavily charged.

Of course, it is quite possible that in a violent, swiftly oncoming storm, the discharge from the pointed rods may be inadequate to prevent a lightning bolt. But if a flash occurs, it will, however, be somewhat weakened in magnitude; and it will probably pass directly to one or several of the rod tips, where a conducting path is already waiting for it, and be guided directly to earth over the multiple cable system, neither tearing its way through walls and leaving fire in its path nor jumping through the air within the house.

IF ONE is caught in the open without any protection from a storm that promises to be of considerable electrical violence, there is little to be done or that need be done in the way of protection. Isolated trees and fences are often hit. One should never seek shelter under them or expose one's self

---

* The detailed phenomena of gaseous ionization are discussed in Part IV, chap. 29.

Where not to be in a thunderstorm

on any eminence. Groups of people and animals seem to be in more danger than isolated individuals. An iron bridge with a lattice work of girders would seem to be a very safe thing to get under for dryness as well as for protection, provided it is well grounded at the ends. However, it affords a large area, correspondingly more apt to be in the path of a bolt. If it is struck, unless its abutments are of metal and go to water, the discharge is very apt to streak to the ground near either end where people usually take shelter. Under normal conditions the area of a storm cloud is so large, compared to the size of a single individual, that if he be seated in a hollow or lies out in the open away from trees and fences, he is never in any great danger. In woods the greater trees themselves form semi-conductors when wet, and are often seared by lightning. It is well, then, to seek shelter under the smaller shrubs and saplings. While there is scientific foundation for the fact that certain locations are more apt to be struck by lightning than others, there is none at all for the many common beliefs that certain objects "attract" lightning because of their material composition. It is true that the electrical tension tends to break down more readily along the edges or pointed tips of conducting surfaces, and this may possibly account for some slight tendency for objects near the edge of water surfaces to be hit. Such a tendency, however, would be very hard to establish.

MAN has not yet been able to duplicate electrical phenomena on the scale of the thunderstorm, and probably never will. However, he has studied in his laboratories the mechanism by which drops of water become electrified in a blast of air; he has satisfied himself that an artificial spray of water falling through the air contains about the same proportion of positively charged drops that occurs in rain; he also has invented various devices to build up and to store very large charges of electricity. Let us look at some of these and see if we cannot get some further experimental verification of the hypotheses we have previously been developing.

CHAPTER 22

Courtesy, Massachusetts Institute of Technology

ARTIFICIAL PRODUCTION OF CHARGE

MOST static electric machines, by means of which large accumulations of charges are built up, are devices that function by a periodic repetition of the process of charging something by induction. The electric "water-dropper" invented by Lord Kelvin is the simplest illustration that we know of such a device. Water drips from a pair of nozzles each within a metal cylinder. The drops are caught in two leaky cups. Each pair of cylinders and cups is insulated, but each cylinder is connected to the opposite cup. Suppose we begin with a tiny — charge on either of the cylinders. Since the water pipe is grounded, an opposite + charge is induced on the first drop. As the drop falls from the pipe, it carries this + charge with it and gives it up to the cup, which thus,

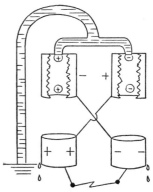

Kelvin's "water-dropper" electrostatic machine

PLATE 44 STEREOPHOTOGRAPH OF SMALL ELECTROSTATIC GENERATOR (WIMSHURST MACHINE) WITH DOLLHOUSE CONNECTED TO SHOW ACTION OF LIGHTNING RODS. A ball connected to one terminal of the machine serves as a "cloud" above the house. The other terminal is grounded to the lightning-rod system.

through the wire cross-connection, adds this + charge to the original amount of + on the opposite cylinder. The next drop in either cylinder has a stronger charge induced on it. It, in turn, gives this up to the cup and connected cylinder. Thus, with the fall of every drop, larger and larger charges are built up. Many thousands of volts* of electrical pressure can quickly be accumulated in this way, and very respectable sparks can be drawn from such a device.

THE mechanical types of electrostatic generators or static machines for producing heavy sparks formerly extensively used by physicists have been for many years regarded as obsolete and of historical interest only. Recently, however, some of these early forms have been redesigned and improved, and from them artificial lightning bolts several score of feet in length can be produced. These tremendous discharges are used in modern research experiments for purposes of atomic disintegration that are described in Part IV. Most laboratories have small static-machines of various types available for demonstration purposes. A photograph of one of these smaller ones (Plate 44), together with some recent gigantic ones (Plates 45 and 46) that have been constructed for research uses is shown by way of contrast.

* The meaning of this term, in terms of work done per charge transported, is deferred to a later chapter, chap. 25.

ARTIFICIAL PRODUCTION OF CHARGE

We can use a small electrostatic generator to demonstrate the correctness of some of our arguments in regard to lightning rods. In the illustration (Plate 47) the dollhouse contains beneath the roof a conductor that is connected to one terminal of the static-machine; the other terminal of the machine supports a ball that represents a "cloud," above the house. A small ball *within* the chimney is arranged to receive

PLATE 45 THE VAN DE GRAAFF ELECTROSTATIC GENERATOR, IN AIRSHIP HANGAR. (Cf. p. 317.)

the "lightning" that flashes between "cloud" and house in the absence of any lightning rods. The lightning rods are there, however, but concealed within the house and arranged so as not to function until a cord is pulled. By means of this cord, however, the lightning rods can be raised above the roof and put into action. When this is done, no more sparks pass from the "cloud," even though the static-machine is operated with the utmost vigor; instead, a slight squeaking and hissing can be heard, and in the dark a little glow is observed near the pointed tips of the rods (St. Elmo's fire [see p. 195]). The continuous discharge of electricity from the points is both audible and visible. As long as it continues, no large amount of charge piles up to spark across.

PLATE 46

ELECTRODES OF VAN DE GRAAFF GENERATOR

PLATE 47 THE EFFECT OF LIGHTNING RODS. (A) No rods. Sparks striking chimney. (B) Rods in position. No sparks. Electricity leaks quietly off the pointed rods. Note the tuft of glow on rods to the rear and brush on forward one. This is suggestive of "St. Elmo's fire."

IF AN insulated set of points, bent to form a little whirligig shaped like a swastika cross (卐), is connected to the terminal of the static-machine, this gadget will be set into violent rotation as the machine is operated (Plate 48). Again we may be able both to hear and to see the electricity as it discharges from the points where it is accumulated in greatest density. Of course, as the electricity escapes from the metal points, it charges up the molecules of air in the vicinity of the points. This air is then repelled from the points, and in turn repels the points backward (action and reaction [Part I, p. 45]), thus setting the device in operation. Such a stream of electrified air molecules repelled from a charged point actually creates a small breeze, the presence of which can be made visible as it blows against a candle flame (Plate 49).

In industry there are many applications of these discoveries and the principles derivable from them. We will take time to mention but one. If points or small wires highly charged are introduced into factory chimneys which normally carry great quantities of smoke or dust out with the heated gases, and thus pollute the air, this smoke and dust is, to a very great degree, precipitated within the chimney (Plate 50). Many thousands of dollars worth of valuable material in the smoke are thus in some instances recovered by these "electrostatic precipitators." In all cases the surrounding community is greatly benefited as to purity of air and general cleanliness, though we suspect the saving of valuable material is often the more influential factor in these factory installations.

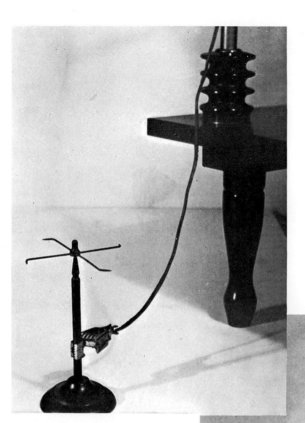

PLATE 48

AN EXAMPLE OF

THE "ELECTRIC WIND"

Leakage of electricity from the points of the suspended "whirligig" imparts momentum to the air near by. This is the "wind." The reaction upon the points (Newton's Third Law) imparts opposite momentum to the points, and on the right the device is seen spinning merrily, when the electric machine is operating.

 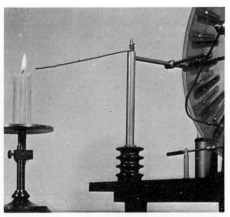

 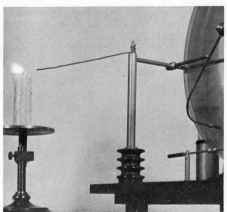

STEREOPHOTOGRAPH SHOWING LEAKAGE OF ELECTRICITY FROM A CHARGED POINT, ELECTRIC "WIND"

PLATE 49

The candle flame burns quietly in the upper photograph when electric machine is at rest. In the lower illustration the machine is in operation, and the escape of charge from the point blows the flame to one side and may blow it out.

STEREOPHOTOGRAPHS OF ELECTRICAL PRECIPITATION OF SMOKE

PLATE 50

Above: Smoke (ammonium chloride) is streaming up the "chimney." *Below:* The static machine is in operation, and its terminals are connected to the metal pipe and to a wire running centrally through the pipe. Smoke particles become charged and condense on wire and inside of "chimney"; very little smoke escapes.

FAR more important for our studies, however, than either practical applications of the principles of electrostatics or the small pleasure some of us may derive from witnessing the performances put on by these myriads of tiny actors, is the knowledge that will be gained in future chapters, as our story progresses, of the actors themselves. For the sake of being able to refer back to them from a subsequent chapter, we must finally tell you about two more experiments with the static-machine which are quite amusing but which also are of significance as giving analogies on a large scale of phenomena that we know occur on a very minute scale.

In the first one, we place within a short glass cylinder, closed at the ends by two metal plates, a lot of little pith balls, made of dried potato and sprayed with aluminum paint or wrapped in thin tinfoil. Placing the axis of the cylinder vertical, we connect the terminals of a static machine to the two metal ends. No photograph can show the lively motion, nor can words describe the rush and clatter that ensues as each pith ball rushes back and forth between the two plates just as fast as it can scamper, each time carrying its little charge acquired by contact at one plate to the other. You at once will understand the reason why it does this, in terms of forces of attraction between unlike and repulsion between like charges. We shall see that the conduction of electricity through solutions takes place in a very similar manner.

The second experiment consists in blowing a soap bubble from a metal pipe connected to one terminal of the static-machine, and then by means of a metal plate, similarly charged, holding it suspended against the force of gravity. If we know the weight of the bubble (which is easily obtained) and the charge on our plate (which also can be determined), the magnitude of the charge on the bubble can be found. Here we are not, of course, especially interested in the charge on a particular soap bubble. But when the experiment is set up on a smaller scale so that it has to be viewed under a microscope, and when the charge is carried by a tiny drop of oil instead of a soap bubble, it will yield very precise information as to the magnitude of the tiny atoms of electricity that we have previously mentioned. This discovery, which we will discuss at length later, has been one of the supreme achievements of our age.

CHAPTER 23

MAGNETISM

LEGEND tells of a Cretan shepherd, one Magnes, who, on being very violently attracted to the earth by the iron nails in his sandals, and by the iron-shod end of his staff, dug around into the ground to discover the cause. He found a mysterious rock which was responsible for this strange power. Whatever may be the facts that gave origin to this story, it is certain that properties that we still call "magnetic" in commemoration of the myth were known to be associated with certain kinds of rock by more than one civilization of antiquity. Pliny relates that King Ptolemy Philadelphos (third century, B.C.) ordered his builder, Timochares, to construct a temple with a roof of this particular kind of rock, in order that the iron statue of Arsinoe, the queen, might remain suspended in mid-air. The story may be fanciful. The hope assuredly was vain. It only goes to show that the relationship of this mysterious power to the metal iron was recognized. The Chinese are said to have been the first to suspend such rocks from a fiber, in order that the definite north-south orientation that invariably took place might be useful in navigation. These rocks, at that time called "lodestones" or "leading stones" from their directive qualities, we now know by the name of "magnetite," a not uncommon iron ore (Fe_3O_4) native to many places.

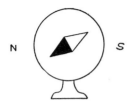

The dip needle

Both Peregrinus and Gilbert, by their careful observations and critical attitude toward commonly accepted traditions in this field, did the first precise and careful work. We mentioned Gilbert in connection with the earliest accurate observations of electrical phenomena. His work *De Magnete*, published in 1600, has been called epoch-making. Galileo referred to him in terms of greatest admiration. Ignorant that Mercator in 1546 had written a letter in which he likened the earth to a huge magnet, Gilbert insisted that this was undoubtedly the case. Into the details of his many observations we will not go. They related to the making of artificial magnets from natural ones, to the recognition of opposite "poles," and to the discussion of the behavior of miniature magnetic spheres in terms of which the phenomena of terrestrial magnetism were discussed.

That the magnetic compass not always—indeed, very seldom—points accurately to the north, a phenomenon, rather inappropriately called the "declination" by mariners, was discovered by the Chinese in the eleventh century. That it dips downward toward the ground if supported on an east-west horizontal axis was observed first by Hartmann of Nuremberg (1489–1564), but the first account of this was not circulated until 1576 by Norman.

IF WE ourselves experiment with magnetic compasses, made of bits of steel* which are magnetized by repeatedly stroking them in one direction with a lodestone or another magnet, we first of all note a property very suggestive of our electrostatic experiments and one which in magnetism is much more obvious. The north-seeking end of one compass will strongly repel the north-seeking end of another. Furthermore, the unlike ends strongly attract each other. In some other respects also the behavior of magnets strongly resembles that of electric charges.

Suppose we take an unmagnetized piece of steel and bring one end of it near the pole of a strong magnet. We find it is attracted, whether this pole be N. or S.† originally. On the other hand, after contact or even close ap-

---

* Steel, which is iron containing carbon, is suitable for making magnets by this method. Pure iron is not, for reasons we will amplify later.

† For brevity we are dropping the expressions north-seeking and south-seeking, and shall hereafter refer to these ends of a magnet simply as N. and S., respectively. Because of the attraction between unlike kinds, we will be required, from our definition of the N.-pointing as N.-seeking, to associate S. mag-

MAGNETISM

proach between the two objects has been established, and
especially if the unmagnetized bar is pounded and jarred
while in this position, it subsequently will be found to possess
magnetic properties on its own account. We will expect to
find that the end brought adjacent to a pole of the strong
magnet will have acquired the type of magnetism opposite
to that possessed by this pole. This process of magnetization by INDUCTION strongly resembles a similar process of
producing electrification. But there is a difference. The steel
bar magnetized in this way RETAINS its magnetic properties
when the inducing strong magnet is removed. Had it been made of pure,
soft iron, however, this would not have been the case, and the parallelism to
the electrical performance would have been much closer.

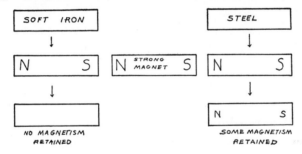

This magnetization by induction is the reason why a magnet will pick
up iron objects with such ease. Whenever such objects come in contact with
a magnetized pole, they become oppositely magnetized where they make
contact, and thus stick closer than the proverbial brother. Steel objects when
subsequently detached carry the imprint of the experience upon them. Pure
iron objects, on the other hand, after being removed are quite as unmagnetized as they were before.

TO MEASURE most conveniently the forces between magnetic poles, we
should, of course, like to isolate them. Here we run into an insuperable
difficulty not encountered at all in the electrical case. If we try to cut off the
N.-seeking end of a magnet and experiment upon it alone, we discover this
cannot be done. Every time we cut a magnet in two, or cut off an end of it,

---

netism with the north geographic location on the earth, whence these forces originate, on the Arctic coast
of Canada, and N. magnetism with the southerly location, where terrestrial magnetism is concentrated
in Antarctica.

we get two magnets. Each piece is a perfect magnet and has both an N.- and an S.-seeking pole of its own.

Because of this, any attempt to determine the law of attraction between magnetic poles always involves experimentation upon four of them, two N. and two S. We find that we can make calculations that roughly represent the forces if we assume, as in electricity, that each combination of two of the poles in air attracts or repels the other with forces that depend on the strength of each pole and inversely as the square of the distance between them. To get a fairly precise result, we cannot assume the poles are exactly at the end of the magnetized bar, however, but at some little distance from it.

Nevertheless, we define unit magnetic pole in a fashion similar to that we used for unit charge of electricity (p. 189), using this Law of Coulomb that seems to apply to these phenomena:

$$f = \frac{P_1 \times P_2}{r^2}.$$

If we make each quantity on the right $=1$, then f will equal 1, and we may say, "Unit magnetic pole is one which, when 1 cm. distant from a similar pole in air, will repel it with a force of 1 dyne." It must be emphasized that here, however, we are talking about something that can never be done even approximately, for, we repeat, we cannot experiment with isolated poles.

Even if we try to find the exact location of either pole in a magnet, we only become confused. We find that there seem to be quite a number of poles clustered about, in the main near the ends of the bar, the average effect of which at one end gives the N.-seeking tendency of that end, or the attractive tendency toward an opposite pole brought near. At the other end a similar rather indefinite clustering of S.-seeking poles is responsible for the net effect.

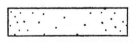
Scattered location of poles within a magnet

To ascribe all of the effect to a point somewhere within the magnet near the end, as we did above, is a convenience for computation but has no foundation in physical reality.*

---

* A quantity called the magnetic moment, M, the product of total pole strength by the length of the magnet, $P \times l$, is a very definite, useful, and accurate measure of the magnet's strength.

MAGNETISM

FURTHERMORE, these magnetic properties, even in those substances that show them most strongly, can be easily shaken out by various methods. Just as pounding seemed to help the process of magnetization by induction, so will pounding upon a magnetized bar, if no other magnet is near, greatly reduce its strength. For this reason any magnet that is to be used for quantitative purposes should be very carefully handled, and not dropped or knocked about.

Now as we have seen, matter, be it iron or anything else, may be jostled in other ways than by mechanical shock. Temperature, we have learned, is a measure of the jostling of the molecules of a substance. Is it possible that we might be able to destroy magnetic properties by heating?

Both pure iron and pure nickel are very susceptible to magnetic influences. Suppose we pick up a piece of nickel by a magnet and then warm it in a flame. With slight heating, i.e., to 350° C., the nickel drops off. Above this temperature it appears quite as insensitive to the attraction of a magnet as anything can be.

The same thing is true of iron or steel—of all common substances magnetically the most susceptible. If we arrange a wire loop of iron (Plate 51) so that it will hang in front of the pole of a powerful magnet, it will of course be pulled up against this pole. We can heat the wire by an electric current through it.* At the moment the wire reaches a dull-red heat (757° C.), it falls away from the magnet. Above this temperature no semblance of any magnetic property remains. As soon, however, as the wire cools below this temperature, it swings back upon the magnet and is as strongly attracted as formerly. Thus it appears that heat motions of the molecules may be sufficiently violent to shake the magnetic properties out of even a strongly magnetic material.

Next we might inquire if there may not be other substances, not at all magnetic at ordinary temperatures,† that might acquire this property at lower temperatures, where there would be less internal commotion. There is but one very common substance that can be used as an example of this. It is not a very good example since at ordinary temperatures it is a gas, and the technique of determining the magnetic properties of gases is quite beyond us at present. The gas is air, or—more strictly—the vital component of the

---

* For reasons that will appear later, we should use alternating current for heating.

† All substances, as a matter of fact, are very faintly magnetic in character, but only to about one millionth of the extent of iron, steel, nickel, and cobalt.

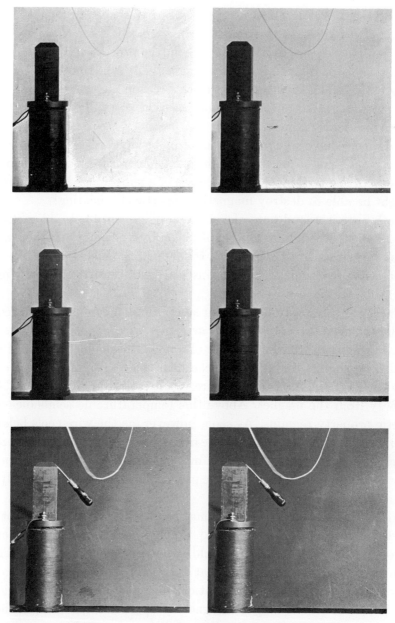

PLATE 51 STEREOPHOTOGRAPHS SHOWING THE LOSS OF MAGNETIC PROPERTIES BY AN IRON WIRE WHEN HOT (737° C.). *Above:* Iron at room temperature. Magnet unexcited. *Middle:* Iron at room temperature. Magnet excited, attracts the iron. *Below:* Iron at about 800° C. Magnet is still excited (note screwdriver), but the iron wire has fallen away from it.

air, oxygen. Its magnetic properties are negligible at ordinary temperatures. When, however, it is liquefied and at a temperature of −182° C., then if a little cup of it is placed between the poles of a powerful magnet (Plate 52) it will all be pulled out of the cup and drawn up all over the pole pieces. The experiment is not of long duration, since the material boils away so quickly; but we have succeeded in getting a photograph of the effect to show you.

PLATE 52   THE MAGNETIC PROPERTIES OF LIQUID OXYGEN. (A) Oxygen in cup between poles of an electromagnet as yet unexcited. (B) On exciting the magnet, the liquid oxygen flows out of the cup and up over the poles of the magnet, forming thereon two caps of the liquefied gas.

BY NOW you probably are wondering what kind of a queer business this phenomenon of magnetism is anyway. Columbus also had ample cause to distrust it, since he found on his voyages that the compass did not swing true at all times. Indeed, we have already observed that the earth's magnetism is far from being a constant and reliable guide for direction. At few places on the earth does the compass stand due north and south; at no place is its direction invariable; and in some rather exceptional locations in Northern Canada and on the Arctic Ocean, and also in the Antarctic continent, it either points exactly the wrong way or else cannot be made to take any particular direction at all.

In some respects the properties of magnetism seem to parallel those of electricity, but the parallel is no sooner made than exceptions to it must be

taken. In most introductions to the subject these difficulties are minimized. However, the fact is that they are of the utmost significance for a true understanding of it. Magnetism is very often treated as if it were a subject quite as fundamental as electricity. We wish to emphasize that this is not true, but is so false as to be utterly misleading. As a matter of fact, magnetism is a phenomenon that is fundamentally electrical. It is a special and very important aspect of electricity, and not an independent phenomenon at all.

A VAST realm of experiments that we shall outline in a later section (chap. 26) shows beyond a doubt that nearly all of the effects of magnets can be duplicated by materials entirely non-magnetic in the ordinary sense of the word, provided only that swarms of electrical charges can be sent streaming through such substances. This of course indicates a close relationship—even a dependence of magnetic effects upon electrical causes. Indeed, without electrical charges moving about there would be no such thing as magnetism.

Furthermore, many of the experiments described above seem to point to the association of magnetic properties with the atoms and molecules of matter: the peculiar way in which the magnetic poles, for example, elude our grasp when we try to pin them down so as to find an isolated individual; the close relation between magnetism and temperature; the very great difference in magnetic properties of a pure substance such as iron and those of the same material even when the merest traces of impurity such as carbon are present. These, as well as several other properties we have not taken space to discuss, all seem to point to a hypothesis that magnetism is first of all a molecular property.

IN PART IV we shall discuss those close relationships between electricity and matter itself which have led us to the conviction that everything material is essentially *electrical*. There, consequently, we will reach the modern point of view that this molecular property of magnetism is merely

MAGNETISM

a manifestation of the motion of electric charges within the atoms and molecules themselves, and thus the entire subject of magnetism is seen not to be fundamental but only incidental to our picture of the physical world.

BEFORE we can continue to pursue this thought, we must pause for a few moments to consider some of the mechanics* of electrical and magnetic forces, treating them independently and as different types of things for simplicity, but warning the reader now not to be misled by what is purely a device of exposition. We must recall, also, that in one of the very first chapters in Part I we said that energy of some form or other was represented in everything that had objective physical existence and that transformations of it were involved in everything that happened in the world of nature.

What we mean by electrical and magnetic energy, therefore, must at this stage in our story be made a little more clear.

* See introductory paragraph, chap. 13.

CHAPTER 24

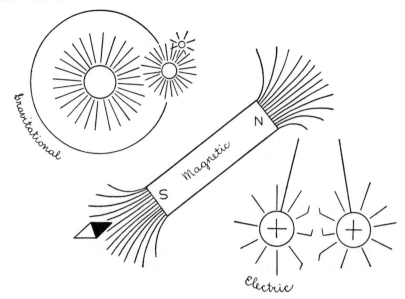

FIELDS OF FORCE

THE chapter heading looks forbidding. The subject, however, is one that we have already discussed at some length in chapters 3 and 5, under the titles of "Falling Bodies" and "The Law of Gravitation." As a matter of fact, you have spent all of your life in a field of force, and have not been unconscious of it either. If you stumble over an edge in the pavement and fall, it is the field of force that is responsible for the tumble. To it you owe your ability to walk, for that matter, since it presses your feet to the pavement, so that, with the assistance of friction in one of its helpful aspects,* you are enabled to establish a reaction with the earth and thus acquire a forward momentum.

A FIELD of force is a region of space where forces are present—nothing more abstract than that.

A GRAVITATIONAL field of force surrounds the earth. The notion is a vector one:† the field at any point has the direction along which a body that has little inertia, or no initial motion, will move upon release to the accelerating influence of the force.

* Note illustration, p. 78. † Discussion of vectors, p. 40 ff.

FIELDS OF FORCE

A barbed arrow, →, would suffice, as formerly, for the picture of the field at any one point; but to look at the region in its entirety, with an arrow attached to each point in it and the barbs all getting in one another's way, would seem somewhat impractical. A simplification suggests itself. Suppose in our imagination we take a small rock, far away from the earth, and let it fall. Suppose, further, that it leaves a trail behind it. Look at this trail. It will be a straight line from whatever point we start, and will lead directly toward the center of the earth. The direction of gravitational force in the region surrounding our planet could thus be represented by the entire system of such radial lines. To represent the strength of the force at any point we might adopt another convention.

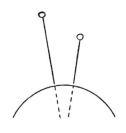

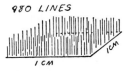

Strength of gravitational field at surface of earth

AT THE earth's surface the force on a unit mass of 1 gm. is 980 dynes. Suppose we imagine 980 such radial lines piercing each square centimeter of the level surface of the ground or water. Looking at them from the point of view of our human size, we would see about a thousand imaginary lines, parallel in our limited perspective and piercing each square centimeter of the ground at our feet. We can imagine, however, that they all meet at the center of the earth, 4,000 miles below.

Let us now take a longer-range view and move away twice as far from the center, i.e., 4,000 miles *above* the surface. Remembering that the surface of a sphere at this distance is four times as great as the surface of the earth at half the radius (Surface equals $4\pi \times radius^2$), we will see that the 980 lines which left 1 sq. cm. at the earth's surface are now expanded to cover 4 sq. cm. at this level; i.e., there are but 245 lines per sq. cm. at this elevation. But this number is the measure of the force of gravity at this point, also (see drawing on p. 27). Eight thousand miles above the surface, i.e., three times as far from the center, there would be only $1/3^2$ or one-ninth as many lines, i.e., 109 per sq. cm., which

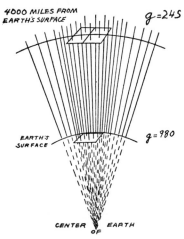

again tallies with the value of the earth's gravitational force (field) at this point.

The reason is obvious, of course. Gravity varies inversely as the square of the distance from the center; and since the surface of any imaginary sphere *increases* as the square of its radius, any given total number of radial lines through any square centimeter of a surface will consequently vary *inversely* as the square of the distance of that surface from the center.

By this convention, that the *number of lines* per normal* square centimeter represents the field strength or intensity—and their direction, the field's direction—we have quite as vivid a picture, and a much simpler one than that given by the barbed arrows formerly used.

AT THE distance of the moon, where the force of the earth's attraction is only 0.271, there will be but 0.271 lines per sq. cm.; i.e., we will have to take 1/0.271, i.e., 3.7 sq. cm. for every line of force. The relative denseness of the lines of force associated with a

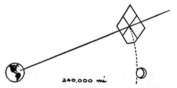

Strength of earth's gravitational field at distance of the moon

given area is thus an excellent device to enable us to visualize variations in force fields, such, for example, as the diminution of the force of gravity as we go from the earth's surface to the distance of the moon. The contrast is between the picture of 980 fine vertical needles thrust through a 1 centimeter square pincushion at the earth's surface, and that of one such needle pointing earthward for nearly four such pincushions at the moon's distance.

Take next a laboratory case, suggested by the measurement of gravitational attraction by the Cavendish experiment (Part I, Plate 5, p. 28), and imagine it set up in its simplest terms, 1 gm. of mass, 1 cm. from another gram of mass. How great is the attractive force? The experiment referred to shows that this force (called G on p. 29) is 666×10^{-10} dynes, i.e., 0.0000000666. Our mode of representation would thus call for this small fraction of a line per each square centimeter of area on a plane at right angles to the line between the two bodies. This spac-

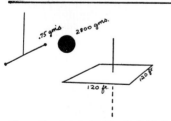

Strength of gravitational field in Cavendish experiment

---

* "Normal" means perpendicular. Note that the square centimeter of area is always taken perpendicular to the direction of the lines of the field.

FIELDS OF FORCE

ing of lines of force would allow but one line for every area of 15,000,000 sq. cm., i.e., for every square of about 120 ft. on a side. These gravitational cases, however, interest us only as stepping-stones to the electric and magnetic ones.

WE WILL now look at the electric *field*, at any region where electric forces obtain, i.e., in which electric charges are attracted or repelled by others in their vicinity. We note at once that in electrical phenomena we must consider both attraction and repulsion; whereas with gravitation there is only attraction.*

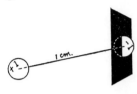

We will fix our thoughts first on two electric charges of unlike sign but each of unit magnitude. According to our definition of unit charge (p. 189), these two charges will attract one another with a force of 1 dyne when placed 1 cm. apart. Taking the negative as fixed, the positive one would then move toward it in a straight line, if free; and since the force is unity, it could be represented by unit vector (barbed arrow) or on terms of our now simplified convention by one "line of force" piercing a square centimeter of area perpendicular to the direction of the force at the location of the positive charge.

Since, however, our negative charge will attract with a force of 1 dyne any unit positive charge placed 1 cm. from it in *any* direction, we will have to associate one line of force with *every* square centimeter of area of a sphere 1 cm. in radius that has the attracting unit negative charge at its center. There are 4π sq. cm. in this area, and therefore our negative charge is visualized as a little spherical pincushion with 4π, approximately a dozen, needles bristling out of it equispaced in all directions.

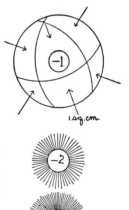

The attractive force on the positive charge is doubled if there are 2 units of negative at the center of the sphere, and in general for Q units of negative there must be $4\pi Q$ lines of force in the surrounding field. The divergence of these radial lines in the electric case, as in the gravitational ones, makes the picture fit at all distances away from it, since the number of lines per normal

---

* There exists in recent speculations about cosmogony something analogous to gravitational repulsion, on a cosmic scale, with respect to the apparent motions of the island universes of stars, the spiral nebulae.

(perpendicular) square centimeter decreases by the inverse square law associated with their solid geometry, just as the electrical forces also follow an inverse square law with distance.

IN THE gravitational case, because of the hugeness of one of the two objects, the earth, we confine our attention more to the attraction of the earth upon the rock rather than to that of the rock upon the earth. Since you clearly remember, however, that there is a double-endedness to all force effects (Newton's Third Law, p. 45), we know that the rock attracts the earth by the very same amount as that by which it is attracted by the earth. In the electrical case we now take up, there seems quite a degree of arbitrariness in speaking of the attraction of the negative unit charge for the positive and of the field of the negative alone. There is likewise a field about the positive in which all other positives will be repelled and all negatives, including the one we have been using, will be attracted. This field will likewise be represented by our pincushion picture of $4\pi\,Q'$ lines of force radiating from any positive charge for every one of the Q' units it contains; i.e., a total of $4\pi\,Q'$ lines altogether.

IF WE have two charges, either of like or of unlike signs, at the same time in our picture, these two fields add vectorwise at every point. The resulting picture of the lines of force is a complicated blended pattern that you may, perhaps, find difficult to visualize. We show you a drawing of it therefore. The figure to the left shows the field around our original $+$ and $-$ charges; each has Q units of charge, instead of 1, and now the separation is not necessarily 1 cm. Originating from each are the $4\pi\,Q$ radial lines, straight for only a short distance or only in one special location, each set blending and

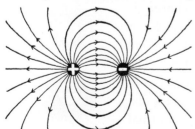

merging with the other, so that there is really only one set after all.*

A tiny $+$ charge anywhere in the figure, if free to move, would follow the path of one of these lines in a direction away from the positive end and toward the negative end.

* The drawing is, perforce, all in one plane. You must imagine a kind of cucumber of lines—a solid bundle between the charges, and lines standing out like hairs on end on the other side.

FIELDS OF FORCE

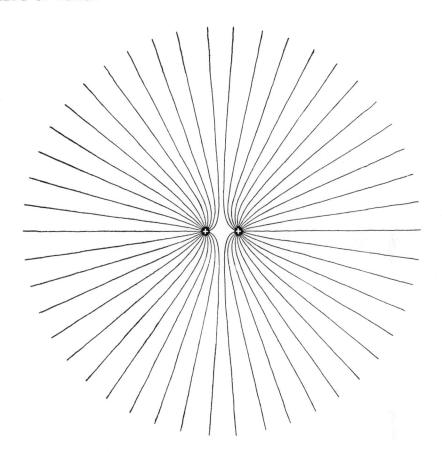

In case we have two equal positive charges there is, of course, a field in all of the space around them also, since a positive would be repelled therefrom. The picture shows the lines of force and their directions; X marks a spot midway between the two charges,* where the repulsion for the tiny + 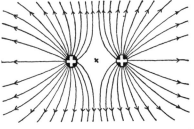 charge that we always use as our field indicator would be exactly balanced in a left-right direction. Our test charge thus would not know which way to go. It will be noted that the right and left components of the forces add to zero all along the perpendicular line bisector in the drawing; and the picture

* Indeed, this represents at one point in the plane surface of the page what is true actually along the entire surface of a plane in space that bisects the line between their centers of the charges.

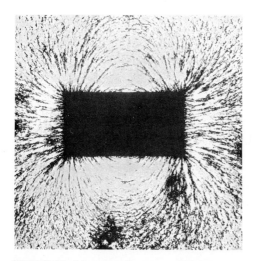

PLATE 53 MAGNETIC FIELD OF A BAR MAGNET. See also Plate 58, p. 237.

indicates the fact, since none of the lines of force pass across this forbidden region, or through the plane it represents if we were looking at it in three dimensions.

If we should now pass to regions quite remote from our two charges of like sign, the shape of the field would be not sensibly different from that of one charge of twice their magnitude. This is, of course, because at a very great distance away from them their actual separation becomes relatively too slight to be appreciable.

At a very great distance away from two equal *unlike* charges we will find no field at all, since at all remote points the attraction of one is exactly balanced by the repulsion of the other.

Let us emphasize again that we are not to think of space around charges as actually having within it *anything objective and real*, resembling a lot of filaments. The *forces are there*, as we can show if we bring up our test charge and investigate. We have used the space as a blackboard, in a sense, and drawn through it in our minds these lines to represent the direction of the forces. By their density we represent the strength or magnitude of the forces.

IN THE magnetic case, we have a similar picture of lines of force surrounding magnets, blending and twisting into complicated patterns as do the forces themselves in any region when we introduce into it a number of poles both N. and S. and of different strengths. In the case of magnets, there is a simple experimental way of verifying our ideas, at least roughly. Fine iron filings, magnetized by induction when sprinkled into a region where there are magnetic forces, tend to arrange themselves in little chains or bridges from one pole to another in the field. These bridges, it can be shown, will take the pattern of the lines of force that we have been imagining. Indeed, it was this phenomenon that first directed attention to the idea of a field of force.

A photographic reproduction of such a pattern is reproduced nearby (Plate 53).

FIELDS OF FORCE

THE earth is a great magnet. Its magnetic forces at any point are variable from place to place, as we have noted (p. 211); furthermore, at any one place, from time to time these forces also change. Another way of saying this is to remark that the earth's magnetic field is neither uniform in space nor constant in time.

We cannot experiment with isolated N. or S. magnetic poles, observing the direction of their motion and thus determining the direction of the earth's field; but small combinations of poles make up every magnetic compass needle. Such a compass sets itself along the direction of the lines of force at every point, the N. pole tending to move one way, and the S. pole the other way at the same time.*

Maps of the earth's surface showing the direction of the entire field at any given time are of great use to navigators and surveyors—indeed, to anyone who relies upon the compass to find accurately his orientation in an unfamiliar spot. The fact that one can find his location and the cardinal directions (N., S., E., W.) far better by observations on the stars is not always to the point, since often the stars for quite long periods are invisible. Data for these magnetic maps are compiled by the various governments of the world, and reproductions of them are to be found in almost any textbook or encyclopedia. The actual field of the earth's magnetism is not usually plotted, but lines are drawn connecting all points that show arbitrarily selected equal deviations of the magnetic compass (declinations) from the true N.–S. direction. Such lines are called "isogonic."

---

* A compass floated on the surface of water shows not the slightest inclination to drift as a whole *in either direction*, and thus we know that in all magnets the total strength of opposite poles or groups of poles is identically the same.

For illustration of use of compass to investigate the direction of a magnetic field see Plate 55, p. 234.

CHAPTER 25

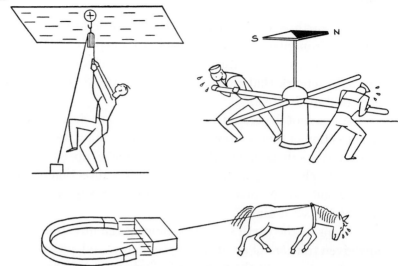

THE ENERGY OF ELECTRIC AND MAGNETIC FIELDS

IN THE latter part of our discussions of mechanics (Part I, chaps. 9 and 10), we emphasized as fundamental the considerations of work and energy. In the field of heat (Part II, chap. 18), we found that, by the identification of temperature with average kinetic energy of molecular agitation, there resulted a great unification and simplicity in our understanding of a vast range of phenomena which included changes between the solid, liquid, and gaseous states of matter as significant of potential energy differences. Indeed, the single unifying thread to be traced in all the different realms of experience that the physical sciences, and especially the more analytic ones of physics and chemistry, include is the notion of energy. In the field of electrical and magnetic discovery—and indeed, directly associated with one of the most important definitions, the volt—energy considerations are of paramount importance.

IF WITHIN the earth's gravitational field of force, and against its gravitational attraction, we lift a weight, we do work upon the latter, and work against the field. This work, equal to the product of the force by the height through which it has been overcome, is also called the measure of the potential energy of the lifted body. Since the weight of the lifted body is due to the mutual attraction between it and the earth acting at a distance through

ELECTRIC AND MAGNETIC FIELDS

the intermediate space, we speak of this potential energy as if it were a property of the space, i.e., as potential energy of the field of force. We make this definition in recognition of the fact that in the absence of friction all of the work expended in a field of force could be recovered by allowing the object to fall back to the lower level from which it originally was raised. If the work goes all or in part to the overcoming of frictional resistance, mechanical energy is lost; but a corresponding amount of thermal energy is produced (see chap. 11). The energy in other cases, however, may nearly all be used up in lifting other objects, i.e., in re-creating potential energy of other objects, or indeed of other forms; or it may all be transformed into energy of motion, kinetic energy. Out of all of our experience and investigation of these natural phenomena has come one of our magnificent generalizations, the Law of Conservation of Energy.

Possibly you wondered at, and probably you were more than a little bored by, our long discourse about fields of force just concluded in the preceding chapter. You must concede the author a foreknowledge of the story and of how its various actors are going to react on one another. His emphasis cannot be felt to be other than significant. It is of the utmost importance now as we proceed that you have images before your eyes of the character and nature of electric and magnetic fields, since we are now about to handle them in the sense of doing work upon them and of letting them return the courtesy.

IMAGINE the surface of a great metal plate scores of square feet in area, highly charged with electricity, so that as you stood upon it your hair would stream out and stand on end, and possibly St. Elmo's fire (see p. 195) would filter from your extended finger tips. Some one well insulated from you hands you a metal ball. As you take it, it acquires its own share of the charge upon yourself and the floor beneath. If you now should let it fall, it is quite conceivable it might fall *up* along the lines of electric force rather than down along the gravitational ones. It is now in two fields: the gravitational one of the earth, and the electrical one of the plate on which you are standing. Whether it falls up or down will depend merely on which influence is the greater, electricity or gravity.

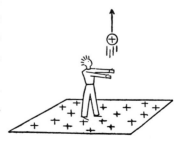

We do not want to press forward to details in this situation merely to make the point that the word "fall" could be used with respect to an electric field as well as with respect to that of the earth, and also that to fall is not necessarily to fall "down" in any sense except "down the field," along the direction of the lines of force. In electrical phenomena, "down" may be *either one* of the two opposite directions along the lines of force, depending upon whether the thing falling is positively or negatively charged. In absence of specification to the contrary, we will take for granted that the object on which the electric force, or field, is acting is a small unit POSITIVE charge. This is a convention merely—unfortunately, as we shall see in later connections, an inconvenient one at that.

To make sure that we are not going to confuse our thinking because of our ingrained acceptance of ever present gravity, suppose now the ceiling of our room is a highly charged plate, negative in sign. Billions of charges are upon it, each with its imaginary lines of force streaming toward it in nearly parallel lines (because we are so close to an extensive charged surface). Our positively charged ball in hand, we may be able to feel the pull with which it is urged toward the ceiling, to "heft it" in the electric field, as we are wont to do commonly in the gravitational one.

We release it—it drops like a shot along the field and it would hit the ceiling with a bang if we did not catch it in time.* We reach up to pick it down, and in "lifting it down" we do work upon it—we give it potential energy in just this same amount. In every sense this matter of doing work against the attraction of one charge, that on the ceiling, for another, that on the ball, is precisely the same as the business of doing work against gravity.

In one rather minor detail, however, the cases are often, but not necessarily, different. With respect to gravity the force is approximately the same everywhere near the earth's surface. Variations exist, of course, depending on latitude and altitude; and there are local variations, but these are relatively very small. In the electrical case the forces vary greatly from place to place. These variations are seldom uniform in rate, and simple averages cannot be taken to find their mean value. More refined methods of

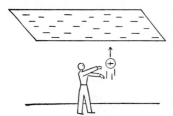

---

* If actually allowed to touch the ceiling, its charge would there be neutralized and reversed, and it would come crashing back. For our argument we want it to remain charged $+$; hence we will suppose actual contact with ceiling has been prevented.

ELECTRIC AND MAGNETIC FIELDS

averaging, those of the calculus, have to be employed. This need not trouble us, however, since the insistence on technical detail and exact rigor of computation is not part of our story.

THE magnetic case is so precisely like the electrical one with respect to energy considerations that we shall not take time for its discussion. Everything that has been said above about electrical fields and work done against them will apply to magnetic fields if you but read N. and S. pole for + and − charge. It is, of course, not possible to obtain uniformly magnetized surfaces of as large extent as electrically charged ones because of lack of conduction in magnetic phenomena. A somewhat Lilliputian scale of experimentation would have to be used magnetically. Another more serious difficulty is the fact that we could not get isolated N. or S. poles to toss about —only pairs that would rotate in the magnetic fields but not move as a whole.*

TO SUM up therefore: When we raise a mass of m through a distance h against the attraction of gravity, we do work—

$$W = m \times g \times h,$$

where g, which is the acceleration produced by the gravitational field, stands also for the field strength itself, since it represents the force acting on unit mass, 1 gm.

Similarly: When we move an electric charge of amount Q a distance h against the force of an electric field of strength E (assumed uniform), we do work—

$$W = Q \times E \times h.$$

And likewise: When we move a magnetic pole, P, a distance h against the force of a magnetic field of (uniform) strength, H, the work done is

$$W = P \times H \times h.$$

In all of these cases the work done against the fields has increased the *potential energy* of the object moved, since the object may fall back to its original position afterward and so give back the work (energy) in some other form

Thus, $$PE = \begin{cases} m \times g \times h & \text{gravitational} \\ Q \times E \times h & \text{electrical} \\ P \times H \times h & \text{magnetic} \end{cases}$$

* See p. 221, n.

Now we define change in potential energy *per unit mass, charge,* or *pole,* in going from one location to another, by the term "change in potential" or more commonly POTENTIAL DIFFERENCE (PD*) between the two locations.

Thus,
$$PD = \begin{cases} PE/m = g \times h \\ PE/Q = E \times h \\ PE/P = H \times h \end{cases}$$

If you will just wait for a moment longer now, we will have something you CAN remember about all this.

Following only the electrical case now, we repeat:

Potential difference between two points, change in potential energy per unit charge moved between these two points, is by definition,

$$PD = PE/Q .$$

Potential energy change, PE, being synonymous with work W done on the charge, may have the latter substituted for it,

$$W/Q = PD .$$

Therefore, by moving Q from downstairs on the left to upstairs on the right, our *fundamental electrical relation* that defines the idea of VOLTAGE or potential difference takes on a slangy aspect,

$$W = PD \cdot Q.$$

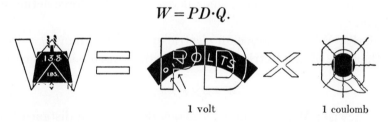

1 volt   1 coulomb

TO MAKE this idea perfectly concrete, suppose that you wish to measure the voltage across the switch that connects the electric wiring in your home to the source of supply brought to you by the local power company. We select the

Measurement of voltage

---

* PD is a symbol for a single quantity, potential difference, just as PE and KE have similarly been used to represent a single idea. The two letters do *not* mean a product of two things P and D.

ELECTRIC AND MAGNETIC FIELDS

switch provided for connecting or disconnecting your wiring from the electric mains simply because the metallic conductors are exposed at that point in the parallel metallic bars of the device.

By means of very sensitive pith ball or electrometer you can detect the actual presence of $+$ charge on one bar and $-$ on the other. With a very delicate balance you would be able to discover the fact that your pith ball with unit $+$ charge upon it is subjected to a force when in the immediate vicinity of the switch, a force that urges it away from the positive side of the switch and toward the negative.

Starting with it as close as possible to, but without touching, the negative side, and measuring this force upon it at every point as the pith ball is carried over to the positive side, you could (with some help, no doubt) calculate the work performed. This work would be the potential difference between the two terminal points of the unit $+$ charged pith ball's journey. The force at every point and the total work done would be much larger, of course, and correspondingly easier to measure if a much larger charge than unit charge were transported on the carrier ball. The potential difference could still be found, of course, in this case as well if the exact value of the charge carried were known also. Dividing the total work by the total charge would give the work per unit charge, the *PD*.*

ONE final comment should be made with respect to the *power* in electrical transformations of energy. Power is the rate of doing work against electric charges in electric fields.

Since work equals potential difference times quantity,
(ergs)   (voltage)   (charge)

$$HP\dagger = W/t = PD\, Q/t$$

power (rate of work) equals work per second, equals potential difference times quantity per second.

Now by the quantity of electricity per second running through an electric circuit we would naturally mean its rate of flow. Quite as if it were

---

* To get the answer in volts (the common practical unit of potential, instead of electrostatic units, the more scientific one), the answer you obtained in the foregoing paragraph will have to be multiplied by 300. If the *PD* is expressed in volts, the work in ergs, the charge will be in units called COULOMBS.

† We use the symbol HP (horse power) for power in general, irrespective of the units employed.

water running through a pipe, we would call this rate of flow the amount of the CURRENT.*

Thus the power of an electric circuit is the product:

$$\text{Power}\dagger = \text{potential difference} \times \text{current} .\ddagger$$

Hence, power in watts equals potential difference in volts times current in amperes.

$$\text{Watts} = \text{volts} \times \text{amperes} .$$

3,000 million charges flowing through per second

IF YOU will now turn back to the pictures on pages 84–85 of the girl first stepping up on the threshold, then walking slowly up the flight of steps, and that of the youth taking the steps three at a time per second, you will see the precise significance of the two electric lamps accompanying the two latter of these pictures. It should be obvious to you why there is no lamp accompanying the first one of the series.

TO THE adage about never being able to get something for nothing, the use of electric power is no exception. Fundamentally it is energy that we must pay for, and energy is sold commercially in units of kilowatt (thousand watt) hours. If you use energy at the rate of so-much-a-second (watts), then of course the total amount you use is the product of its rate of use times the time it is used. It seems rather roundabout to divide energy by time to get power, and then to multiply it by time to get energy again. A practical reason for this seemingly circuitous behavior has to do with the size of the demand put upon the power companies for energy. This varies enormously during the day and night, and also seasonally.

* In the analogy to flow of water in a pipe, the voltage corresponds to the pressure.

† A practical unit of power, the watt, was mentioned in Part I, chap. 12; a horsepower, HP, is 746 watts.

‡ The practical unit of current is the ampere. In the next chapter there will be more about this. We will ask you to note at the moment only that it is the current produced by the passage of three thousand million electrostatic charges per second through the conductor.

ELECTRIC AND MAGNETIC FIELDS

When you apply for electric service the distributing company wants to know first what will be the probable maximum of your *rate* of consumption. To enable you to make very large withdrawals over short intervals means that a large power supply must be available to you; and to get it there requires more expensive, heavier equipment. Of course a large demand for a short time may take no more total consumption than a small demand of long duration. The meters, therefore, record on dials the total energy, much as miles are recorded by the odometer or mileage register on a car. A small electric motor within the meter spins at a rate proportional to your rate of use of energy at any moment. If all the lights, ice machine, electric stove, flatiron, etc., are running simultaneously, this little motor turns over at a lively clip, and its total revolutions mount up swiftly like the dimes on a taximeter. If, however, very little current is being used, it crawls around quite slowly, and in the course of a month no very great total is recorded. Unlike a taximeter, fortunately, there is no charge for waiting. When all switches are off, inspection of the meter (where part of the motor is visible

A   B

PLATE 54   Illustrating Comparative Costs of a Toaster and a Lamp. (A) Electricity to toast ten slices of bread daily amounts to 3 kilowatt hours per month. (B) The lamp, used 6 hours per day, consumes over 9 kilowatt hours per month.

as a horizontal revolving aluminum wheel or disk) should show it motionless. If it moves even slowly under these conditions, all switches being off, an inspection and accounting for possible overcharge by the utility company should be demanded.

IT IS of practical importance to remember that, although an electric toaster, a flatiron, or an electric heater or sun lamp may be rated high in watts—several hundred or a thousand, even—they seldom are in use for very extended periods. When they are, the bill mounts up of course.

Half a dozen incandescent lamps, however, even of as little as 15 or 25 watts consumption each, turned on at dusk of a long late fall or winter evening and left burning continuously for six or seven hours, may cost in the long run much more than the high-powered appliances that are used for short periods only (see Plate 54).

Quite often the service company is blamed for defective equipment or for overcharging for one reason or another, when it is the little unnoticed use of small amounts of power continuously over long intervals that is entirely responsible.

With this brief pause to comment on some of these homely matters of household economy, let us now mount Pegasus again and travel a little farther into the interior of this magic land of electricity, first to look at those strange interrelations between electricity and magnetism that arise when electric charges begin to move and flow in wires or when magnetic fields begin to move and sweep across electrical conductors. Only when equipped with this knowledge can we hope to approach the more mysterious domain in which the true nature of matter begins to be revealed to us in terms of its fundamental electrical and magnetic character.

CHAPTER 26

CHARGES IN MOTION

THE idea of *electric current*, electric charges in motion, has already been defined in the last chapter. The total quantity of electricity (Q) passing through a conductor in a given time (t), divided by the time of passage is the definition and the measure. Mathematically expressed,
$$I = Q/t.$$

ELECTRIC currents were called at first "Galvani currents" from the name of their discoverer, Luigi Galvani (1737–89). He was not a physicist, but a professor of anatomy and obstetrics at the University of Bologna. About 1780 he associated spontaneous contractions of the muscles in the severed legs of frogs with the effect of an electrostatic machine operating in the vicinity. His essay on the "Force of Electricity on the Motion of Muscles" was published in 1791. He also found that strips of dissimilar metals in contact with nerve and muscle, respectively, at one end and in contact with one another at their other end produced similar contractions. Certain selections of dissimilar metals were more potent than others, and the foundations for the chemical electric battery were laid.

Allessandro Volta (1745–1827), professor of physics in the University of Pavia, whose name is cherished in the name given our practical unit of potential, the volt, was another distinguished Italian experimenter of the same era. Of electrostatic experiments he was a master, inventing an electrometer, electrical condensers, and electrophorus before Galvani's discoveries were published. He added greatly to the latter's work in systematizing and extending knowledge in this field of enterprise. Electric batteries, as they came to be developed, were originally called "Galvanic" and "Voltaic cells" in acknowledgment of their direct outgrowth from the work of these two men.

NO MOTION picture, even with the most skilful animation, could ever portray on the screen what we visualize in our imagination as this flow of current*—myriads upon myriads of atoms of negative electricity, atoms as much smaller than material atoms as the sun is smaller than the space filled by the solar system—these myriads of infinitesimal things swarming, tumbling, rolling helter-skelter, through and between the vibrant trillions of thermally agitated atoms of the solid metal of the conductor. Suppose we could catch in a bucket the number of these electric entities that pass through the filament of our study lamp in 10 sec. (i.e., 4×10^{19}). So great a number is this that if we should set the entire population of the United States to counting them—every man, woman, and child, working twelve-hour shifts throughout the twenty-four—they might now be approaching the end of their task only if they had started the job about 5,000 years ago.†

CURRENTS from electric batteries and from certain types of dynamos are called "direct currents." In these the flow of electric charges is continuous in one direction all the time. More commonly our homes are provided with "alternating currents," where the direction of flow reverses about one hundred and twenty times every second, and the moving myriads sweep first one way and then the other. Our sense of touch is able to detect this phenomenon, provided the field between the electrified terminals is great enough so that a potential of 50 or 100 volts exists there. Care must be used, however, lest too much electricity pass through our rather poorly conducting bodies painfully to shock our nerves and to cause our muscles to go into powerful and uncontrollable contractions.

* The intricate structure of the analysis which has enabled us to discern the elements of the picture cannot be presented in detail except in technical treatises where the language of mathematics formulates and carries on the logic of the argument. A simple account of some of the broad lines of evidence for the picture that we have will be presented in Part IV.

† This number is of the same order of magnitude as the number of molecules in 1 cc. of gas at standard temperature and pressure. See Part II, p. 171.

CHARGES IN MOTION

Interrupted or alternating currents can be felt at considerably lower voltages than direct ones; but if the current strength is not limited to a very small amount by one means or another, either type of current is swiftly destructive to the living tissues so traversed and equally fatal to the continued existence of the individual. Not infrequently electric shocks so slight as to be quite harmless in themselves will interfere with the automatic periodic action of involuntary muscles, and so stop heart action or respiration. Prompt application of the usual methods of resuscitation then frequently restores the individual. In any case of unconsciousness produced by electric shock, resuscitation exercises should be promptly applied.

FROM the point of view of physical science there is one outstanding and fundamental phenomenon accompanying the motion of electric charges that gives us a unique and absolutely certain test for the existence of this motion.

MAGNETIC FIELDS OF FORCE EXIST IN THE VICINITY OF MOVING *ELECTRICAL* CHARGES

Magnetic fields where there are no magnets seems quite odd to you, perhaps, although by now you are probably somewhat *blasé* to novelties of this sort.

WE SHOULD like to retain, if possible, a certain lightness of touch in our exposition of the somewhat more technical aspects of electricity that form the remainder of this subject. This will be difficult, if not impossible, to do. We are dealing more and more with phenomena quite removed from direct sensory contact in the ordinary activities of life. Although the principles are embodied in a thousand and one devices that surround us on every hand, they form, as it were, only the skeleton of the physical structure that surrounds our life in an electrical age. In the explanation of these principles by means of the more simple examples, the only ones we could hope to make clear in a brief survey, it is necessary to divest these familiar forms of so much clothing, fat, and incidental outer tissue that, in looking at the articulation of the skeleton beneath, we will find it difficult to relate it to the homely aspects of daily life with respect to which our non-technical reader has quite as much, if not more, background than the trained specialist.

As we lose certain opportunities in one direction, perhaps we may hope to gain a very real advantage in another. If you can bring yourself to this reading with a fixed determination to read it thoroughly and slowly, to ponder upon it, to re-express it in your own mind, and thus make it part of yourself, you will soon find your interest growing, not directly as does your understanding, but at a swifter pace, perhaps as the square or the cube of the latter. Give us but the stirrings of a genuine interest: that is the most that we could hope or desire. The marvelous way in which the physical world is built, the almost unearthly beauty of its design, and the unerring precision of its laws of operation never fail to grip and hold the attention of the beholder or to fire his imagination.

WE OWE to Hans Christian Oersted (1777–1851) the discovery of the magnetic effects of electric currents. Many others during the preceding twenty years since Galvani and Volta (see p. 231) had paved the way for the development and use of electric batteries of considerable power, and had sought to find some connection between electrical and magnetic phenomena. The mutually exclusive character of these two types of fields of force had been recognized for a long time; yet many had attempted to find some common factor of interrelation, since there were so many striking and obvious similarities in the behavior of static electric charges and magnetic poles. That a magnetic compass should happen to be brought near a wire in which electric current was flowing became much more probable after the discoveries that resulted in the bringing of electric batteries into the laboratory.

Oersted was a professor of physics in Copenhagen when this pregnant combination of current in a wire and a compass needle in close proximity occurred before his trained and watchful eyes. He made few false moves in his subsequent investigation, and soon had recognized and published (in 1820) many of the facts we are now about to relate.

The magnetic fields that exist in the vicinity of straight conductors carrying electric currents (charges in motion) are quite different in shape from those associated with bar magnets. The lines of force are *circular* and are in planes at right angles

PLATE 55 Circular Magnetic Field around a Wire Carrying a Current Shown by Compasses. Negative charge in the wire is moving upward toward observer.

CHARGES IN MOTION

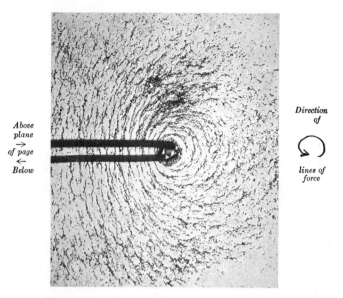

Above plane → of page ← Below

Direction of

lines of force

PLATE 56 MAGNETIC FIELD IN SPACE AROUND A WIRE CARRYING A CURRENT AS EXHIBITED BY IRON FILINGS. To the right, influenced only by current in conductor passing from above to below, the field is circular. To the left, where the effect of the wire "leads" going toward the center (above) and away from the center (below) is present, the condition of a current in a loop (Plate 57) is approached. The field is less curved and not truly circular here.

to the wire (Plates 55, 56). If, however, the wire is wound into a cylindrical helix, a sprinkling of iron filings about it shows the magnetic field to have taken the same form as that which it assumes about a bar magnet of approximately the same dimensions as the helix. Within the helix, however, there is open space, and the parallel lines of field in this region cannot be duplicated, for obvious reasons in the case of the bar magnet, since we cannot sprinkle our filings throughout the corresponding region within the solid material of the bar. Other aspects of analysis, however, sustain our conviction that a similar state of affairs, with respect to field form, exists, even within a magnetic material, although we cannot directly demonstrate it.

THE direction of a field about a straight wire may be determined by leading the current parallel to a pair of bar magnets free to rotate, and by confining the current to one end of the magnets only, thereby limiting its effects to one kind of pole only, let us say the N. one (see

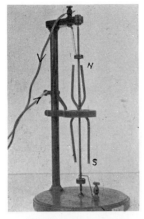 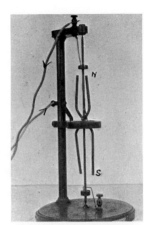

No current flowing in axle

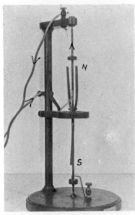 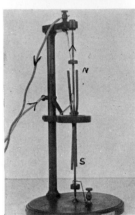

Momentary current in axle has shifted position of magnets slightly. Shows direction of rotation

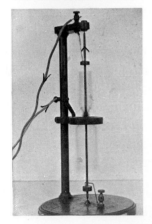 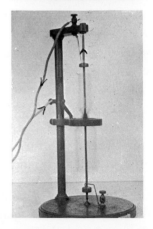

Effect of continuous current, i.e., continuous motion. The isolation is accomplished by passing the current upward (in sense of electron flow) through the upper half only of the axle to which the magnets are attached

PLATE 57 N. POLES ISOLATED IN FIELD AROUND A CONDUCTOR CARRYING CURRENT

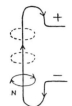
Plate 57).* If we grasp the wire in the *left* hand, our thumb pointing in the direction of flow of the *negative* electricity, i.e., away from the negative terminal of battery or other source of supply, and toward the positive, our fingers will encircle the conductor in the direction of the magnetic field, i.e., will point in the direction of the motion that a free N. magnetic test pole will take.

Returning next to the helical coil of current-carrying wire, Plate 58, it is obvious that if such a coil be floated or suspended on a pivot or a thread, like a compass needle, it likewise will swing into the direction of any surrounding magnetic field, such as that of the earth, in this case pointing in a magnetically N.-S. direction. If we are not familiar with the points of the compass, however, in the location where we perform the experiment, we must know the direction of flow of the current through the coil, apply our left-hand rule to its loops, and so determine from which end an N. pole would be repelled. This end will be the N.-seeking end of the coil.

PLATE 58 Magnetic Field of Current Flowing in a Helix (a Solenoid), Shown by Iron Filings. The similarity to form in Plate 53 is clear. The parallel lines of force in center of coil are also visible in this case.

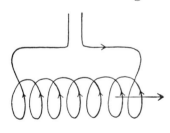

If we have but a single loop of wire carrying a current, it will act as if it were in effect a magnet of infinitesimal length—a magnetic shell, as it were—with an N.-seeking face on one side and an S.-seeking one on the other. As we shall see later in Part IV, we have good reason to believe that within the atoms of matter are circulating negative electric charges, often so many of them, circulating in so many different directions, that the tiny magnetic fields accompanying their motion

* Our convention for flow of current is that of direction of motion of negative charges (electrons). This is exactly contrary to the convention developed historically before the phenomenon was understood, which is still found in most textbooks. Reference to this fact was made on p. 182.

are almost entirely neutralized and have no sensible residual effect. In strongly magnetic substances like iron, however, a very definite residual field remains. Thus the atom behaves like a tiny magnet. These are more or less free to move around and line themselves up parallel to fields applied from the outside. The sum total of their combined effects is what gives iron and steel and other magnetically susceptible materials their gross magnetic properties. Although the magnetic elements themselves are atomic or molecular, as evidence in a former chapter showed, this molecular or atomic property is thus seen to be not a primary one but simply a result of circulating electric charges in the atom itself, a point of view mentioned earlier but not justified until now.

THE idea that currents of electricity and their accompanying magnetic fields were indicative of charges in motion was accepted as a working hypothesis for quite a while before definite proof of the fact was offered. Indeed, it was seventy years after the discovery of a field around a wire before Henry A. Rowland, of Johns Hopkins University, in 1889, by whirling insulated static charges on a plate, not dissimilar to that of a static machine (p. 198), was able to show the existence of a magnetic field in the vicinity of the path of motion of these charges and thus prove by a crucial experiment the correctness of this hitherto hypothetical identification of current with moving charge.

ANOTHER very important series of phenomena results when current-carrying coils of wire are placed in magnetic fields that originate elsewhere than in themselves. One case of this sort has been referred to, that of the coil that acts like a compass needle and points in a N.-S. direction. The force to which it is subjected arises, of course, from the reaction of its own magnetic field to the surrounding field of the earth. The latter is, after all, a comparatively weak field. However, if we hang a small coil of wire, having many turns on a very light and easily twisted suspension, between the poles of a strong magnet and then send a current of electricity through the coils, we have a device that is exceedingly sensitive for the detection of current. The current may be very feeble; but if the

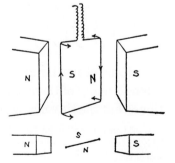

CHARGES IN MOTION

coil through which it flows is made of fine wire, very small and with many hundred turns, the feeble magnetic field in each is added to that of every other one and the total is correspondingly increased. The smaller the coil the more its field is concentrated. The outside magnets likewise are especially designed so that their field is very strong and very concentrated around the coil. The coil is suspended so that its field, when current is flowing, is at right angles to that of the magnet. Even then the force between the two fields—(of magnets and of coil) which rotates the coil so that its lines tend to become parallel with those of the magnets, i.e., N. pole of coil to S. of magnet, and vice versa—may be very small; but if the suspending fiber holding the coil is made excessively fine, as it was, for example, in the Cavendish experiment (p. 28), there is practically no resistance to the rotation of the coil at first and the opposing twist, even after the fiber has been slightly twisted, remains extremely small. Electric currents of a few hundred thousandths of an ampere or even less not only may be detected but also may be measured with high precision by such a device, which is called a GALVANOMETER. We represent galvanometers in our drawings by the symbol ⌀.

IN ITS most simplified and diagrammatic form a galvanometer may be represented by a straight section of a wire like the axle of a pair of wheels. Suppose it is lying at right angles to a uniform and vertical magnetic field. When a current, in our usual sense of flow of negative (electrons), is led in through the rails and wheels and flows through the wire, its field reacts with the field already present and alters the picture of lines of force in a manner none too simple. For us this detail is not important, but the resultant effect on the wire we ought to have clearly in mind. The wire is urged in a direction *at right angles to both the magnetic field and that of its own length.*

The force upon it depends only upon (1) its length in the field as it crosses it at right angles (or the perpendicular component of its length to the field if it does not), (2) the strength of the field it is in, and (3) the intensity of the current flowing through it.

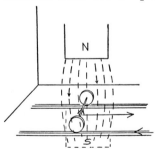

Motion produced by current through a conductor in a magnetic field

Force, f, = current × length × field strength = $I \times l \times H$.

Since force, length, and field may be independently measured, this fact gives us a method for arbitrarily defining electric current in terms of its mechanical magnetic effects. One system of units, the electromagnetic, from which our practical electric units—volt, ampere, watt, coulomb, etc.—are derived, has its origin in this phenomenon.*

AN IMPORTANT instrument for the measurement of current, using the field of a coil to move a magnet, is the TANGENT GALVANOMETER. A compass needle that points N.-S. in the earth's magnetic field, is at the center of a coil of wire, the plane of which is also N.-S. The field due to the electric current in the latter is at right angles to that of the earth, i.e., E.-W. in direction. When current flows in the coil the compass needle sets itself in a position along some line between the N.-S. and E.-W. lines, whence from the "tangent" of the angle so determined the relative strength of the two fields may be found. In this instrument the current-carrying coil is fixed and the magnet is suspended and deflected out of the N.-S. line by forces emanating from the coil. In the other types of MOVING-COIL GALVANOMETERS the magnet is fixed, the coil suspended, and the latter moves. Three of these are photographed in Plate 59.

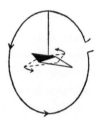

"Tangent" galvanometer

WE HESITATE to remind you yet once again of Newton's Third Law of Motion regarding the simultaneous and equal character of action and reaction. Force between coil and magnet, each pulling equally and oppositely upon the other—which will move? The answer may be neither, or both, or either one. It depends, obviously, upon which is more solidly attached to the earth, whose recoil is always insensible. The one that moves is the one not so solidly attached.

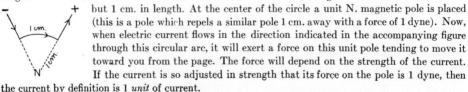

* Unit current in this system of units (the electromagnetic system) is defined as follows: we arrange a circuit in the form of the arc of a circle of 1 cm. in radius, using of the entire circle an arc of but 1 cm. in length. At the center of the circle a unit N. magnetic pole is placed (this is a pole which repels a similar pole 1 cm. away with a force of 1 dyne). Now, when electric current flows in the direction indicated in the accompanying figure through this circular arc, it will exert a force on this unit pole tending to move it toward you from the page. The force will depend on the strength of the current. If the current is so adjusted in strength that its force on the pole is 1 dyne, then the current by definition is 1 *unit* of current.

ELECTRIC MOTORS of practically all types function as a result of there being a force exerted by a magnet on a conductor through which a current of electricity is flowing. The magnets are usually of highly permeable iron or steel, whose native susceptibility to magnetization is excited and held constant at its highest possible value by having wound around them coils of wire carrying a current: the magnets are, in other words, electromagnets. They are in most cases fixed in position in a circle within a heavy frame. At the center of the latter, in a position to be entirely surrounded by the electromagnets, is mounted the ARMATURE, a series of coils of wire through which

 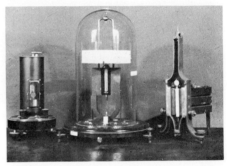

PLATE 59 THREE COMMON FORMS OF MOVING-COIL GALVANOMETERS. *Left:* High sensitivity type; *center:* lecture-table type; *right:* common laboratory type for student's use. (Stereophotograph.)

electric current also may be sent. The magnetic field of these armature coils (also very often reinforced by being wound on iron [see Plate 60]), by its reaction against the fields of the surrounding electromagnets, furnishes the driving force of the motor. One of the simplest types of motors was designed by Barlow in 1823 (see Plate 61).

EXCEPT for size and general ruggedness of build, there is only one significant point of difference in the mounting of the coils of a galvanometer and of a motor. In the former the coil is attached at one end mechanically by a suspension, a spring, or some other elastic constraint, so that, as it is twisted by the forces of the magnetic fields, this twisting of one end of the support calls forth a reacting force in the support which increases as the twist increases, and finally brings the coil to rest in equilibrium between the mechanical stress in the support and the electric stress between coil and magnets. In the motor, however, the coil is free to rotate continuously in either direction. As it does this, of course the relative positions of the wires carrying the current alter with respect to the field, and some of them momen-

Courtesy of Museum of Science and Industry, Chicago

PLATE 60 AN EARLY FORM OF MOTOR BY JACOBI, 1834. The stationary, "field" magnets are to the right, the moving coils of the armature on the left. These coils are separate, and being wound on iron cores are virtually electromagnets just like the field magnets. A switching device (commutator) on the axle to the right of the coils shifts connections to armature coils, so that there are always two like poles, armature and field, near by and repelling, and two unlike poles near by and attracting in such a sense that both effects rotate the armature magnets in the same direction. At the extreme right is a replica of the old type of "galvanic cell" or battery used by Jacobi.

tarily in location to receive the greatest force move out from those positions and others take their places. As this happens ingenious switching devices, called "commutators," automatically, as the coil is in motion, switch the current successively through those particular loops on which the field reacts most strongly. With alternating-current motors commutators may not be necessary except in starting, as the variation of the current in both magnets, those of coils and those of fields, may so synchronize with the rotation of the motor that what results is quite the same as what is produced by a switching device on a direct-current machine. In one way or another a motor is always designed so that a practically constant force is acting on its armature coils at all positions.

CHARGES IN MOTION

ACTION and reaction here likewise are present as always when forces are involved. The force that spins the armature in one direction would spin the field magnets in the opposite direction if the armature were clamped and the fields were free to move. In quite a number of types of motors this is indeed the case. Similarly, if both armature and field magnets

Courtesy of Museum of Science and Industry, Chicago

PLATE 61 BARLOW'S WHEEL, 1823. In this early form of motor the essential principles are clearly shown. The field of the magnet is directed toward us. The current flows from the axle at center of wheel toward the end of whichever spoke is pointing downward and is thereby in contact with the mercury in the small tank to which the other end of the battery is connected. If the flow (of negative) is downward, this current, being in a field at right angles, suffers a force at right angles to both these directions, i.e., to the right, so that the wheel spins in a counterclockwise sense. As one spoke moves out of the field, it also leaves the mercury cup, and the current is then switched to the next spoke coming in.

were mounted so as to be able to rotate freely, both would rotate in opposite directions and would acquire speeds inversely as their respective rotational inertias—Newton's Second Law again.

WHEN all is said that might be said about them, motors are, after all, but servants created by man's ingenuity to perform useful jobs and to serve practical and physical needs. There is little they can teach us about the nature of the world; they are patient, dumb, blind, sturdy, tireless creatures, designed to transform energy efficiently and to work silently, powerfully, and quickly.

Not so the galvanometer to which we now return in the closing paragraphs of this chapter. Frail and delicate in construction, puny and of feeble strength, a breath of wind may blow away its tiny coil and shatter the hair-fine suspending fiber; yet this same instrument has been one of mankind's most sagacious and inspired teachers. The faintest whispers, coming from almost the innermost recesses of the atoms themselves, it can hear and interpret for us. Unafraid in the face of death-dealing amounts of electric current from great power stations, its delicate finger can measure their strength as well as the potentials from which they are hurled with equal rapidity and precision.* Messages from the stars the galvanometer translates also (Part II, p. 123).

Frequently in the future we shall refer to some of these attributes or describe others germane to our context there. Briefly now we will comment only upon the enormous range of electric current strengths that the galvanometer can adapt itself to measure.

Small as are the currents that may be measured directly by galvanometers of high sensitivity, some of those involved in modern physical investigations, amounting to as little as a few billionths of an ampere, are below the range even of these instruments. By means of certain kinds of radio tubes these currents may be amplified. A local source of supply is drawn upon, and from it the original current is built up by a definitely known and constant factor until it comes within the range of the galvanometer sensitivity. In quite an opposite way, by devices known as SHUNTS, which divert all but a tiny fraction of the current elsewhere, the sensitivity of a galvanometer may be reduced and much greater currents than the instrument could tolerate without becoming a mass of melted wreckage may thus be scaled and measured.

IT IS due more to the great usefulness and wide range of adaptability of galvanometers and other current-measuring instruments than to any other single thing that the last fifty years has witnessed such prodigious strides in the understanding, control, and widely varied and disseminated use of electricity. Similarly, we owe to this little instrument the lion's share of the credit for the swiftly increasing degree of our understanding, not

---

* Nearly all voltmeters and ammeters are in reality only galvanometers arranged in special circuits and appropriately calibrated.

CHARGES IN MOTION   **245**

only of the gross aspects of electrical phenomena, but of the more subtle implications as well. Three instruments—the analytical balance for weighing, referred to in Part I; the galvanometer for measuring electric currents; and the spectroscope (see later, Part V)—separately for the most part, but sometimes in conjunction, have done more to further mankind's understanding and control over the forces of nature than all other inventions put together. This unrecognized and unsanctified little trinity of devices might well be taken together as the symbols of the modern scientific age.

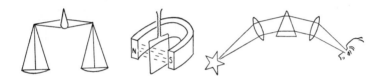

CHAPTER 27

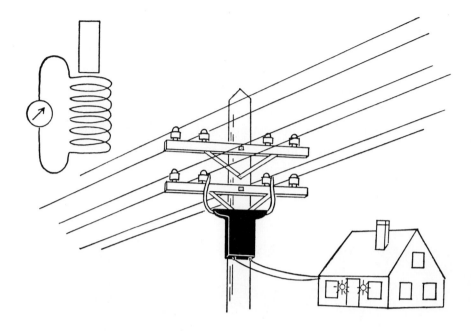

MAGNETIC FIELDS IN MOTION

FEW men have been recorded in the entire history of science who will be remembered and respected for their achievements more than Michael Faraday (1791–1867). As a bookbinder's apprentice not long out of his teens, this blacksmith's son learned first of physics and chemistry from his perusal of books sent in for binding. By shrewd perseverance he obtained his first chance to acquire more knowledge of these studies through experimental opportunities by obtaining what was little better at first than a bottle-washer's job in Sir Humphrey Davy's laboratories in the Royal Institution. Here he became Davy's successor as director in less than twenty years. His earlier years were very fruitful and productive, notably in the field of chemistry, where his discovery of benzene was fundamental. His investigations of alloys and the invention of new kinds of glass were pre-eminently practical, but it was not until he became the director of the Royal Institution in London at the mature age of forty that his most penetrating researches were made.

MAGNETIC FIELDS IN MOTION

INTENSELY interested in the magnetic field surrounding iron and other magnetic materials, but like many of us non-mathematical, either by talent or by training, he was the first to visualize and name the lines of force which all succeeding generations of students have found so useful in describing and picturing the various complexities of magnetic phenomena. The work of Oersted and Ampere revealing the presence of magnetic lines of force about moving charges, the phenomenon of electromagnetism, Faraday repeated and verified in every detail. At once he became obsessed by the question: Since electricity in motion produces magnetism, why should not magnetism produce electricity? Magnets and coils were always in his pockets or near his person; daily in his laboratory, and continually when forced to be away, he pondered on the problem, turning over magnet and coil, visualizing the imaginary lines of force, speculating on their properties and reactions and their possible relations to the corresponding lines of an electric field.

One day in the laboratory when he had one coil of wire connected to a galvanometer and a rather heavy current of electricity was flowing through another coil near by, the elusive phenomenon appeared and was caught by his all-observing eye.

AT THE MOMENT OF STARTING OR OF STOPPING THE CURRENT in *one* coil a slight deflection of the galvanometer, significant of newborn electric current in the OTHER coil, appeared. The phenomenon is called ELECTROMAGNETIC INDUCTION.

WITH this observation was created the beginning of the vast field of electrical engineering that was to come: great cities, still more extensive rural areas lighted by electricity; industry, formerly concentrated in mining areas close to the coal supply necessary for running of steam engines, now released to become diffused and spread into smaller units; the energy of fuel, converted either directly or through the medium of steam into mobile and silent electrical energy. The giant that was to perform men's manual labor for them in the new culture that was arising not only took on more colossal stature, but was at the same time provided with the winged sandals of Hermes. In flight this Colossus was to be as silent and invisible as his godly prototype.

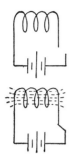

Charges in motion create a magnetic field

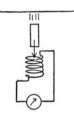

A moving magnet creates a current (motion of charge)

Where one man has struggled blindly in the dark and found the doorway, the light admitted by his opening of it makes its location seem much more obvious to those who come later. Electromagnetism and the induction of electricity by magnetism are perfectly reciprocal effects. Moving electric charges engender and maintain magnetic fields—MOVING magnetic fields engender and maintain the circulation of electric charges in closed conducting circuits in their vicinity. Electric charge at rest is unassociated with any magnetic property. Magnetic lines of force call forth no electricity unless they move across a conducting path.

It makes no difference what may be the origin of the magnetic field, or whether it moves in response to the motion of some physical object with which it is associated or not. A few simple experiments quite similar to those of Faraday may serve to make this clear.

Plate 62 A shows a simple coil of wire attached to a rugged, and not especially sensitive, current-indicating galvanometer. A bar magnet completes the equipment. The galvanometer shows no current whatever: the magnet RESTS with respect to the coil; but any movement of the magnet near, or especially through, the coil produces a large momentary deflection (Plate 62 B). As one end of the magnet is thrust into the coil, the deflection is in one direction. As it is withdrawn the deflection is in the opposite direction; but it is then in the same sense as if the OPPOSITE pole of the magnet had been inserted, and oppositely directed to the deflection produced by the withdrawal of the opposite pole.

We called attention above to the fact that whether the magnetic field had its origin in a permanent magnet or in the current flowing in another coil made no difference whatsoever. In Plate 62 C and D we have two coils: one connected as before to the galvanometer, and the other to a battery. As before, the galvanometer shows no current except when one coil is moving with respect to the other. Reversing the direction of the current in the battery's coil or reversing the direction of motion of this coil reverses the direction of the current through the galvanometer.

A

B

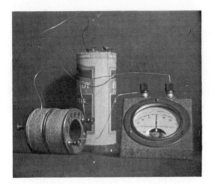

C

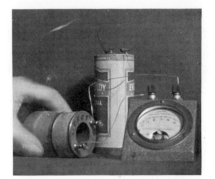

D

E

ELECTRO-MAGNETIC INDUCTION

(Time exposures all. Moving objects, hand, magnet, coils or needle of galvanometer *when in motion* have not registered clearly.)

PLATE 62

IT SEEMS strange to us now that Faraday hit upon neither of these two experiments at first. His discovery of the effect was in the following wise. In Plate 62 E are our two coils again. The one associated with the battery (which from now on we will call the PRIMARY) is not connected to it at the moment, but the wire is held in the hand and may be placed in contact with the binding-post terminal of the wire on the coil. The other coil, the SECONDARY, is connected to the galvanometer in the usual manner. Both coils are fixed in position, and neither of them will be moved throughout the course of our experiment. Let us watch the galvanometer as we now CLOSE THE CONTACT. It gives a sudden jump. As we break the contact, it gives a jump in the opposite direction. None of the apparatus has moved, only an electric contact that starts or stops current in the primary coil has been operated. We could do this far away in another room or in a distant city, running long wires to maintain the conducting attachments to the rest of the apparatus, and the effect would still be obtained. Something MUST HAVE MOVED, of course—the magnetic field or, if you prefer, the lines of force.

Before the switch was closed, there was no current through the primary coil, there was no magnetic field surrounding it. On closing the switch, the current began to flow; and, as it did so, magnetic lines opened out from each loop of wire in the coil like ripples spreading from the splash of a stone in a pond. Unlike ripples, however, as soon as the current reached its maximum full strength in the primary coil these lines ceased to move and remained like frozen ripples, forming their customary pattern in the region around the primary coil and extending far beyond its immediate confines into, through, and even beyond the space occupied by the secondary coil. In moving out into this position these lines passed across—or cut, as we say—all the loops of wire in the secondary coil, quite as effectively, indeed, as if they had had their origin in permanent association with a bar magnet and had moved up and across the secondary by the actual motion of the bar magnet near it.

When the battery switch is opened, the magnetic lines, quite unlike our ripples, retrace their paths, converge back upon the turns of wire in the primary coil, and vanish therein. In making the return journey they recross the turns of wire on the secondary coil in the opposite direction and thereby excite a momentary counter current in it.

Faraday observed this effect in a sense accidentally, although he was expecting something like it to happen. He quickly and correctly interpreted it and satisfied himself that in general the motion of magnetic fields

MAGNETIC FIELDS IN MOTION

ACROSS conductors produced currents in the latter as long as the fields were moving. For the effect to be a maximum, the field must move at right angles to the direction of motion of the conductor. If it moves at any other angle, only the component (see pp. 41, n., and 61) of its motion at right angles to the conductor is effective. If the field moves parallel to the conductor, there is no effect at all.

We hardly need point out that *relative* motion between field and conductor is all that is essential in the process of generating a current. If the coil is at rest on the table and the magnetic field is swept across it by motion of nearby magnet, or other current-carrying coil, or if both coils are at rest and moving lines engendered in one sweep out and cut the other, just the same result is produced as if the magnetic lines of force are motionless, and we move the generating conductor so that it cuts across them.

PLATE 63 A MAGNETO. The so-called "low-tension" type, meaning that only low voltages (10–12 v.) are produced by it. The crank turns it easily until the button is pressed and the lamp connected. Then it is more difficult to turn. Why?

IN DEVICES called MAGNETOS (Plate 63) we have coils of wire arranged to rotate within and across the lines of force of the magnetic field of strong permanent horseshoe magnets. When the terminals of the moving coil are connected to a lamp, a path is provided for the circulation of the charges through it and back through the coil, and the lamp is lighted. Here we have in essence the modern power-house and the use of electricity in regions remote from it for illumination.

IN SUCH a machine, if the rotating coil of wire is connected through a switch or push button with the lamp, comparatively little effort will be required to whirl it in the magnetic field, UNTIL THE SWITCH IS CLOSED, and the closed path through coil, lamp, and back to coil is completed, so that the electrical charges can CIRCULATE therein. Before this is done, i.e., on an open circuit, positive and negative charges simply pile up electrostatically on either side of the break in the circuit. Practically no work is done upon them to accomplish this; and, except to overcome friction in the mechanical bearings of the machine, no work is required to drive it. Just as soon as the circuit is closed, however, and the electric charges have a path through which they can circulate, much more effort is required to turn the crank of the magneto. The electrical charges, in plowing through the rather resistant material of the lamp filament, suffer opposition to their motion; work has to be done to drag them through, or push them through if you prefer; the work is converted into heat at this point to so great an extent that the poorly conducting filament is heated to incandescence. Thus we have a lamp.*

THE energy sent out by our little magneto in the form of forces moving electrical charge may be transformed into many other different forms and in many different ways. We may run a toy motor and produce kinetic energy with it; the motor may in turn wind up upon its pulley a thread to which a weight is hanging, and thus store up some potential energy against the force of gravity; we may pass the current through conducting solutions as well as through conducting wires, and not only heat them but produce chemical changes therein, which will store up for us some part of the work we have expended as chemical potential energy. It was in this connection that Faraday made another equally famous series of discoveries, which we will examine in considerable detail in Part IV.

DYNAMOS and electric generators function in accordance with the same principle as does our little magneto. Instead of permanent magnets their magnetic fields come from more powerful large electromagnets (see Plate 64) that are sustained often by diverting some of the current produced by the machine itself for this purpose. These generators are driven by powerful steam or oil engines or, where water power is available, by great turbines that are, of course, simply very efficient water wheels. Dynamos ordinarily generate alternating current, since their coils, in rotating, pass first in one direction and then in the opposite one across the lines of force. By means of commutators (see p. 242) quite like those used on motors, the current can be switched into a unidirectional form upon leaving the coils.

* The current through the lamp of course creates its own magnetic field around filament, wire leads, and especially around the coil of the magneto. This field opposes the permanent field of the magnets, and the work necessary to drive the current through the lamp is done against the forces of interaction of the two fields.

Courtesy of Museum of Science and Industry, Chicago

PLATE 64   A REPLICA OF AN EARLY EDISON DYNAMO

Danger! Keep off

The modern counterpart in our machine world of that kind of electric generator that corresponds exactly to Faraday's first discovery and our third experiment (p. 250) has NO MOVING PARTS AT ALL. Silent and motionless, it is the more deadly to the ignorant inquisitive mortal who sometimes tampers with it. These devices we have scattered everywhere in city and in country. Usually they are put in inconspicuous locations and always out of reach of the casual passer-by, who, like yourself, has probably never noticed them. They are usually hung on poles. The pole sometimes is primarily for their support, but usually has a much more obvious purpose of supporting wires which the layman supposes to be for telephone or telegraph purposes. There is not much difference between the looks of the wire that carries the tiny currents engendered by the sound of your voice in front of the telephone transmitter and that which may be carrying several hundreds or thousands of horse-power to run a factory near by; but a world of difference is *there*, nevertheless, for contact with the latter wires may swiftly kill the person who touches them.

NEVER CLIMB POLES to investigate apparatus that may be contained in black boxes mounted thereon. Never, for any purpose, expose yourself to contact with wires that may have blown down unless you are absolutely certain whence and whither the wires lead and that no other wire, which may have something to do with a power transmission line, has accidentally become connected to them or in contact with them.

WITHIN these black boxes on poles or elsewhere around our homes in regions where people have electric light, are electrical generators, usually called TRANSFORMERS. The latter name is apt. They generate electricity in one of the coils, electricity you use in your home, from the surge back and forth many times a second of electricity in the other coil, electricity that may be sent to them from many miles away by the power company. From what we have told you above about the production of electric current in one coil by the starting and stopping of its flow in an adjacent coil, you can now suspect, we hope, that it is the secondary coil that is connected to your house and the primary one to the lines of the power company.

MAGNETIC FIELDS IN MOTION

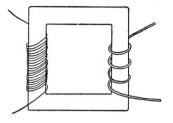

THE reason it is necessary for every home, or at least every neighborhood, to have its own special generator or transformer is a practical and economic one if the supply of electric energy is rather far away. It usually is far away, since we prefer not to live in the vicinity of power-houses. For reasons too technical to spend time upon here, it is far more economical to transmit electricity at high and dangerous voltages. These high-voltage currents are then transformed or "stepped down" to comparatively harmless low-voltage supplies for the use of the community.

Within each transformer are the primary and the secondary coil, wound on iron to make their magnetic fields more effective, and in very close juxtaposition.* Instead of interrupted current from a battery, there comes from the power-house an alternating current that starting from zero rises to a maximum value in about 1/240th of a second, declines again to zero in the next 1/240th second, reverses direction, and rises to a maximum in a third interval of equal length, and falls back to zero again in a fourth. The cycle of change is then repeated over and over again every 4/240ths or 1/60th of a second. The current is thus called "60-cycle alternating."

As the current alternates in the primary coil, so do the magnetic fields associated with it sweep out and back and then in the opposite sense, out and back. In doing this they move across the secondary coil and generate another quite separate and independent 60-cycle alternating current in it.

THE relation between the voltages and current strengths in primary and secondary windings depends on the relative number of turns in each. This will be better understood if, once again, we look at a few simple laboratory experiments instead of taking a commercial transformer to pieces. Two very simple cases are shown in Plate 65. The large coil is connected to the house-lighting alternating current (AC) circuit. The other one, since it is smaller, can intercept only a small portion of the oscillating field of the former. Nevertheless, it captures enough power (energy per second; $PD \times Q$ per sec.; $PD \times I$; volts $\times$ amperes) in one case (Plate 65a) to light a small lamp and in the other (Plate 65b), a larger one. In Plate 66 we have another simple

* In the sketch the two coils are shown separately to avoid confusion of one with the other.

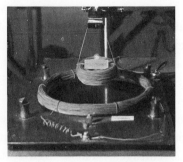

PLATE 65a A Simple Transformer (Stereophotograph). The large coil is the primary connected to the lighting circuit. The small coil is the secondary, suspended in the air above the primary. The small lamp is lighted at center of this coil.

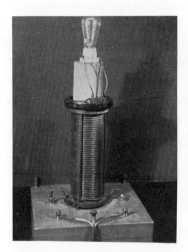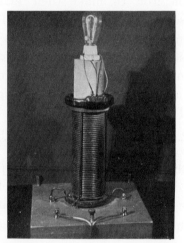

PLATE 65b An Iron Core Transformer. The large coil is here wound on an iron core, made up of a bundle of strips or wires ("laminated"). The strength of magnetic field is thus greatly augmented. The secondary coil, resting on top of the other, generates sufficient energy to light a full-sized lamp. (Stereophotograph.)

case. The primary consists of relatively few turns connected to the lighting circuit. The secondary, however, consists of but ONE turn, a solid ring of copper. Whereas 100 volts and 50 amperes are utilized in the primary, only a very small fraction of 1 volt is generated in the ring *but* many *thousands* of amperes circulate through it. What an enormous current it carries can be seen from the fact that, although it is of large cross-section and of excellent conducting material, it becomes red hot in about half a minute.

MAGNETIC FIELDS IN MOTION

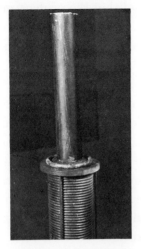 

Transformer where secondary is a copper ring

Secondary ring of this transformer at red heat owing to circulation of current in it (Stereophotograph)

PLATE 66 Transformer Giving a Very Large Current

This is an extreme case of a step-down transformer. The ones in common use within our homes are usually quite harmless, since their primary high voltage is that of the lighting circuit, 110 volts. Their low-voltage output is usually from 4 to 15 volts. Their most common use is that of a toy transformer for running electric trains and toy machinery, motors, miniature incandescent lamps for Christmas trees, etc.

The power-company's transformers outside our houses are also of the step-down type. The danger involved in contact with them lies, as we

Shocking

have seen, in the fact that the primary current is sent to them at a high voltage of several thousand. The reduced voltages of their secondary coils are usually only one or two hundred volts. Such potentials are in general harmless upon the more common types of accidental contact. However, it sometimes happens that someone will make an unusually perfect contact over considerable body area with the house supply. This is most apt to happen in bathrooms when one is in contact with water in the tub and then with wet hands grasps an electric heater, a curling iron, a lamp, or some other device the insulation of which has become defective. With good contact over a considerable body area (connection with water pipes and ground forming one terminal) sufficient current may pass through the body to produce uncontrollable contractions of the voluntary muscles and thus render the person helpless. Involuntary muscles like those of the heart are sometimes affected, and unless prompt action is taken death may result. Electric fixtures correctly installed in bathrooms should be entirely out of reach from the vicinity of the bath tub.

STEP-UP transformers are a much less common type. They are usually to be found only around laboratories and industrial plants, where they should be accessible only to those technically trained in electrical matters. Many types of technical experiments require high voltages, and transformers that will increase the electric fields, and hence the voltage, many thousand times are not uncommon in the laboratories of technicians. The production of X-rays is one of the most common needs that can be served only by the use of electric currents driven at these high electrical pressures. X-ray equipment is usually installed with adequate protection not only for the patient but also for the technically trained physicians and nurses accustomed to its use.

ONE very important mechanical effect is involved in connection with currents induced in coils by moving magnetic fields. We have already (p. 238 ff.) discussed at some length the forces that act on conductors carrying current when these find themselves within magnetic fields. Apropos of the

MAGNETIC FIELDS IN MOTION

magnets, we commented in a footnote on the reaction of the field of the induced current with the field inducing it. There is not the slightest mystery about this, of course; the mechanical force that arises from the reaction between the magnetic field *produced by the current* and the inducing magnetic field, of some exterior origin, is no different from the reaction of the fields of two bar magnets which pull them together or force them apart. This is Newton's Third Law of Motion cropping up again.

That there should be mechanical forces between primary and secondary windings of transformers, then, would seem obvious. In each coil there is a current flowing; around each coil there is a magnetic field—in rapid motion, it is true—but nevertheless in no way surrendering its fundamental characteristics. Each field reacts mechanically on the other field just as each also reacts electrically upon the coil producing the other. The direction of the forces so produced is always one opposed to direction of the motion of the field (in one of the coils) that resulted in the production of the current responsible for the other field (in the other coil). Here the Principle of Conservation of Energy finds illustration.

Take as a specific case that described on page 248 where we have a steady current from a battery flowing in one coil. When we bring this near, and thus sweep its field around, the other coil connected to the galvanometer, the current induced in this other coil sends out its own field in a direction such as to oppose and resist our effort as we move the two coils together.

If currents and fields are fairly strong, the force is readily felt and the sensation is strange indeed. It feels as if you were trying to move something around in a sticky, viscous fluid like molasses which clings and opposes every motion, and resists the more strongly the more effort is put forth. In the absence of anything visible the phenomenon seems quite weird to one who has never before encountered it.

The magnitude of the forces involved in these mechanical reactions between primary and secondary coils, or in general between magnetic fields of whatever origin, can be seen in the experimental transformer referred to before on page 256 in which the secondary was simply a solid ring. If this ring is resting around the top of the iron core of the primary coil before the current is turned on, it can be readily tossed up 6 or 8 ft. by the reactions of the magnetic fields when the current is turned on (Plate 67).

PLATE 67 FORCE OF MAGNETIC FIELDS. Secondary coil (ring) tossed several feet into the air as current (A.C.) is sent through primary coil, showing magnitude of force between original field in magnet and the field of the current induced in the ring by field of magnet. Stereophotograph and double exposure.

It doubtlessly has occurred to you by now that the field created by the induced current in the secondary coil will, in cutting back across the conductors of the primary coil, engender in it another current where already there is one current flowing. The existence of this current is no deterrent to the creating of another one, a *second* secondary current in the primary coil. Just as gross material objects can respond simultaneously to the effect of a variety of forces and move in a manner which represents the sum of all of the component effects, so can these electric charges, which themselves are the quintessence of all things material, respond to different forces simultaneously.

Details of these effects are too complicated and not of sufficient general interest to ask you to apply yourself to a complete exposition. One of the most important results, and the only one we shall mention, is that, although all of the voltage supplied by the power company is available and ready on its side (the primary) of the transformer, and although a free and open conducting path of wire in which it may circulate runs therein from one side of the power line to the other, no current to speak of flows across and through this primary winding until you, within the house, begin to withdraw some

MAGNETIC FIELDS IN MOTION

from the secondary coil. Precisely as you tap and utilize the energy pdQ from the secondary, does energy PDq filter through the primary side. We have used lower case and capitals in the expressions for energy to reiterate that large PD, potential, voltage, is furnished by the company necessitating relatively small quantities of electricity q carried across to do a given amount of work. At the lower voltages, pd, at which you use the energy, a larger charge Q must be carried to do the same amount of work.

Power relations, as we have said, are more commonly employed; dividing the energy by the time we have, in similar notation, that only as you use power, pdI, from the secondary side of the transformer does power from the company, PDi, filter through its primary side.

THE forces against which work must be done to furnish this supply of total energy, or, as power, to deliver a given amount of it per second, are the mechanical forces that must be overcome in the generators at the power station. Just as you can feel the force resisting you as you move a primary coil up to a secondary one and thereby create in the latter a large current whose field reacts to and opposes your motion, so does the generator "feel" the drag upon it which it must overcome if it is to maintain this current which many miles away you are demanding of it as you withdraw energy and power. In large generating stations the effect of any one individual user is quite negligible at any moment. However, as many thousands of users begin to increase their demand as dusk comes on and to add many hundreds of thousands of lamps to the circuits, very well designed methods of absorbing and carrying these "peak loads" are necessary.

ONCE again we have to pause with this, to you perhaps, far too close approach to matters technical. It is necessary, however, for us now and again to penetrate a little way into the deep and dark recesses of the forest that you may have some vague impression of its style and magnitude. We are approaching closely now those regions where until recent years no man had ever penetrated. Here there has been found, after a little clearing of the underbrush was made, a soil so rich and fertile for imagination, that at first we could hardly believe in the reality of the strange and wonderful plants that began to flourish in it. The little excursions we have made into and through the outlying regions have developed, we hope, at least a slight amount of taste for this type of woodcraft. Without it we will be hopelessly in the dark. Others, perhaps, equipped with trained techniques, can explore and return and describe what they have found. However, to see with one's own eyes, even though dimly, is far better than never to have seen at all. Such glimpses as we hope to make possible for you will enable you perhaps as time goes on to develop your own eyesight and appreciation from firsthand experience, and to share our own conviction that in the vast book of nature the truth may be far more strange than the wildest flights of human imagination.

PART IV
ELECTRICITY AND MATTER

Courtesy of Science Service

PLATE 68 SIR J. J. THOMSON

CHAPTER 28

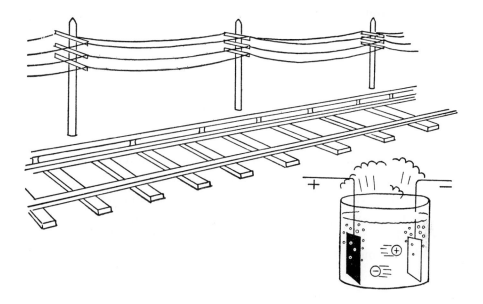

CONDUCTORS, SOLID AND LIQUID

AT LAST we approach new country. No historical preamble is needed to show the growth of these new ideas about electricity and matter slowly germinating in the minds of men of former generations. We are to deal now with the achievements of our own age, the twentieth century. Black and hopeless, perhaps, have been many of its pages as the men of our times have written them: what its true perspective will show when seen from the remoteness of future centuries yet unborn, we cannot know except in one respect. Its early decades have witnessed the bringing to light of the most remarkable array of astonishing truths about the nature of the physical world that has ever been disclosed. Scientific investigation in the realm of physics seems suddenly to have come into man's estate in our own period of history.

That other generations, rather than our own, may reap the wind or whirlwind of these advances in fundamental knowledge may be true. That may be something for us to be thankful for. That they may understand things that still seem dark and confused to us is certainly our loss. But the joy of uncovering the treasure and of being among the first to see its marvel has been ours and ours alone. Some of those men whose names will never be forgotten in the vanguard of this quest have been and are still our own teachers, friends, and associates. They are a numerous company too, for in our times, as never before, have the industries as well as universities collaborated in giving the scientific investigator encouragement, appreciation, and—what is still more important—freedom and support.

NO SINGLE generation, of course, is without its deep obligation to the past; and even in the development of the new physics and chemistry of which we are about to tell, we must pay high tribute to at least one great figure of the past whose experiments definitely foreshadowed and whose mind dimly saw what was to be clearly seen almost a century later.

Michael Faraday was, as we have seen, the author of the discoveries that resulted in our being born in an electrically minded world. He was also perhaps the only man of his century who saw reasons to suspect what we now know to be tested truth, that ELECTRICITY, like matter, is of ATOMIC CHARACTER.

In previous pages we have more or less naïvely adopted a language that implies that this is so. The fact is common knowledge, and such language no longer creates surprise or is challenged. But so far we have not presented any evidence for it, and this now becomes our task.

WHEN we come to look at the details of the conduction of electricity through material substances, solid, liquid, or gaseous, we begin to find a great many phenomena that are of the utmost significance not only with respect to the light they shed on the nature of electricity but equally with regard to the information they give us of the close relation of it to the nature of matter itself.

In chapter 20 we pointed out that all substances conducted electricity to a greater or lesser degree, that the best conductors were the metals, and that certain non-metallic materials, like sulphur, glass, amber, and oils or resinous substances, were such poor conductors that they were excellent insulators.

If a large charge from a static machine, transformer, or any other source, is allowed to accumulate in two opposed electrodes, a strong electric field is set up between these two locations. Air, ordinarily an excellent insulator, does not facilitate the transfer of an accumulated negative charge from the

CONDUCTORS, SOLID AND LIQUID

terminal where it has been piled up, even though every part of this charge is being repelled thence by the combined effects of all of the rest of it with which it is associated and even though it is all being attracted at the same time from across the gap by a region 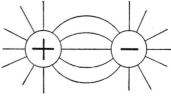 positively charged in like amount.

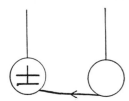

IF A conducting wire is led between these two charged regions, along this easy path the negative charge at once rushes; a magnetic field is sent out around the path while the charge is in transit. In a very small fraction of a second all the charge passes across, the current and its associated magnetic field die out, and all electromagnetic phenomena disappear. If, however, a powerful source of electricity is continually renewing the supply of negative to one terminal, then the initial rush dies down not to zero but to a condition of equilibrium where the amount leaking off through the conducting path is no greater than the supply continually maintained, and a steady current results with its associated constant magnetic field.

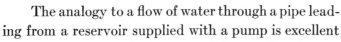
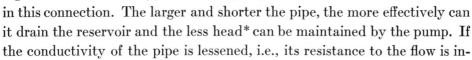

The analogy to a flow of water through a pipe leading from a reservoir supplied with a pump is excellent in this connection. The larger and shorter the pipe, the more effectively can it drain the reservoir and the less head* can be maintained by the pump. If the conductivity of the pipe is lessened, i.e., its resistance to the flow is increased, the more will the pump catch up and create a greater pressure. Of course, as the pressure is increased, the outflow through the pipe increases and, within rather wide limits, the greater becomes the current through a given pipe. Thus equilibrium is again established when the pump builds up a sufficient head so that the pipe removes the water just as fast as it is supplied.

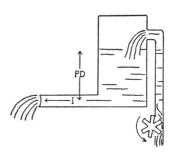

* By "head" we mean pressure-head, height of water above the level at which it flows through the pipe.

In other words, the flow of a given pipe depends on its cross-section and its length and also upon the pressure under which the water is driven through it. If the pressure is doubled, so is the current (quantity per second) through the pipe. The constant ratio between pressure and current defines what we mean by the RESISTANCE of the conducting pipe. This quantity, of course, depends also on the physical dimensions of the pipe. The shorter it is and the larger in area, the correspondingly less the resistance and the larger the current from a given pressure.

Use the words "voltage" (or "potential difference") for pressure, and replace the pipe by an electrical conductor, and think of a current of electricity instead of one of water, then every word of what we have said fits electrical conduction of direct currents.* Since, however, there are specific differences in the electrical conductivity of different materials, there must be added something characteristic of the material for the story to be complete. A silver wire, for example, under the same potential difference will conduct nearly twice as much current as a copper one of exactly the same dimensions.

Thus we have by definition: resistance, $R = PD/I$, a relation discovered by Georg Simon Ohm and published about 1827, known as Ohm's Law.

The specific resistance of a material, R', equals the resistance between the faces of a 1 cm. cube of the material, so that from the physical dimensions the resistance of a wire l cm. long and r cm. in radius is $R = R' \dfrac{l}{\pi r^2}$.

FOR all metals, except an occasional alloy, the resistance increases with the temperature, and at very low temperatures becomes extremely small.

PURE water and all other non-metallic liquids are extremely poor conductors of electricity. Pure water is very difficult to obtain. Ordinary distilled water conducts several hundred times better than the purest water that has been made. The slight conductivity of distilled water, and the con-

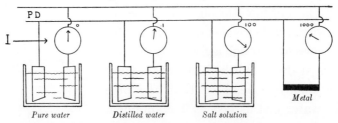

* Some modifications have to be made for the case of alternating currents, modifications that become progressively more drastic as the rapidity of the alternations increase.

CONDUCTORS, SOLID AND LIQUID

siderable conductivity of ordinary water, is due to certain substances dissolved in it, notably acids, salts, and bases.* The solution of many common substances like alcohol, or sugar, in water do not render it conducting at all; but any of the three groups mentioned above, if dissolved in appreciable quantity, will make water readily conducting. It never becomes as good a conductor, however, as some metals.

ELECTRIC currents in their passage through metallic wires produce no change in the material of the wire. They always produce a certain amount of heat. In solutions, however, very fundamental changes in the material are brought about. For one thing, there is in general a tendency for the current to remove from the solution the substances that are responsible for the conductivity. This effect is not always accomplished on account of secondary reactions that prevent it; but in general *chemical change* is produced by the passage of current through a solution (ELECTROLYSIS). The chemical changes are quite different at the different poles, and may or may not involve the material of the electrodes, i.e., the terminals which are immersed in the solution to lead the current in and out of it. The terminal through which negative charge is led in is called the CATHODE. The other is the ANODE. There is also the best of evidence that, during the passage of current, material is transported *through the solution.*

As evidence for this take the case of the electrolysis of copper permanganate,† a deep purple-colored liquid. If this liquid is placed in the lower half of a U-tube, with a layer of colorless dilute nitric acid above it in each arm, near the top of which the electrodes of platinum are inserted, we can observe the transport of material through it by the color changes produced.

Copper permanganate contains copper and manganese combined with oxygen, and the association of the two latter elements is rather close. During the passage of the electric current a blue-green zone of color appears in the nitric acid surrounding the negative electrode (cathode), and a reddish deposit forms on the cathode terminal of platinum. Copper salts are blue-green in solution, and the deposit can be shown to be copper. A pinkish color appears meanwhile in the solution around the positive pole (anode), and a

---

* Bases are substances that react with acids to form salts and water.

† For an excellent treatment in more detail of these phenomena see H. I. Schlesinger, *General Chemistry* (Longmans, Green, 1931).

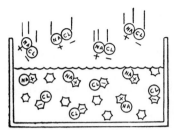

blackish deposit may be formed on the anode plate. Manganese salts are pink in solution, and the deposit is a compound of manganese.

Thus we see that the current has transported material each way through the solution, and that this transport of material represents a breaking-up of the compound originally present, since the different constituent parts of it have been taken out of the solution, which of course becomes more dilute thereby.

Many different kinds of experiments, that we cannot detail here, lead us to believe that in the very process of going into solution, the acids, bases, and salts, that act as ELECTROLYTES (render solutions conducting), are separated into two constituents, one atom or group of atoms being quite as much or more strongly attracted in the liquid condition by the molecules of the solvent than they were by the other group of atoms in the solid compound in which they were originally introduced. Thus, after such a material as common salt (NaCl, sodium chloride) is dissolved, the sodium atoms and the chlorine atoms are quite freed from one another. Because of the strong attraction of the water molecules for them, they are pulled apart and separated from one another; then by the agency of the kinetic thermal agitation of the liquid they become diffused quite uniformly throughout the entire body of the latter.

IF THE atoms or molecular groups of these dissolved substances were electrically charged in some way, then one might easily understand why they migrate through the solution in the presence of the electric field applied by inserting charged electrodes. The negatively charged groups would pass to the positive anode, and the positively charged ones to the cathode. This hypothesis looks attractive—let's adopt it!

BY NOW you begin to smile, perhaps, at our notable weakness in trying to frame some artificial kind of an explanation for everything that comes along that we do not understand. What evidence have we that sodium and chlorine atoms individually possess a charge? When combined they certainly give no evidence for it any more than does the copper permanganate of our experiment. Now, of course, we would not like to have you think that in forming a tentative hypothesis we would be too hasty in jumping to conclusions. We ask

CONDUCTORS, SOLID AND LIQUID

you therefore simply to accept this hypothesis as tentative, to see what further implications it might have and how it might be tested. Furthermore, we will keep alert to find if there are any other phenomena in the passage of current through solutions which independently would seem to justify such a hypothesis.

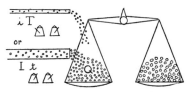

Total charge determines weight of substance deposited, Faraday's First Law

Michael Faraday in this field of study did yeoman service. His three laws of electrolysis are as famed as are his other discoveries that we recounted in the preceding chapter. The fundamental experimental fact we must admit. It is: The passage of electric current through conducting solutions in many cases removes from the solution one or more of the constituents of the dissolved substance.

Faraday discovered, in addition, that: **I. With respect to any given material deposited out (or coming off as a gas) the amount by weight depends solely upon the total QUANTITY of electricity passed through the solution, and is independent of the current strength or solution concentration.**

A strong current for a short time gives the same result as a weak current for a correspondingly longer time, since $Q = I \times t = i \times T$.* A strong solution of the electrolyte gives no more than a weak solution, provided the same quantity of electricity passes through. This would assuredly seem to be good evidence that any given atom or molecular group is always associated with a given amount of electric charge, since the same total weight, i.e., number of atoms or molecular groups is always associated with the same total quantity of electricity.

II. When different substances are compared, a large number of them can be found, the weights of which, deposited out by the same quantity of electricity, are proportional to the relative weights of the different atoms or molecular groups themselves. The entire group of substances of which this is true contains, incidentally, only and all of those substances that have the same chemical valence, or combining power with respect to hydrogen.

This was a very remarkable result. To find that an entire group of substances (which previously had been associated together because of a striking

---

* As heretofore we use again a trick of type. $I \times t$ means a large current flowing for a small time; $i \times T$, a small current for a correspondingly longer time.

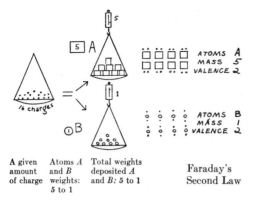

A given amount of charge | Atoms A and B weights: 5 to 1 | Total weights deposited A and B: 5 to 1

Faraday's Second Law

similarity of chemical properties and behavior) were all, without exception, associated with, or carried by, one and the same quantity of electricity in electrolysis was of the highest importance, not only for the study of electricity, but for that of chemistry as well. An electrical significance was given to the idea of chemical valence.

Faraday's third law gave added weight to this and imparted yet another new idea: **III. When different groups of substances of different chemical valence are compared, the amounts deposited out by the same quantity of electricity, as measured by weights, are found proportional to the weights of the atoms or molecular groups involved, DIVIDED by 1, 2, 3, 4, ... , etc., the INTEGERS, and by that very one of those integers which also stands for the valence of the substance involved.**

The atomic theory of electricity really began with the interpretation of this observation. It is equivalent to saying that the smallest charge that can be observed, indirectly, of course, in the flow of electricity through solutions,

Same given amount of charge | Atom C weight 2 | Total weight same as B | Thus half as many atoms

Faraday's Third Law

is the one associated with the hydrogen or silver or any other univalent atom.

LET us recapitulate more concretely:

An electric charge (or quantity) of 96,494 coulombs will be found to release 1.0078 gm. of hydrogen from an acid (sulphuric) electrolyte. A charge of 96,494 coulombs will likewise be found to deposit out 107.88 gm. of silver, from a silver nitrate solution. These numbers of grams, 1.0078 and 107.88, also represent the relative weights of one atom of hydrogen and of silver, respectively, so that in these two weights there will be found equal numbers of atoms of hydrogen and of silver. This number is, by the way, about 606,400,000,000,000,000,000,000; i.e., 6.064×10^{23}!

The charge on one atom then will be

$$\frac{96{,}489}{6.023 \times 10^{23}} = 0.000{,}000{,}000{,}000{,}000{,}000{,}160; \text{ i.e., } 16.0 \times 10^{-20} \text{ coulombs.}$$

The electrostatic unit of charge we used in chapter 20 was a much smaller quantity than the coulomb, 3×10^9 times smaller. In terms of the electrostatic unit, the charge associated with one atom of hydrogen or silver as described above is obtained by multiplying by 3 and knocking out nine zeros. Thus the charge on one atom is $0.000{,}000{,}000{,}480$ of an electrostatic unit; i.e., 4.80×10^{-10} (e.s.u.).

If elements like copper or bivalent iron salts are used for the conducting solution, a quantity of electricity equal to exactly TWICE the 96,494 coulombs will be required to get an equal number of atoms of these metals, i.e., to get weights proportional to their atomic weights. This means, of course, that each one of them will have to be associated with and carry through the solution TWO of the 4.80×10^{-10}ths of an electrostatic unit. Similarly, iron in a ferric instead of a ferrous solution, trivalent in character, or any other trivalent atom or molecular group will carry THREE such charges; and so on.

NEVER in any kind of a solution or under any kind of an electrical process has any *smaller* amount of charge made its appearance. Never has any other fraction or multiple of this amount been observed except 2, 3, 4, 5, or some other *exactly integral* multiple of it. In certain other entirely different experiments that we shall discuss in the next chapter, every single integral multiple of this number up to nearly 200 has been observed.

This is our first bit of evidence that electricity comes in "packages" of a definite amount, that electricity is ATOMIC in character.

A FEW names should be mentioned. The 96,489 coulombs have been given a special name, the "Faraday"—the amount of electric charge that, when passed through an electrolyte, deposits a gram molecular weight, weight in grams equal to the relative weight in question (referred to oxygen as 16.000 or hydrogen as 1.0078), or one-half, one-third, or one-fourth this amount, depending on the valence of the atom in the molecule in the solution. So convincing seems this evidence for the existence of these charged atoms and molecular groups in electrolytes, and so plausible is it to imagine them migrating in the presence of the electric field, positives to cathode, and negatives to anode, that we call them IONS (Greek for travelers). The substances which in solutions give birth to them (the electrolytes) have also another name, IONOGENS. The solutions are said to be IONIZED to the extent that the "ionogen" has completely "dissociated" into its component ions. Complete ionization, for reasons we shall not go into, takes place only in solutions as they approach infinite dilution, and therefore, having a negligible number of ions left, cease to be conducting altogether.

The hypothesis of ionization outlined above was first proposed in 1857 by the great German physicist Rudolf Clausius (1822–78), known better perhaps for his extensive and fundamental contributions to the kinetic theory of gases. His hypothesis was subjected to many tests and elevated to the dignity of a theory largely through the work of a prominent Swedish physical chemist, Svante Arrhenius (1859–1927) of Stockholm. If we had no group of phenomena except those of conducting solutions, we might possibly have some misgivings about accepting as fixed and final so fundamental a conclusion about the nature of electricity.

ATOMS of ordinary matter we have never seen and never shall see directly with our eyes;* the case is far worse with respect to these ultra-atomic-dimensioned atoms of electricity.† The next six chapters, and parts of several others that follow, will deal almost exclusively with a great variety of different kinds of experimental evidence and interpretations that all, from one angle or another, converge upon this same conclusion as to the atomic character of electricity. The implications are even more far reaching, to the effect that matter itself is essentially electrical in nature; and our entire physical world, its processes, its changes, our own physical selves, and even our thoughts seem to become reducible in simplest terms to interrelations between fields of electric and magnetic forces.

The last of these conclusions certainly lies beyond the ken of physical science at the present time. With respect to the others we shall proceed with the evidence and let you decide for yourself.

* This statement contains something more than mere pessimism as to the possibilities of human accomplishment or prediction as to the future, both unworthy of a scientific point of view. It refers to the fact that, to the best of our scientific knowledge, now quite extensive in these matters, it is impossible to form images, and so to "see" objects that are of smaller dimensions than those of light-waves themselves.

† In round numbers atomic diameters are of the order of 10^{-8} cm; the charge on these atoms, however, is an entity ten thousand times smaller, 10^{-13} cm, lost, as it were, somewhere within this atom, like a moth-miller flying about within a cathedral.

CHAPTER 29

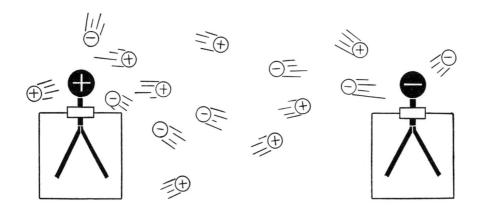

THE CONDUCTION OF ELECTRICITY BY GASES

IN THE light of what has been said before you may feel the desire to take exception to this title. Gases, you have been led to believe, are nonconductors. The leaves of our electroscopes and electrometers are insulated from their cases by sulphur or by amber, and certainly they are not supplied with a ground connection by the circumambient air. Electricity does not flow across the air gap of our lighting-circuits' switches when they are open and heat the filaments within the globes. The power companies certainly could not send energy at the rate of thousands of horse-power to remote communities and have any left at the destination there if the miles and miles of air around its transmission lines were normally conducting for the flow of charge. You are quite right. Neither air nor any other gas is *normally* a conductor.

The abnormal conductivity that air sometimes acquires may be produced by such simple means that it is surprising that this fact was not discovered until comparatively modern times. Suppose we try a simple experiment with our electroscope. We will use the same one as formerly and will alter the arrangement in no way except by attaching a short length of wire to the insulated knob and leaf system. The other end of the wire we will fasten to a bit of sulphur, so that the insulation is perfectly maintained.

We may now charge the electroscope either by contact or by induction (see chap. 20) and with either sign that we prefer. If the insulation is good

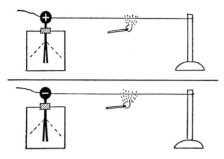

and the weather dry so that no moisture condenses on the insulating surface, the charge will be held for hours; no leakage can be detected by any ordinary methods. The air, while not quite perfectly nonconducting,* is so nearly so that we can, for the present at least, assume that it is.

Now we will light a match and hold it so that the flame lies below but does not touch the wire attachment. Almost instantly the leaves collapse.

The virtue that the flame of the match possesses is shared by a candle flame, or that of a bunsen burner, or gas stove. Indeed, a white-hot poker held below the wire produces almost as rapid a collapse of the gold leaves. Whether we use a negatively charged electroscope or one charged positively there is no variation in phenomena. An electroscope charged with either kind of electricity is instantly discharged by something associated either with fire or with any sort of incandescent heat. That the discharging entity resides in the air rather than in the flame itself seems already evident. This point may be investigated with a little more complexity of arrangement.

Suppose the insulated wire be connected between the gold leaves at one end and a short terminal of wire passing through an insulating movable plug inside of a tube either of metal or of glass. By means of a gentle suction let air from the immediate vicinity above a flame be drawn through the tube, with a velocity approximately known. The conductivity of the air is all present in the air immediately after being drawn away from the flame, since after entry into the tube it still discharges the electroscope. As one moves the electrode farther and farther away down the tube, however, the conductivity gradually disappears, although the gas may fall only slightly in temperature and be unchanged in its constituent chemical character. The conductivity is a physical, and not a chemical, effect. Indeed, the conductivity may be filtered out of the air by drawing it through a plug of glass or cotton wool after it leaves the immediate vicinity of the fire.

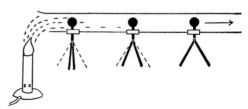

* See later, under "cosmic rays," p. 281.

CONDUCTION OF ELECTRICITY BY GASES

WE MUST now, according to our custom, begin to make hypotheses again. The discharge of an electroscope *negatively* charged can be understood only on one of two alternative assumptions: either that negative leaks off or that positive leaks on and neutralizes. If it is positively charged, then either the positive leaks off or negative leaks on. Which one of these alternative points of view we should adopt would seem at first like Hobson's choice. 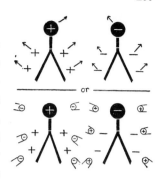 Our only past experience with the details of conduction has been with metallic solids, where whatever it is that happens leaves no trace or evidence behind; and with conducting liquids, where we have a most excellent working hypothesis as to the existence *of ions*, charged atoms or molecular groups, of both signs, which by their migrations in a field neutralize the charges on the electrodes that cause the field, or at least would do so if a fresh supply of charges were not furnished continually to maintain it.

Let us be guided by this past experience. It is always the part of common sense to do so. Using a hypothesis that we formerly found useful, let us assume that the flame in some mysterious manner produces ions, of both signs, negatively and positively charged molecules of nitrogen and oxygen (that comprise 99 per cent of the air) in the immediate vicinity—billions upon billions of them of course. Toward the wire, if it is positive, the negative ions will swarm and neutralize its charge (see drawing at the beginning of this chapter). The positive ions will do the same if the wire is negatively charged. So far the hypothesis works.

If the ionized air is left to itself for a time, the individual ions, positive and negative, will seek out one another under their mutual attractive forces, in spite of the disturbing and mixing up that is going on continually owing to the thermal agitation. The air will ultimately regain the uncharged condition that it originally had. Its conductivity is not inherent originally; it is something that has been added by the processes of combustion.* Given

 a few seconds in the absence of outside field, the much greater freedom of path which gas molecules enjoy in their kinetic motions over molecules of a

---

* Gases produced as products of combustion can be shown to play no fundamental rôle in the effect. They may be, and are, ionized just as the air is; but the ions disappear in time by neutralizing one another, and the gaseous combustion products are normally quite as perfect nonconductors as the air itself.

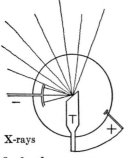

X-rays

substance in the liquid or solid state (see chap. 18, p. 164) will enable them much more rapidly to seek one another out and to become neutralized if oppositely charged. The filtering effect described above holds back the gas and swirls it in all directions through the interstices of the wool, and thus gives more opportunity for neutralization of the opposite charges. Suppose we accept our hypothesis tentatively and see if we can find other processes than flames which will also produce temporary conductivity in gases.

There are, we find, several agents rather common in this modern age that serve not only equally well but better, and there is one other that does the trick but not nearly as well as fire. These agents are known to us at present merely as names —words in our vocabularies, but words still replete with mystery. They are: X-rays, radioactivity, and ultra-violet light. Accept them for the moment for what you think they are, quite uncertain as to whether you really know what they mean or perhaps quite certain that you do not.

An X-ray tube in operation far away across a large room, 30 or 40 ft. from the charged electroscope with its wire lead, will discharge it almost as

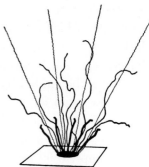

Ionization agents associated with radioactive materials

quickly as you could by grounding it with a wire. Indeed, if you shut off the tube and recharge the electroscope, at once it will begin to discharge again from the residual ionization that still remains in the air. X-rays are a very powerful ionizing agent.

It is quite the same with radioactive substances, which, formerly very rare and expensive, are now becoming much more abundant in kind and in amount as a result of the atom bomb investigations. These produce strong ionization also, when near the electroscope wire.

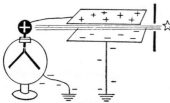

Ultra-violet light is not nearly so effective. Among the most powerful sources of it are electric arcs in mercury vapor inclosed in quartz tubes. If, to the other end of the wire attached to the knob of the electroscope, we connect a broad metal plate, and have near it

CONDUCTION OF ELECTRICITY BY GASES

another similar plate connected to the earth, and if we then direct the light from the lamp between the plates, taking care that it does not shine directly on either,* there will be sufficient ionization produced in the region between the plates to cause an appreciable leak of the electroscope.

WITH all of these agents, apparently equal numbers of ions of both signs are produced. Their mobilities of drift in the field can be measured. This mobility is the velocity acquired in falling down a field whose potential varies by 1 volt per cm. The mobility depends on the nature of the gas, and slightly on the sign of the ion. In hydrogen, mobilities are about 7 cm. per sec.; in carbon dioxide, about 1.

We have seen in the previous chapter how it was possible to measure the charge on ions in solution by dividing the total charge carried by the number of the atoms involved in transporting it. The most interesting studies with

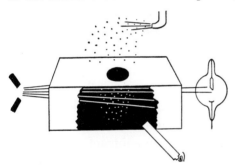

respect to gaseous ions are connected with the successful attempts that have been made to measure their charge. The names of J. S. Townsend, J. J. Thomson, H. A. Wilson, and R. A. Millikan, several of them familiar today as Nobel laureates, are associated with this work. Millikan's method is simplicity itself.

An atomizer blows into the air a fine spray of minute droplets of oil, some of which, settling down, fall through a hole drilled in the upper of two condenser plates (parallel metal plates that may be charged). A strong beam of light illuminates the space between the plates from left to right, and in its radiance these little drops may be seen if one has a low-power microscope directed across the beam toward a black background on the other side.†

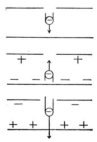

Thus the oil particles may be watched as they slowly settle under gravity, their constant terminal velocity being

---

* To avoid the photoelectric effect (see next chapter, p. 284).

† This is essentially the same method used to observe the motions of colloidal particles in liquids cited in chap. 18, p. 170, as evidence of molecular agitation. Oil droplets or particles of tobacco smoke, made visible in the same way, show the Brownian movements in gases, especially at somewhat reduced pressures.

quite small because of their minute size (see relation on p. 23, n.). If, when they are between the plates of the condenser, an electric field is created in this region by charging the two plates oppositely, the velocity of the little oil drop usually is greatly altered. This can happen, of course, only if the drop *also is charged;* and it usually is, since it becomes electrified by friction as it is blown out of the atomizer.

Suppose it were negatively charged; then if the top plate of the condenser were made positive when the drop was in sight between the plates, its downward motion might be stopped altogether and it might now even drift upward, like our imaginary pith ball beneath a charged ceiling. If the charges on the plates were reversed, it would at once acquire a swifter terminal velocity downward, since now the force of the electric and gravitational fields upon it would be in the same direction and would add.

From the ratio of the two rates of motion in either case, and the masses obtained from Stokes' Law (p. 23, n.), the charge on the droplet may be measured.* By suitable switching arrangements the strength of field may be readily altered or reversed, and in this way the little oil drop may be made to travel back and forth, up and down, with the field opposing gravity and then with gravity alone. No matter how the drop was originally charged, in the course of time the charge will suddenly be found to alter. There are always a few gaseous ions in the air.† One of them probably collided with the oil drop and left its charge upon it, thus adding to, or subtracting from, the charge that was already there.

If an ionizing agent (for example, a beam of X-rays) is shot for an instant through the space between the plates and myriads of ions produced, the little droplet will change its charge often faster than one can observe it. Every time the charge on it changes, remember the speed of the drop changes if the field is on.

* The relation is very simple:

$$\frac{V_g}{V_\epsilon} = \frac{mg}{(E\epsilon - mg)},$$

where V_g and V_ϵ are the velocities under gravity (g) alone and under the combination of gravity and electric field, E, ϵ is the charge on the drop and m its mass. If the same drop on two different trips past the microscope has different charges on it, then this difference

$$\epsilon - \epsilon' = \left(\frac{mg}{EV_g}\right)(V_\epsilon' - V_\epsilon)$$

is equally simple to find from the difference between the speeds on the two occasions. For any single drop, and with a constant field, E, the factor in the first parenthesis is constant.

† Cf. under "cosmic rays," later in this chapter, p. 281.

FROM such experiments Millikan announced in 1910 that, in all of the many hundreds of cases that he and his students had observed, the differences in speed, reduced to differences of charge, were all exact integral multiples of a certain smallest difference that was frequently observed. This smallest difference corresponded, within the limits of accuracy of observation, to a change of charge of 4.77×10^{-10} electrostatic units. Thus, they found, first, that: THE CHARGE CARRIED BY IONS IN GASES NOT ONLY CHANGES BY THE SAME CHARGE AS THAT CARRIED BY IONS IN SOLUTIONS BUT ALSO NO SMALLER CHARGE OR CHANGE IN CHARGE EVER APPEARED. Subsequently, by many thousands of patient observations Millikan and his pupils watched and waited until they had observed the tiny oil drops carrying not only just one of these charges but 2, 3, 4, 5, 6, 7, ... , and every integral multiple of this amount to almost 200. Never in the course of all this work did they find, nor has any one else found, any authentic case of any other small-sized charge that did not correspond exactly to some one of these whole numbers multiplied by this fundamental one. No intermediate fractional charges ever appear.

The marvelous confirmation of the same conclusions, by precise work in this field of conduction of electricity by gases, supported by the same numerical values that resulted from the study of liquid electrolytes, has placed the atomic theory of electricity on a solid foundation ever since the experiments were published.

TWICE we have referred to cosmic rays in footnotes (on pp. 276 and 280) for the purpose of taking exception, slightly, to some statement we have made. Not very long ago the public imagination was fired by a series of cosmic-ray experiments, the outgrowth of thirty years of unheralded, patient work that also centered on the leakage of charged electroscopes due to ionization of the air.

Since 1902, when Elster and Geitel in Germany found that electroscopes in air always leaked by a slight amount, the air has been known to be not perfectly nonconducting. So slight was the leakage in cases where known ionizing agents were absent that for many years the actual existence of the leak, apart from well-known accidental causes, was even in doubt. Indeed, it is extremely difficult to find a place on earth where slight traces of radioactivity—in rocks, in sea water, on land or over ice, on seas or lakes—do not produce slight ionization, and hence slow leakage. Experiments first on mountain peaks and then in balloons have shown finally that a cause of

Cosmic rays

ionization outside the earth exists, and that some mysterious agent coming to us, either from extremely high altitudes above the earth, or perhaps even from interstellar space, filters down through our atmosphere and produces now and then a few ions within our electroscopes. This agent goes by the name of the COSMIC RAY. To measure such feeble ionization taxes the limits of modern apparatus and technique. A large number of the most careful investigators all over the world, representative of most of the great nations of the earth, independently and co-operatively, have for the last several years been making every effort to find, not only consistent and concordant results, but also an interpretation with which all can agree. We shall refer in Part V, page 435, to these phenomena in more detail.

HAVING presented evidence that is today unquestioned as to the existence in electrical phenomena of facts that show electricity to be of atomic character, we shall next proceed to see if we cannot find out still more about the nature of these electrical atoms and their more intimate connections, if they have any, with the atoms of ordinary matter. So far we have seen them only as mysterious, tiny charges, caught and entangled, as it were, upon ordinary atoms and molecules as flies or dust specks might be caught on baseballs or dumb-bells smeared with the adhesive that comes on fly paper. We know much more about them than we have yet disclosed. To tell you about these things we shall have to pick out of the vast material of experimental studies that physics has now collected two seemingly unrelated items, both discovered many years ago, neither understood with full realization of its significance until modern times. Our next chapter deals with them.

CHAPTER 30

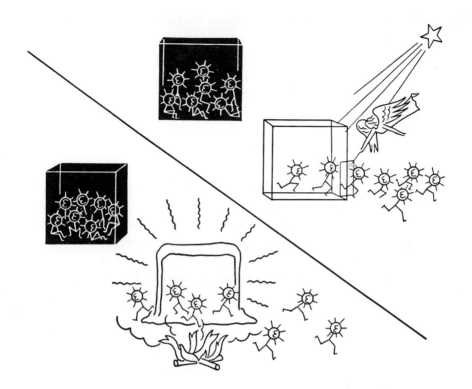

PHOTOELECTRIC AND THERMIONIC EFFECTS

BACK in the eighteen-eighties, experimenters, drawing sparks from static machines and from other sources, not infrequently noticed that sparks seemed to be able to jump a greater distance in the light than in the dark. If the two electrodes were separated just far enough in ordinary light so that the potential of the supply of energy was insufficient to enable a spark to pass, the flashing of the light from a brilliant spark from another source near by upon these electrodes would invariably produce a leap of the spark between the first set of terminals. Electrical influences between the two spark gaps were carefully studied and eliminated from any part in the story. The effect was one of light upon electricity, a PHOTOELECTRIC EFFECT.

NOW let us tell you another. Suppose we take our simple electroscope, unscrew the knob at the top, and replace it with a plate of zinc. Near by we will set up our ultra-violet mercury lamp, used in experiments of the preceding chapter to show that its light ionized the air. We charge the electroscope negatively, either by contact or by induction, and then let the light shine upon the zinc plate. The electroscope leaves slowly collapse. If we remove the zinc, polish it well with sandpaper, and then replace it, the discharging action of the light is more vigorous.

We now recharge the electroscope, this time *positively*, and repeat. NOTHING HAPPENS. Negatively charged again, the leaves wilt the moment the light falls on the plate; positively charged, nothing happens whatsoever.

Quite similar effects can be obtained using other metals than zinc, and with all of them the same unidirectional character appears.

IN TERMS of our ionization hypothesis, set forth in the preceding chapter, we might assume that in these experiments we have been producing ions of only one sign, positive ones that drift to the plate when the latter is negative, but no negative ones simultaneously, which by their presence would neutralize the charge on a positive plate when in its vicinity. This at once gets us into difficulties. There are many other experiments on gaseous ionization; and all of them attest to the fact that equal numbers of both positives and negatives are invariably produced when either one kind or the other is formed. Indeed, our explanation above seems to be at variance with everything we have formerly learned about the existence of equal charges of opposite signs always being present when electrical phenomena appear.

What shall we do? You will recall that when we started our interpretation of gaseous ionization on page 277 we noted that we had to make a choice of one of *two* alternative assumptions with respect to the discharge of a negatively charged electroscope. They were *either* that the positive charge leaked on, *or* that the negative leaked off. The former choice (and the corresponding one with respect to a positively charged electroscope) took care of the phenomena of gaseous ionization to the queen's taste. But, as we have seen, it brings difficulties in its train with respect to photoelectric phenomena. Let's try the other hypothesis and see where we come out.

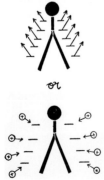

PHOTOELECTRIC AND THERMIONIC EFFECTS

IF THE action of the light causes negative charge to leak off from a metal plate exposed to its radiance, but has no effect upon positive, this point of view at once associates itself with the fact that in metallic conductors, as we have frequently observed, it is only negative that is free to move about within and on the surface of the metal. Moreover, it is quite obvious that negative could *not* leak off if positive charge were in excess on the conductor, because the mutual attraction between these unlike kinds would prevent. On this hypothesis, furthermore, the implication would be that the gas surrounding the conductor had little, if anything, to do with the phenomenon. Possibly it would work in a vacuum just as well as in air.

CONSEQUENTLY, we design and build a vacuum tube. It cannot be all of glass, since ultra-violet is not transmitted by ordinary glass. Quartz is transparent to these rays, however, and it can be attached to glass in an air-tight manner. We seal in the zinc electrode, well polished as before, arrange the quartz window so the light can shine through and strike the zinc, seal the tube onto a vacuum pump line, evacuate, test for the quality of vacuum (which we make as perfect as possible—about a millionth of an atmosphere pressure), seal off our toy, and now, connecting the electroscope to a lead-in wire that goes to the zinc plate, turn on the quartz lamp and see what happens. Everything that happened before, happens now! *No change in the results has been produced by removing the surrounding gas.* Thus, then, we evidently do not have a case of gaseous ionization here at all. Consequently, we could hardly expect it to conform to a theory designed to describe the behavior of gaseous ions.

There is still another implication of our new hypothesis of negative leaking off a plate even in vacuum under the action of ultra-violet light. It provides us with that *summum bonum* always desired by the puzzled scientific worker, an *experimentum crucis*, a crucial experimental test by the results of which a hypothesis either stands or falls.

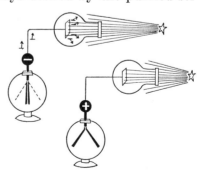

In terms of the idea that light releases negative only from a negatively charged plate where the electrostatic field is already there to assist in the process, and cannot pull it out against an opposing field of excess positive charge on the plate, it is quite

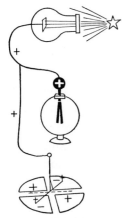

logical to argue that, if the plate originally were neutral, then likewise negative would come off unopposed, at least for a time. In other words, a neutral plate would become positively charged by the light action. Ultimately, of course, the excess positive charge left behind might prevent the escape of any more negative.

We have a convenient apparatus at hand. We start in by grounding the leaves to earth, to assure neutrality; then we insulate the plate-leaf system, turn on the lamp, and watch. Alas! the leaves show *no* divergence, no acquisition of positive charge through loss of negative—not enough, at least, to affect our instrument. We really hoped to get an answer here; we are going to die hard; but, nevertheless, we must play fair and not deceive ourselves. Maybe our electroscope is too insensitive. Light is a tenuous affair whatever it may be; its force to detach negative may not be very strong; and if it is not, a very weak positive field set up will prevent the escape of all but a very little negative charge initially. Electroscopes are all right for fields of thousands of volts potential, but cannot indicate only a few volts. We get hold of an electrometer, but we may have to spend weeks setting it up and getting it adjusted, for it's a delicate thing at best.

WE ARE fully rewarded as we now repeat our test. We observe that the moment the light begins to shine on the uncharged plate in the attached electrometer needle system, the needle begins to move. The acquisition of a positive charge is indicated, but its amount is very small—one that builds up a potential of about 5 volts only before the action ceases. Our hypothesis has stood its crucial test; a new phenomenon is hereby attested. LIGHT shining on a piece of metal gently removes a little of the negative that is always there associated with the immobile positive. The touch of the light is feathery almost beyond belief, as if it were an insect with wings of gossamer that flew by and dusted off the surface with the faintest contact.

Further experiments in this field we will not detail. Not only metals, but a vast array of substances, exhibit this photoelectric effect. With metals like sodium and potassium, ordinary visible light is quite effective. X-rays that lie far beyond the ultra-violet produce the photoelectric effect intensely.

PHOTOELECTRIC AND THERMIONIC EFFECTS

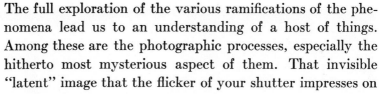

The full exploration of the various ramifications of the phenomena lead us to an understanding of a host of things. Among these are the photographic processes, especially the hitherto most mysterious aspect of them. That invisible "latent" image that the flicker of your shutter impresses on the film was a phenomenon that for generations defied analysis, either chemical or physical. How it could carry in its shadowy existence minute delineation of detail that was to be revealed only by subsequent development was a great puzzle. It turned out to be a purely photoelectric effect in silver haloid grains* and gelatine. Through the action of organic reducing agents these delicate electrical impressions are embodied in visible metallic silver.

There is an entirely new and modern science of photochemistry, based on photoelectric actions, that covers hitherto unexplained effects ranging from explosions to the production of vitamins.

The equation by Einstein that describes these processes will always be remembered in connection with his name. This indeed may be a far more significant contribution from him than the theory of relativity, for which he is also justly famous. For that matter, through their various ramifications, those early experiments in which a flash of light released the negative electricity stored up in electrodes and initiated a spark that otherwise would not have occurred have done quite as much toward the building of a new science of physics and chemistry as any other single thing.

As for practical uses and devices depending on these principles, they are almost too numerous to mention. Photoelectric CELLS, as they are called, consist of some active photoelectric material, combined with radio tubes or other devices to magnify the rather feeble currents that they normally give, and thus work electric relays and so turn on or off much larger supplies of electric power. They are utilized for opening doors, for counting people that go through a passage, for sorting cigars, or almost any kinds of things

---

* Photographic emulsions, films, plates, and papers usually consist of a fine precipitate of halogen derivatives of silver (combinations of silver with chlorine, bromine, and iodine) thrown down within gelatine, then coated on the film.

PLATE 69 Apparatus to Demonstrate the Edison (Thermionic) Effect. A, Galvanometer; B, incandescent lamp with wire sealed inside of bulb; C, variable resistance to control current in lamp.

that show differences in size or color. There are photographic exposure meters of guaranteed precision that measure light intensity. There is television, too. Most of these developments are comparatively recent. There would seem to be here an almost virgin field for the play of inventive ingenuity. Whatever the eye can see (and indeed many things that it cannot see) these unwinking photoelectric eyes can see far better, since they never need sleep and never become fatigued.

We shall return to a more complete interpretation of these phenomena later, after a brief glance at some experiments of an entirely different nature whose applications in engineering and the art of electrical communication have brought us radio and television—achievements that our forefathers could not even have imagined.

THE late Thomas Alva Edison, in the early days of his experiments with "the light in a bottle," noticed a phenomenon that took place within an evacuated bulb containing an incandescent carbon filament. The phenomenon was an electric current. It was not that current which, in passing from the electric mains, generated heat in the filament and made the latter luminous. It was a quite different one—one that in part took place across a region that was as free from all material substance as it is possible to get such a region, *a current through a vacuum.*

The experiment is one that anybody who can blow glass a bit, and who has access to a vacuum pump and who can borrow a rather insensitive galvanometer, can repeat for himself. Since perhaps this lets you out, we have

PHOTOELECTRIC AND THERMIONIC EFFECTS

supplied a photograph (Plate 69) of the experiment and a drawing of the very simple circuit involved. The heating current from battery or generator is applied in the usual manner. The lamp bulb has had a hole pierced in it on the side in which a wire has been sealed, and then the bulb has been re-evacuated of all the air possible down to a pressure of about one-millionth of an atmosphere again.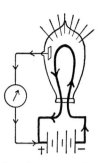

To the wire sealed in the bulb, another wire connection has been made which leads through the galvanometer and thence to the positive side of the battery, or to where this positive terminal is attached to the socket near the lamp filament. In order to control the brightness of the filament, a slide wire resistance, not shown in the drawing but seen in the photograph (Plate 69), is connected in series with the battery and filament. We start the experiment with the resistance all "cut in," so that the lamp glows but feebly; then gradually reduce the resistance; and finally, when it is all "cut out," the lamp shines with its normal brightness.

As the brilliance of the filament is slowly increased, quite suddenly the galvanometer needle begins to show a deflection which quickly mounts up to several thousandths of an ampere. This is not large compared with the quarter- or half-ampere passing through the filament; but nevertheless, it means that something like 10,000,000,000,000,000, i.e., 10^{16}, of our natural units of atomic electric charge are passing through the galvanometer every second; and this is a fact that must be reckoned with.

THE phenomenon is called the EDISON EFFECT, sometimes the THERMIONIC EFFECT. By using special kinds of filaments—tungsten containing borium; or platinum containing exceedingly thin deposits of oxides of calcium, barium, or strontium—upon it, we may increase the effect enormously and also make it occur at temperatures only slightly above red heat. Currents of several amperes passing across the highly evacuated space between the filament and the sealed-in wire (often called the "plate" because it usually has attached to it a metal plate inside the bulb) have been observed.

No explanation in terms of ionization or of the existence of gaseous ions would seem possible, at least with high vacuum inside the tube. Further-

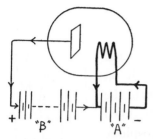

more, the effect, like the photoelectric, is unidirectional, i.e., it does not take place at all if the plate is connected to the negative, instead of to the positive, terminal of the battery. Frequently the potential between plate and filament is more conveniently altered and controlled by a separate battery inserted between filament and plate.

Undoubtedly you have already recognized in what we have been describing the beginnings of a radio tube. The two batteries are those two usually designated as "A" batteries, used to heat the filament, and "B" batteries, used in connection with a lot of other gadgets for some mysterious reason to make a radio set out of the tube or a combination of several of them. Large tubes of similar construction but with larger plate surfaces within, called "kenotrons," are used to "rectify" alternating current, i.e., to convert it into direct without the medium of any commutators or moving machinery of any sort.

It was Preece and Fleming who first saw the "valve" possibilities of devices involving this effect of Edison's. They invented the early arrangements that made practical use of them to transmit a current while it was flowing in one direction and not in the other. Fleming likewise was the first to put such tubes to work, detecting the presence of the electromagnetic waves discovered earlier by Hertz and used first in the art of communication telegraphically by Marconi. Some of these things will be referred to later in other connections. At the moment we need, first of all, an interpretation of the phenomenon, an understanding of it.

With apologies to Prof. Foley

IF WE have such a tube built on a rather large scale and with but a small bit of filament or a small spot of emitting oxide on it, with some traces of gas left in the tube, we will observe luminous rays traveling out in all directions from the filament. These are not especially directed toward the anode

PHOTOELECTRIC AND THERMIONIC EFFECTS

(positive) plate. If we assume that these rays are not light but only the paths of streaming electricity, made visible by impinging on the gaseous matter in the tube, we would class them (by definition) as the paths of currents and expect them to be deflected by a magnet as a current in a solid conductor is.

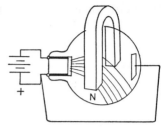

This expectation turns out to be justified. If a magnet is brought near, the path of the discharge bends away at right angles to the magnetic field. The sense of the deflection is the same as if negative electricity were leaving the cathode. There is, of course, no stiffness to the path of this current as there is when it is confined within the solid matter of a wire. If we arrange our discharge tube to lie within a magnet especially designed to give a uniform magnetic field at all points, our discharge winds itself around into a perfect circle. We shall return to this fact a little later (p. 304) to do some quantitative measurement.

With any appreciable amount of gas left in the tube, however, to make the path of the current visible, our experiments are apt to be somewhat erratic, and we are not certain that interpretations involving gaseous ions are shut out. The experiment of Edison's in very high vacuum was ideal, but we "see" nothing but a galvanometer deflection. Nevertheless, we can trust our galvanometer to "see" for us. Indeed, it is a far more reliable witness than our eyes, for it not only detects, but it also measures, the strength of this mysterious current. It shows us that it increases enormously with the temperature of the filament. In the case of tungsten the current is increased 100,000,000,000 (i.e., 10^{11}) fold as the filament is heated from 1,000° to 2,500°, Absolute, in temperature.

THE Edison effect we see has several points of remarkable similarity to the photoelectric effect. Among them are:

1. Both effects take place in a vacuum; consequently, an appeal to hypotheses of gaseous ionization for their interpretation would seem futile.

2. Both effects are unidirectional, occurring only when the active agent, illuminated plate or heated filament, is negative.

3. Hence, the alternative interpretation we used for the photoelectric effect would seem also to apply to the Edison effect, viz., that in the latter

the light (or heat) in the filament is responsible for the release of negative charges therefrom, which, repelled by the excess negative there, at once leave the vicinity.

With these thoughts in mind let us now describe to you some other old experiments that for many years puzzled physicists. These, likewise, will be found to have many points of similarity with those we have just been describing.

CHAPTER 31

CATHODE RAYS

IN THE preceding chapter, the phenomena of the photoelectric and thermionic (Edison) effects seemed to give evidence that streams of negatively charged particles under certain circumstances might be shot out from negatively charged electrodes, or cathodes. Such streams are aptly called "cathode rays."

This name was not a new one in physics at the time attention began to be focused on these two particular experiments. A quite different group of experiments involving neither ultra-violet light nor incandescent filaments had brought to light some rays that were already known as cathode rays. Curiously enough, also, these experiments were in connection with studies in the conduction of electricity by gases, especially when the pressure was considerably reduced below atmospheric pressure; i.e., they were experiments within sealed tubes partly evacuated.

SUPPOSE we take a tube, about 3 ft. long, insert at the ends electrodes (e.g., aluminum plates with connecting wires running through the glass), connect this tube to a common vacuum pump that will exhaust it to about a thousandth of an atmosphere, and, starting with the tube at atmospheric pressure, attempt to pass an electric current through it by applying an electric field from some source of potential across the terminal electrodes.

We find that, as usual, the air is an extremely poor conductor. If we have a sufficiently large source of potential like a great static-machine, we may be able to send a miniature lightning bolt crashing through the air within the tube. Here, however, we have a momentary current of great magnitude, a kind of current that we do not want. The resistance of the tube, $R\ (=PD/I)$, is very great, since under a strong electric field, which means a great PD, practically no constant current, I, can be detected.

We will now set our pump slowly into operation and reduce the pressure of the air within the tube, all the while maintaining many thousands of volts potential difference across its terminals. By the time the pressure has been reduced to about one-eighth of an atmosphere, the space within the tube suddenly "breaks down" and a long and narrow filament of luminous color like a glowing snake appears in the tube, each end resting on one of the electrodes. Simultaneously a milliammeter (ammeter whose full-scale deflection is only a fraction of 1 ampere and whose smaller divisions read in thousandths of an ampere) indicates for the first time a current of electricity flowing through the tube. We nowadays call the phenomenon an arc, or "vacuum arc," although the vacuum is far from being good.

As the evacuation proceeds, the writhing, snakelike arc straightens out and rapidly grows in diameter, soon turning into a luminous glow of a terra-cotta red that fills the entire space within the tube. The resistance now is rapidly diminishing, and with a fixed source of potential the current would now be mounting into amperes and the tube soon might melt as the heat from the arc increases. Usually such sources as will supply the high potentials that are required to "break down" the tube and start the current flowing cannot be expected also to maintain a very large current. Since the current is thus limited by the limitations of the generator and cannot increase to a large value, the greatly lessened resistance through the tube results in the potential, PD, swiftly falling to keep pace with the declining $R(=PD/I)$. The circumstances are helped by the fact that, once started, this arc will maintain itself on a much smaller PD than that which is needed to get it going.

We continue the evacuation of the tube. The next change that we notice is the development of an intensely blue tuft of light surrounding the negative electrode (cathode) and a dark space between it and the

CATHODE RAYS

reddish glow that extends down to the anode. This red glow is called the "positive column" for brevity of reference, and the glow around the cathode the "cathode glow." The space between is named the "Faraday dark space," in honor of its discoverer.

As the evacuation proceeds, the positive column recedes toward the anode, under certain circumstances breaking up into segments or "striations," alternate sheets of dark and light perpendicular to the axis of the tube. The Faraday dark space expands between the receding positive column and the expanding blue negative glow. Then we begin to notice another dark space—at first close to the surface of the cathode. This accurately follows every contour of this electrode and grows larger within the

cathode light as the pressure becomes still lower. This second dark space is named after Sir William Crookes, who spent many years in the study of these phenomena.

The resistance of the gas, meanwhile, has been gradually declining, but at a less rapid rate. Finally, it reaches a minimum as the evacuation proceeds to a pressure that depends considerably upon the size and shape of the inclosing tube. By this time the positive column has almost disappeared and we have about reached the limit of the capacity of our simple vacuum pump. If a better one is now put into action to continue evacuation we find the resistance to the discharge begins *to increase* and by the time the Crookes dark space has expanded to fill most of the tube, practically all of the gorgeous colors have faded, the current has fallen so as to be barely detectable on our instrument, and the source of high potential has again been built up across the terminals where we again can hear it sputter and hiss. Just before the tube gets entirely dark again, the little light that is left comes *not from within the tube* but from a faint and spotty *fluorescence of its glass walls.* Finally, all action once more ceases, and the resistance becomes once more almost infinitely great.

WE HAVE described these phenomena in considerable detail, not that you should remember them all, but that you may better realize that in every electric sign of the glass tube type you are looking at the positive column of some glowing gas excited electrically by current passing through conducting gas, in a tube filled to a pressure equal to that of a mercury column 1 cm. high, or somewhat less. (Atmospheric pressure = 76 cm. mercury [see chap. 13, p. 98].) These positive columns may be made of all the colors of the rainbow, depending on the gas content of the tube. They are yellow for helium, red for neon, blue-green for mercury vapor. Green, purple, and violet shades are all possible. Important as these glowing vapor arcs have become for industrial uses, they are far more important in what they teach us about electrical phenomena.

WITHIN such a vacuum arc, and especially throughout this positive column, there is maintained a most intense ionization of the gas. The relatively low resistance of the tube is due to a vast multitude of these carrier ions, positive and negative, which by their motions in the field swiftly transport their charges and constitute the mechanism of conductivity. Compared to the numbers of ions formed by the flame in our experiments in chapter 29, the numbers in these tubes are legion. The luminosity of the gas is due to phenomena attendant upon their collisions with one another that cause recombinations of oppositely charged ions that occur therein. At higher pressures the phenomena responsible for the origin of these ions and their maintenance are too complex to detail.

Ultimately, as the pressure is reduced their number diminishes; the resistance of the tube increases as the current-carrying quality disappears from the gas; the luminosity fades.

IT WAS as this stage of increasing resistance set in that the early observers noted the presence of what they called CATHODE RAYS emanating from the region around the cathode. The faint fluorescence of the glass just before the tube becomes finally dark gives us a clue. Suppose we put some solid material like zinc sulphide, which is known for its fluorescent powers, within the tube. Not until the evacuation has proceeded to the point where the Crookes dark space has expanded and extended to the location of the sulphide does it show any very great fluorescence. The cathode rays are evidently confined to the Crookes dark space. At all pressures below this, and until the

CATHODE RAYS

tube goes out completely, the sulphide shines with a gorgeous brilliance. When the tube goes out, it goes out all at once; and at the same moment we find that the flow of current through it has also stopped. The rays that excite the fluorescence are themselves invisible. That they 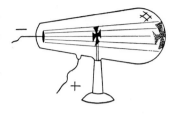 proceed in straight lines from the cathode may be shown by ingenious contraptions in tubes of special design, with opaque shadow-casting screens that can be shifted in position between cathode and a fluorescent wall within the tube.

Not only do these rays proceed in straight lines, with little reference to the lines of force of the initial field which is greatly modified by the discharge,  but these invisible rays proceed from the cathode in a direction which is initially, at least, at right angles to its surface. If this surface is hollowed out so as to form part of the surface of a sphere, these rays become focused at the center of the sphere. A bit of platinum foil inserted in this focus does not fluoresce. It becomes RED HOT. These rays, then, might be supposed to have mass associated with themselves since they carry energy. Independently, in other ways, this can be verified; if directed against a lightly suspended target the rays will impart *momentum* to it as well as heat (Plate 70).

Finally, and most important, if a narrow beam of cathode rays is shot through a mica slit beyond which lies a long fluorescent screen intercepting their path, and so indirectly rendering them luminous, a magnet brought near this beam will cause the latter to bend aside just as if a current of electricity were flowing along this path as through a wire, with negative charges traveling away from the cathode (Plate 71).

Indeed, current-detecting instruments of somewhat greater sensitivity than the one we used in the Edison effect will show the

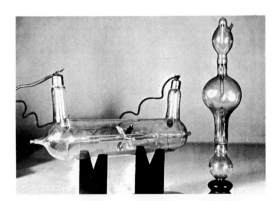

PLATE 70 Mechanical Systems in Vacuo, Operated by Cathode Rays

PLATE 71

MAGNETIC DEFLECTION OF CATHODE RAYS

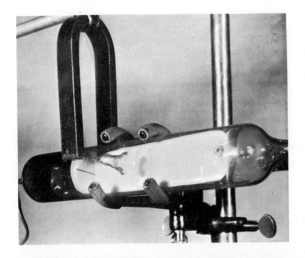

A Tube to show magnetic deflection of cathode rays. Note slit and fluorescent screen.

B Tube in operation; cathode-ray stream photographed by light it causes on the fluorescent screen. No magnet near—stream of rays is straight.

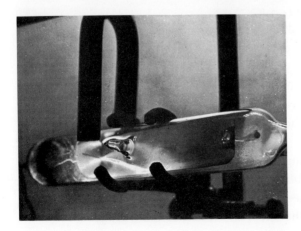

C Tube in operation: cathode-ray stream deflected downward by transverse field of the horseshoe magnet.

CATHODE RAYS

passage of a current between the electrodes in this sense. Also, an electrometer less delicate than the one we used in the photoelectric effect to detect the positive charge taken up by the plate under the action of light will show here the rapid acquisition of negative charge taken up by an insulated target interposed in these streams of cathode rays. Thus we show the latter to consist of moving negative charges.

One very sensational fact in addition is connected with their production. Just as soon as they make their appearance in any kind of discharge tube, the tube at the same moment BECOMES AN X-RAY TUBE. To this point we shall return later.

FOR a moment now let us take stock and see how these characteristics of the old-time cathode rays compare with those we have ascribed to the more recently discovered rays emanating from the heated filaments of incandescent lamps in the Edison effect, and "dusted off" the surfaces of metals, negatively charged in vacuum, by the delicate touch of ultra-violet light impinging thereon. We find the following characteristics in common:

(1) Occurrence is best at the lowest stages of evacuation, i.e., high vacuum

(2) Negative "rays" leave the cathode, carry −charge, suffer magnetic deflection in the proper sense

(3) The phenomena are unidirectional; the rays are never found coming from the anode

The following properties found in the earlier forms of cathode rays may be established also for those originating in the thermionic and photoelectric effect:

(4) They cause fluorescence of various minerals and of glass

(5) They travel in straight lines and cast shadows

(6) They convey momentum, i.e., can exert force

(7) They carry energy, transformed into heat where they strike

(8) They cause X-rays to emerge from any target that they strike.

At least one precise and quantitative test of the properties is needed. The momentum and energy transfer by these rays is so prodigious in amount compared with that which might be carried by ordinary light that we wonder if the term "rays" may not be unfortunate. Cathode "streams" might be more suggestive of the idea of individual particles. Indeed, why should we not expect them to be particles from other evidence we already have of the atomicity of electricity? Here are perhaps the negative electric atoms, *isolated from matter* in comparatively matter-free space, to be dealt with on their own account.

We will describe the quantitative test and give these entities their own proper name in the next chapter.

CHAPTER 32

ELECTRONS: THEIR MASS AND VELOCITY

WE COME now to the discussion of something that has become almost a household word. That all material substances are composed of electrons, that they are the building-blocks that enter at least as one of the constituent parts of all matter, is very well known. In these days of high-powered education, probably you were told about them when you studied general science in the grade schools. If you have ever thought about such things as the nature of the physical world, you have undoubtedly come to take electrons quite for granted.

Of course, with respect to them, as with respect to many things said to be true by scientific men, you may share a very common feeling to the effect that scientists will believe in electrons only until some new idea appears in the fevered brains of those eccentric individuals, the research investigators, especially that class of them that sit and spin the theory of what all things are. You think that later, perhaps, you may learn that it was not electrons that were at the bottom of it at all, but some other queer conceit that then will be regarded as the essence of everything.

"One theory today, another tomorrow," however, is not the slogan of natural science. We have tried so to arrange our words in the many pages that have preceded this chapter that by now you understand this just as well as we do. You have been taken frankly into our confidence, you have witnessed experiments with us, have shared our provisional hypotheses, the testing of them, the tossing of them away when they did not click with experience; and, perhaps, you have felt just a bit of the thrill that came when one of these hypotheses seemed to be unexpectedly effective, to contain something more than we put into it.

WE SUMMED up at the close of the last chapter the experimental facts that are associated with what were there called "cathode rays" because of their obviously close physical relation to a negative (cathode) electrode. Only qualitative properties, however, were discussed. At the outset of this story we agreed to try to keep it on the plane of common language, to use tables as little as possible and equations still less.

But it was necessary in the case of Millikan's oil-drop experiments to show you the integral multiples of a certain number that resulted from them in order to give you the cogent and ultimate reason for believing that electricity is atomic, and to present some idea of the order of size of the natural unit of charge, and how we count what is uncountable. Now once again it becomes necessary to describe the quantitative experiment that is our ultimate reason for believing that these entities in cathode rays have *mass* and *velocity* as well as negative charge.

Did you ever swing a heavy object around your head at the end of a string? If you did, or if you now will, just to please us, you will find that it pulls on the string with a force that increases very greatly the more times a second you make it go around; and also you will note that the force increases notably as you increase the length of the string. Naturally, the larger the chunk of material the more it pulls; and you associate this not with its size as such, but with its mass (inertia), as usually determined by its weight.

The name, CENTRIFUGAL, of the force is well known to you also. But here you may have been puzzled sometime or other by someone's saying that this word was not the right one exactly; that this force should be called CENTRIPETAL. Both words are right. Their respective usage depends entirely on whether one adopts the point of view of the stone at one end of the string or that of your hand at the other end, usually a matter of indifference in casual conversation among laymen. In chapter 7, on action and reaction, we laid great stress on the dual aspect of force, whether it be tension in a string or any other kind. That is why there are two equally good words for this phenomenon. The tendency of the stone is "centrifugal": inertia pulls it tangentially to the curve and tugs outward on the string in that sense. It is prevented by your hand at the other end, which continually tugs at the center in the opposite direction holding the stone in its circular orbit, making it accelerate the *direction* of its velocity toward the center continually: a center-seeking or "centripetal" force is acting on it.

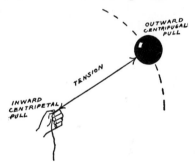

We will not trouble you with the analysis of this situation which results in our exact knowledge of how the pull in the string depends on mass, speed in the path, or number of turns per second of the stone about your hand, and length of the restraining string, for which omission you will, we hope, express your gratitude by accepting our statement of the results as correct. Of course, we both understand that neither should we be dogmatic nor should you accept anything without proof.

The centripetal force is either

$f_c = 4\pi^2 m n^2 r$, where m is the mass of the stone, n the number of turns per second, and r is the radius of the circle.

or

$f_c = \dfrac{mS^2}{r}$, where m is the mass, $S*$ is the *speed* along the circular path, and r is the radius of the circle.

*This is the sort of thing we had in mind when, in chap. 1, p. 7, we took great care in our definition of velocity to point out the distinction between velocity and speed. If, for example, we suppose the stone to be whirled at a constant rate by a string of fixed length, its *speed* will then be constant but ve-

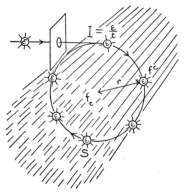

Let us now apply our knowledge of centripetal force to the case of a stream of cathode rays whirled around in a circle by a uniform magnetic field perpendicular to their path. Experimentally we can do this quite readily. We are using our imaginations, of course, when we so boldly assume that an argument that applies to a stone on a string will apply and will have the same meaning in the case of the cathode streams, which we have guessed might be particles, also. The justification of such use of imagination cannot be determined until we see where we come out and whether or not it is possible to make certain tests or checks, or find some independent way of reaching the same conclusions.

The *centripetal* force of the field that holds the stream of cathode rays in a circle, since the stream constitutes an electric current, is:
$f_c = IlH$,* I being the current they represent; l, the length of their path; and H the strength of the magnetic field that they are in.

The *centripetal* force required to hold the particles in a circle, *assuming they are particles with mass and speed*, is:
$$f_c = \frac{mS^2}{r}$$
, as indicated in general in the preceding paragraph.

These two expressions are really identically the same, on the basis of our assumption, italicized above.

Thus,
$$\frac{mS^2}{r} = IlH .$$

Focus attention now on one of the particles. Call its charge ϵ. Its mass, m, is the m we have above. The part of the current that the motion of *this charge* contributes (since $I = Q/t$) is $I = \epsilon/t$; thus
$$\frac{mS^2}{r} = \frac{\epsilon l H}{t} .$$

---

locity will not, since the direction is changing. There is an acceleration, of course, if velocity changes. This is due to the force which you exert in holding the stone in its orbit. The acceleration has the direction of the force, i.e., is toward the center. This is also to be seen from the fact that the speed in the path does not change. A radial force has no component along the path to change the speed, since the path is perpendicular at all points to the radius.

* Cf. chap. 26, p. 239.

ELECTRONS

But l/t in this expression, the distance covered by the particle divided by the time taken, is the speed, S, of the particle; so that

$$\frac{mS^2}{r} = \epsilon SH .$$

Canceling S's and solving for ϵ/m,

$$\frac{\epsilon}{m} = \frac{S}{Hr} . \qquad (1)$$

Whence did the particle get this speed, S? How can any *charged* particle acquire speed? Only by falling down a field of force, provided in this case by the high potential apparatus which generated these streams of cathode rays. From what we have learned in several other places, the kinetic energy acquired comes from the work done on the particle by the electric field, which is by definition (see p. 226) $W = PD \; Q$, here, $W = PD \; \epsilon$. Thus, we have another relation,

$$\tfrac{1}{2}mS^2 = PD \; \epsilon . \qquad (2)$$

We will now rewrite these relations in close juxtaposition for a better look at them.

$$PD \; \boxed{\epsilon} = \tfrac{1}{2} \boxed{m} \boxed{S}^2 .$$

$$\boxed{\epsilon}/\boxed{m} = \frac{\boxed{S}}{Hr} .$$

Unknown entirely are the values of the quantities in boxes $\boxed{\epsilon}$, $\boxed{m}$, $\boxed{S}$, charge, mass, and speed of particle, respectively. We can from our experiment measure all the others; viz., r is the radius of path around which the rays are bent; H is the field strength of the magnet we use to swing them, and PD is the voltage used to excite the experimental tube which produces the rays.

Can we not, perhaps, get these boxes open—get the unknowns expressed in terms of the knowns? If you are good at algebra, you might try to multiply, divide, add, subtract, and perform all the legitimate operations you can think of on the foregoing two equations and see if this is possible. However, if you are good enough at algebra to do this intelligently and correctly, *you will not even try to do it*, for in that case you already know that to solve for three unknown quantities in terms of known ones you must have at least *three* simultaneous equations (three relations that are independently true) and that three is also sufficient for the purpose. Here we have given you only two equations.

From this conclusion as from all conclusions, proved rigorously by sound mathematical skills, there is no escape. We must accept and be content to find the value only of two boxes, one containing some one of our three unknowns, the other containing the other two as if they were one. Suppose we solve for $\boxed{S}$, and for the ratio $\boxed{\epsilon/m}$, which we might call the "specific charge" i.e., the charge per unit mass or charge per gram.

If we do this, $\boxed{S}$ turns out to have values, depending on our experimental arrangements, of anything between a few hundred miles per second up to almost 186,000 miles per second—a very interesting fact, indeed, but one of no great significance for our story at the moment. However, our other box, $\boxed{\epsilon/m}$, comes out to have ALWAYS ONE AND THE SAME VALUE.* Irrespective of any experimental arrangements we may choose to make; ϵ/m is always 531,000,000,000,000,000, i.e., 5.31×10^{17} electrostatic units per gram mass of particle, or 17,700,000, i.e., 1.77×10^7 electromagnetic units of charge per gram mass of particle.

OF COURSE, we *can* take one step farther if we wish to *assume* that these particles in cathode rays are our fundamental electric atoms themselves, i.e., are those whose charge is 4.77×10^{-10} electrostatic units. This amounts to supplying another equation to the two we had; but this was not a proved relationship originally, only a "hunch,"† a guess as to what one of the two things boxed up together might be. If this assumption is correct, then the *mass* associated with this charge would have to be

$$\frac{0.000,000,000,477}{531,000,000,000,000,000} = 0.000,000,000,000,000,000,000,000,000,9 \text{ grams!}$$

I.e., 9×10^{-28} gm.

This *is* a small amount of mass! The lightest atom of matter is hydrogen, and it has only the twenty-fourth negative power of 10 in its abridged form of

---

* Provided velocity of the cathode rays is small compared with the velocity of light. When it is not, the effect predicted by the theory of relativity that mass depends on velocity is observed.

† Subsequently quite independent experiments proved this guess to be correct.

expression ($M_H = 1.66 \times 10^{-24}$ gm.). If we are on the right track here at all, we are impelled to the conviction that we are dealing with entities about 1,846 times less massive than the lightest weight we know anything about. With respect to size, and spatial dimensions likewise, although the arguments are considerably less convincing, we visualize something 100,000 times smaller, i.e., 10^{-13} cm. in diameter instead of the 10^{-8} cm., which many different lines of evidence show to be about the size of the atoms of matter.

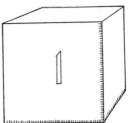

One billion (10^9) subdivisions on each side divide the block into tiny volume cubes that are each 10^{-27}th of the whole

We have emphasized already that the value of ϵ/m is independent of the types of experiments we make to determine it. Six of the most careful determinations lie in value between 5.29 and 5.33, $\times 10^{17}$. In these the sources of the cathode particles were: oxidized filaments of metal, such as we might use in radio tubes or in the Edison effect; discharge tubes containing gases at low pressure with altogether different metals from the preceding ones for electrodes; radioactive sources (which also emit such particles); and metals of all sorts illuminated by ultra-violet light. Taking for granted that we had a right to associate with ϵ in this ratio, ϵ/m, the atomic charge which these same experiments suggest and for which many other types of experiments give evidence, we can then say more simply with respect to the *mass* of cathode particles that, whether they originate in heated filaments of carbon, tungsten, or platinum—pure or associated with any kind of impurity—in vacuum tubes, or in any of the metals from the lightest to the heaviest, under the photoelectric action of light of any quality, or whether they come from solids, liquids, or gases subjected to the action of powerful electric fields, or from radioactive materials—THE MASS OF THE CATHODE PARTICLE IS ENTIRELY INDEPENDENT OF ITS ORIGIN.

> The name **ELECTRON** was given to these entities first by Johnstone Stoney. The evidence we have just recited has been accepted as showing that, whatever they may ultimately prove themselves to be, they are constituent parts of all of the atoms of all material substances; they constitute something more fundamental than matter itself, because they enter into the construction of all matter; they are one of the basic materials of atomic architecture, just as wood, brick, and stone are the basic materials of which human edifices are built.

WE WOULD not have you think that, after the thirty or more years during which electrons have definitely been studied and subjected to thousands of experiments and tens of thousands of pages of reports of technical studies, our convictions now rest entirely on the simpler earlier experiments that we have detailed. There are innumerable other phenomena that yield to our understanding in relatively simple terms on the basis of these hypotheses of the atomic character of electricity and the electrical character of matter. Many of these other phenomena would seem quite inexplicable to us still in any other terms. This does not mean that we have, as yet, much more than a superficial acquaintance with this little entity, the electron. We know quite exactly the values of its charge and of its mass. We know exactly how many of these entities are contained in the outer regions of all atoms of the elements of the periodic table which is given to us by the work of the chemists* (see table, p. 309).

WE KNOW how to handle and use electrons and how to make them serve innumerable industrial and scientific purposes. We can even finance ambitious and expensive expeditions and enterprises for the purposes of scientific and industrial research and developments, placing the utmost

---

* The "periodic table" is an arrangement of the chemical elements in such a manner that elements that have similar chemical properties lie in vertical columns called GROUPS. As one goes from the lightest element, hydrogen, in the upper left-hand corner where it is sometimes placed, and reads from left to right and down, as one does a page of print, with one or two exceptions the elements are in order of increasing atomic weight. The elements on the left side of the table are "electropositive"; i.e., when they occur as ions in conducting solutions they migrate toward the negative cathode terminal. Electronegative elements are on the right side of the table. As ions in solutions these appear in the region of the positive anode electrode. Elements in the middle are rather indifferent electrically, being neither conspicuously electropositive or electronegative.

The so-called "zero group," all of which are gases, are likewise neither electropositive or electronegative in character, but their weights compel their being placed at the extreme left (or right) of the table.

The horizontal rows are called SERIES. A series begins with an inert gas, the next member of it is strongly electropositive, the next less strongly so, the next is but weakly electropositive, and the next quite indifferent in this respect. Then follow a very weak electronegative, a more strongly electronegative, and finally an element with intense electronegative qualities. The next element is another inert gas, and so begins the next series.

Not all series are of the same length. The one above described was one of the short series. The first two are like this, beginning with He and Ne, respectively. The third, fourth, and fifth series are much longer. The sixth is incomplete, since the heaviest element known, uranium, is reached before the end of the line is finished.

The complete interpretation of the meaning of the periodic table was not found until the detailed knowledge of how atoms were built up out of electrons was acquired from experiments such as we have been describing and, of course, from many others.

The next table (p. 310) gives the electron arrangements for some of the lighter elements.

0	I	II	III
2 He 4.00	3 Li 6.94	4 Be 9.02	5 B 10.82
10 Ne 20.18	11 Na 23.00	12 Mg 24.32	13 Al 26.97
18 A 39.94	19 K 39.10	20 Ca 40.07	21 Sc 45.10
36 Kr 82.9	37 Rb 85.44	38 Sr 87.63	39 Y 88.92
54 Xe 130.2	55 Cs 132.81	56 Ba 137.37	57 La 138.92
86 Rn 222.0	87 AcK 225.95	88 Ra 225.95	89 Ac 227

	IVa	Va	VIa	VIIa		VIIIa		Ia	IIa	IIIa	IV	V	VI	VII
											6 C 12.00	7 N 14.00	8 O 16.00	9 F 19.00
											14 Si 28.06	15 P 31.03	16 S 32.06	17 Cl 35.46
	22 Ti 47.9	23 V 50.96	24 Cr 52.10	25 Mn 54.93	26 Fe 55.84	27 Co 58.94	28 Ni 58.69	29 Cu 63.57	30 Zn 65.38	31 Ga 69.72	32 Ge 72.60	33 As 74.96	34 Se 79.2	35 Br 79.92
	40 Zr 91.22	41 Cb 93.1	42 Mo 96.0	43 Ma 96.?	44 Ru 101.7	45 Rh 102.91	46 Pd 106.7	47 Ag 107.88	48 Cd 112.41	49 In 114.8	50 Sn 118.70	51 Sb 121.77	52 Te 127.5	53 I 126.93
58–71 Rare Earths	72 Hf 178.6	73 Ta 181.5	74 W 184.0	75 Re 187.?	76 Os 190.8	77 Ir 193.1	78 Pt 195.23	79 Au 197.2	80 Hg 200.61	81 Tl 204.39	82 Pb 207.22	83 Bi 209.0	84 Po 210.0	85 Unnamed but identified
90–96 *	90 Th 232.15	91 Pa 234.?	92 U 238.17	93 Np	94 Pu	95 Am	96 Cm							

Hydrogen (H) is not included in the chart, because it fits into none of the columns.

* Alternate position for elements 90–96 according to Seaborg.

Periodic Table after Muskat

Chemical Element	Numbers of Electrons in Consecutive Shells, Which Are Called— (As One Proceeds Outward from the Center)			
	Innermost K shell	L shell	M shell	N shell
H........	1			
He.......	2			
Li........	2	1		
Be.......	2	2		
B........	2	2+1		
C........	2	2+2		
N........	2	2+3		
O........	2	2+4		
F........	2	2+5		
Ne.......	2	2+6		
Na.......	2	2+6	1	
Mg.......	2	2+6	2	
Al.......	2	2+6	2+1	
Si.......	2	2+6	2+2	
Ti.......	2	2+6	2+6+2	2
Kr.......	2	2+6	2+6+10	2+6

ELECTRON ARRANGEMENTS IN ATOMS

The simplest part of the table here given is far too complex for us to attempt to interpret it for you. The only item of importance is for you to get the impression of how regularly the electron structure is built up in consecutive concentric regions or "shells." The innermost "K" shell contains 2; the "L" (when filled), 8; the "M," 18; etc.

reliance on our control of these inconceivably tiny entities, the use of which is essential to success.

We do not know what electrons are; but we interpret everything else in terms of them. To know what they are, would be to know what electricity is, to know what matter is. The business of "knowing" anything is a very complex and confusing matter, after all. We use this word and all other words in the rather loose language of common speech, to which the language even of the physicist is much closer than to that of the psychologist, philosopher, or dialectician. That these ideas are of great importance to those engaged in the pursuit of these other specialties more technical than ours we are, of course, aware. Our function is, however, not to forget for very long that we began as *experimenters* with natural phenomena, and that we must continue to experiment, as wisely and understandingly as possible, of course, if our experiments in the future are to be as significant as they have been in the past. The control we gain in this way over natural forces has been in the past perhaps of more significance to mankind than have been the new interpretations that have resulted with respect to the nature of the physical world.

THAT matter contains other building-blocks than electrons is certain. We shall, in the next chapter, review some evidence as to what some of these other entities may be.

CHAPTER 33

POSITIVE RAYS, PROTONS, NEUTRONS, AND ISOTOPES

IF WE carry on further experiments with evacuated electrical discharge tubes in the light of what we have already learned about them, we discover some new and interesting phenomena. We may use unheated electrodes as cathodes and work at gas pressures where the Crookes dark space (see p. 295) extends several inches from the cathode, or we may use a heated cathode to emit electrons more freely and then work at lower pressures and voltages.

If the cathode is a metal plate perforated with holes and if it is placed so that a considerable space lies behind it in the tube, we will observe that, in addition to the dim luminosity of the gas between it and the anode to the front of it, caused by its electron emission and the ion production in that direction, streams of luminous glow also extend from the holes out into the region behind the cathode. These rays, discovered by Goldstein in 1886, were called by him CANAL RAYS, from their association with the holes (German, *Kanallen*) in the cathode.

The earliest studies seemed to show no deflection of these rays by magnetic fields. Later, however, these luminous bundles were found to suffer slight deviation in the *opposite* sense from electron (cathode ray) streams. There are two possibilities; either positive flowing away from the cathode or negative flowing toward it! The latter seems an impossible

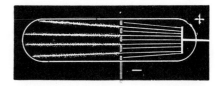

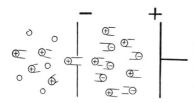

Formation of positive rays by ions

interpretation. Now, since these rays appear only when there is enough gas in the tube for there to be a considerable amount of gaseous ionization, and since this ionization involves the presence of both plus and minus charged atoms or molecular groups, a very simple conclusion would be that these rays are due to *plus ions* proceeding *away* from the cathode instead of negative proceeding toward it from some mysterious origin behind.

Positive ions formed in front of the cathode would fall toward it. Many would strike its surface. Some, however, would hit the holes in it and, having acquired a very respectable velocity, would fall right on through and proceed some distance beyond before being stopped by collisions with other gas atoms there or by the now reversed pull of the cathode on them. In their collisions with other molecules, light is produced by mechanisms we shall take up later in Part V. If molecules are present in sufficient numbers, this light will be intense enough to see. All this turns out to be verifiable by careful analysis of the phenomenon.

By special arrangements of electric fields in combination with magnetic ones at right angles, simultaneous and independent deflections can be produced and measured. The deflected particles, if too few to render their paths visible by light incidentally sent out, may be caught on photographic plates placed in their path, and the positions where they strike rendered visible by subsequent development of these plates.

As a result, sections of parabolic arcs appear on these plates, from the geometry of which the number (n), of elementary charges (ϵ), carried by the particles, and their masses (M) may be found. Only integral multiples of ϵ appear, which is to say that n is always 1, 2, 3, 4, or some whole number. The masses, however, depend entirely upon the nature of the substances in the tube, and turn out to be identical with the masses of the atoms present.

BY THIS means we have made these streams of positive ions give account of themselves. The account tallies in the main with what we would expect to find, but it contains some items of surprise, nevertheless. When we experiment with hydrogen, n is never anything but 1. No hydrogen atom ever has but one unit positive charge, i.e., can never lose more than one negative charge; and thus we can assume that this is all it has to lose. This singly charged positive ion of hydrogen is in a sense the *alter ego* of the electron. It has consequently been given a special name, PROTON (primitive substance).

The common element hydrogen, then, consists when electrically neutral, of two parts—this proton and an electron. They are equally but oppositely charged; but they are vastly unequal in their respective masses, the proton having 1.6608×10^{-24} gm. of the total mass of 1.6617×10^{-24} gm. and the electron $0.0008994 \times 10^{-24}$ gm. The ratio of the masses from the most recent deflection experiments of this sort, such as we discussed on page 304 ff., is thus,

$$1846:1.$$

One other important aspect of this method of analysis of matter by study of positive rays is its great sensitivity. In mixtures of different substances the trails which correspond to each of them may be clearly differentiated on the photographic plate and the charge/mass ratio determined so accurately that the masses (or relative atomic weights) may be found to within about one part in a thousand.

THE greatest surprise in this subject, however, came to the early workers,* Thomson and Aston, in the case of neon gas. We have met this substance before as one of the most common gases used in electric vacuum arcs for advertising signs. It is one of the very rare constituents of the atmosphere, and one of a group of elements that are chemically very inert (see p. 309, col. 0), and never found combined with any other substance. Its atomic weight relative to hydrogen the best methods had previously shown to be 20.2 times as heavy as this lightest of all elements.

Thomson and Aston, in swinging positive ions of neon through their electric and magnetic fields, found TWO parabolic curves instead of one—indeed, that neon was indubitably a mixture of two different kinds of atoms, one kind having a mass of 20 and the other a mass of 22. The amount of the heavier kind was 10 per cent and of the lighter, 90 per cent; thus the atomic weight of the mixture was a WEIGHTED average of these two kinds obtained as follows:

$$0.9 \times 20 = 18.$$
$$0.1 \times 22 = \underline{2.2}$$
$$20.2\dagger$$

* Among other names distinguished in this general field are those of Soddy and Dempster.

† Later experiments have disclosed a third kind of neon, mass 21, only one-quarter of 1 per cent of the total. Hence the figures above are approximate. A more accurate value of the atomic weight is 20.183.

This discovery was of the utmost importance because of the light it shed upon the periodic table of the elements so fundamental in both chemistry and physics.

It had long been known that if one chose the weight of oxygen as standard and made its weight arbitrarily 16, the weights of many other substances relative to it came out to be very close to other whole numbers. Carbon, for example, was 12.010; nitrogen, 14.008; sodium 22.997; etc. Indeed, it was this very circumstance that led the chemist Prout (many years ago) to propose the idea that hydrogen (weight 1.008) entered into the composition of all other things. Many elements like neon, however, did not have even approximately integral atomic weights; for example, magnesium is 24.32, chlorine is 35.457, and copper 63.57.

Positive ray analysis has shown that all of these latter substances are not homogeneous assemblages of atoms all alike in weight, but mixtures of several different kinds. Magnesium has three kinds of atoms—24, 25, and 26 in relative weights, respectively; chlorine has two kinds of atoms, one 35 and one 37; copper, likewise, two, 63 and 65.

Element	At. wt.	Isotopes
B	10.82	11, 10
Ne	20.183	20, 22, 21
Mg	24.32	24, 25, 26
Cl	35.457	35, 37
K	39.096	39, 41, 40
Hg	200.61	202, 200, 199, 201, 198, 204, 196

SOME OF THE ISOTOPES

Not to be distinguished in any other way than by their slightly different masses, these various kinds of the same element are called ISOTOPES of one another. As the name indicates, they all occupy the "same place" in tables of the elements, since these tables are arranged on the basis of chemical properties and in chemical properties the various isotopes of any element are identical. We shall meet them again in the next chapter.

BECAUSE of these facts, and also because we can obtain protons (positive hydrogen ions) from many other substances that are not hydrogen at all (by experiments referred to later, p. 328), there has resulted the revival, on modern grounds, of the old hypothesis put forth many years ago by Prout that all other atoms were made up of hydrogen. Of course, our more modern knowledge would make us say of positively charged hydrogen ions. Indeed, so universally does this entity crop up in all sorts of experiments that it is undoubtedly another one of the building blocks of matter and is really entitled to have its own special name of PROTON.

Thus, the study of positive rays in discharge tubes has been found to yield several very important fundamental facts about the nature of matter. That protons are another type of atomic building-block there is no longer any doubt, since the atomic weights of atoms are really very close to integers when hydrogen is 1.008. This charged particle, that goes by its chemical name of hydrogen when associated with a companion electron, is a most fundamental entity and quite entitled to share a leading position with the electron. However, as we shall see in the next paragraph, still another fundamental particle of mass almost identical with the proton, but UN-CHARGED electrically, more than divides the honors with the proton in contributing mass units to the nearly integral weights of individual isotopes. This particle is the NEUTRON, brought to public fame by the ATOMIC BOMB.

The experiments that revealed this new fundamental particle, the NEUTRON, were first correctly interpreted by Chadwick in England in 1932. It had been found earlier in Germany and in France that when high-speed particles from natural radioactive substances (see chap. 34) impinged on certain light elements, particularly Li, Be, and B, an unusually penetrating radiation resulted. This was first supposed to be of the same nature as X-rays, then was ascribed to high-speed protons. Chadwick in 1931 conclusively established the fact that the "radiation" consisted of particles of proton mass but without the proton's charge. He named them NEUTRONS. Later work has established that they are present, together with protons, in the deep interior of all atoms. These two entities, indeed, form the building-blocks of atomic nuclei—those central massive cores of atoms, whence originate all the phenomena of radioactivity that will be discussed more systematically in the next chapter.

POSITIVE RAYS and the discovery of neutrons, then, have clearly revealed to us a new type of precise atomicity, that of the isotopes from which atomic weights are derived. Indeed, if atoms are to be regarded as being built up of more fundamental entities, we might expect that the mass of any one would be the sum of the masses of its constituent parts. It will surprise us, no doubt, to learn that when complex entities like atoms are built up of protons, neutrons, and electrons, the masses resulting are not always precisely the sums of the masses of the separate parts! Here again are facts revealed by precise measurement which clearly foreshadowed by many years the arrival of the atomic age we have just entered upon.

Under the label of phenomena of a quite different sort in the next chapter we shall find these new acquaintances bobbing up again.

CHAPTER 34

RADIOACTIVITY AND TRANSMUTATION

FROM earliest antiquity mankind has cherished two dreams of achievement. To fly in the air was the ultimate in the way of sport and adventure; to transmute base metal into gold was the *summum bonum* of the avaricious. The twentieth century has accomplished widespread conquest of the air; and it has also witnessed artificial transmutation of all the elements. Although the manufacture of gold out of base metals has not been, nor is it likely to be, accomplished, results of even more dire importance, undreamed of a decade ago, have been accomplished under the urgency of war. Here we pause a moment and, stepping out of our character as physicist, as a layman we ask: Are the leaders of society in matters economic and social prepared, or making plans, at least, for the consequences of these scientific advances in physics and chemistry which make another war with atomic weapons unthinkable?

Less than twenty years ago, although transmutation of one chemical element into another was well known and observed wherever radioactive processes were being studied, we would have said such processes were as inviolate from human handling and control as was a distant spiral. We observed with awe and wonder, but we never dreamed that we could learn to

RADIOACTIVITY AND TRANSMUTATION

design and build apparatus that could set quite similar processes into operation. Not only has this been done by now in the case of the entire 92 elements of the periodic table, but at least four *new elements*, all of them heavier than uranium, have been created. Marsden may have been the first, in 1914, to effect transmutation, artificially, of one element into another but what was taking place was not recognized by him or by the scientific world in general at that time. Sir Ernest Rutherford in 1919 and in 1921, and in collaboration with Chadwick in 1924, and subsequently a considerable number of workers have been doing these things advisedly, and in accord with predetermined design and plan. Fermi, formerly of Rome, Bohr, of Copenhagen, and Urey, at Columbia, are but three of many recent contributors. The great Van de Graaff electrostatic generator (see p. 199) was one of the first high-voltage machines designed and built by men of the Massachusetts Institute of Technology for the purpose of attempting the disintegration of atoms with potentials never before available for human use. The invention of the cyclotron by E. O. Lawrence, of the University of California, and of the betatron by Kerst, of the University of Illinois, following in rapid succession, have provided men with the means for hurling atomic projectiles and electrons, respectively, at other atoms with energies so great that the smashing of atomic nuclei proceeds apace and in measurable, even tangible, quantities. The construction for war purposes in recent years of great radioactive "piles" for the purpose of controlling self-perpetuating radioactive "chain reactions" for the production of new elements (like the now famous plutonium) to provide enormous supplies of atomic energy for the production of atomic bombs is one of the latest chapters in this amazing story.

RADIOACTIVITY was discovered by H. Becquerel quite by accident in 1896. Stimulated by the discovery of X-rays by Roentgen the previous year, Becquerel was searching to find whether there were any substances other than glass which, by being rendered fluorescent by ordinary light instead of cathode rays, would emit these mysterious "X" radiations. Uranium salts, known for their fluorescent properties, were being used and were left for some days entirely in the dark with a photographic plate near by. With the usual care of a trained experimenter the equipment was retested before experiments were resumed; the old plate was to be replaced with another, but, before it was thrown away, it was first developed, and on it was the earliest record we have of the radiations from radioactive materials.

The details of subsequent events make up a long and complicated story—a story full of false leads, misinterpretations, disagreements, discouragement, and mystification. These radioactive rays or emanations were far from being simple things; the background of achievements in related fields, with which we are familiar from our perusal of the last few chapters, was then an unwritten page in the human record. Persistence, a most important attribute for investigators, ultimately won. Three distinct types of rays or emissions of some sort were disentangled from one another, and called, for want of any more descriptive terms, by the names of the first three Greek letters—alpha, beta, gamma.

Alpha (α) rays* produce intense ionization in the region through which they proceed, but they are those most easily absorbed by matter through which they travel. Their range is only a few centimeters in air, and metal sheets 0.01 cm. thick stop them completely. They carry positive charge but suffer such extremely slight deflection in a magnetic field that they were originally thought to show none at all. This deflection is only about 1/7,400th of that shown by electrons, and in the opposite direction, of course, since their charge is positive. It is about one-fourth of that shown by streams of protons (positive H ions) and in the same sense as the deflection of the latter. Alpha rays have a speed of about 15,000 km. per sec.; the range of variation of speed is not large (1.42–2.07 $\times 10^9$ cm./sec.).

Beta (β) rays produce much less intense ionization, but penetrate much farther into solid substances (e.g., through several millimeters of aluminum) without being absorbed. They carry a negative charge, suffer the same deflection as do electrons; indeed, they are not to be distinguished from electrons originating in any other way except by their prodigious velocity. This, in the case of some of them, approaches closely the fastest speed that occurs in nature, 186,000 miles per second, the speed of light.

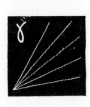

Gamma (γ) rays are chiefly distinguished by two things that place them in quite a different category. They carry no electric charge whatsoever, and they are of tremendous penetrating power, passing through most solid objects as light does through glass, and going through even such substances as iron or lead in thicknesses of several centimeters. They are rather

* The thumbnail sketch shows in the wavy line the characteristic trail of an α particle on which fog has been precipitated in a device called a "Wilson Cloud Chamber" (see pp. 322–23). The figure

indifferent in their capacity to ionize gas, but this is really more on account of the very small amount of their absorption by it than of ineffectiveness of the rays themselves. As far as can be determined, the **velocity of gamma rays is no different from that of light itself.** Indeed, these radiations are a kind of super X-ray. They are light waves of extreme minuteness. We shall discuss them later, therefore, in their proper context in Part V.

MOST significant for our present study of the group of radioactive radiations are the alpha rays, which, because of their obviously corpuscular characteristics, are usually called "alpha particles." Electric and magnetic deflections made by methods such as we have indicated previously show that they carry a DOUBLE positive charge and have a mass that is 7,400 times as great as electrons and FOUR times that of hydrogen, almost, but not quite, exactly. If we allow them to stream for months on end into a small sealed and evacuated tube, with walls thin enough for them to penetrate, an astounding thing results. In the course of time HELIUM GAS is found to be inside the tube. Indeed, these alpha particles are positive helium ions—helium atoms that have lost both of the two outer negative electrons they possess in the neutral state. We can produce these helium ions artificially in tubes of helium gas, as positive rays of helium and with modern acceleration devices, like cyclotrons, speed these up to reach and slightly overlap the prodigious velocities (thousands of miles per second) with which they are emitted naturally by the heavy radioactive substances like uranium, radium, and thorium.

What does this mean? Do these heavy elements contain helium as such? No evidence has ever been found that they do. Indeed, the evidence all shows that after the helium comes forth from the radioactive atom the latter is no longer what it was before the ejection of this alpha particle. It has quite different physical and chemical properties. **The case is clearly one of the *transmutation* of something like a heavy, chemically rather active, metal into two inert elements that are gaseous at ordinary temperatures and pressures.** This is exactly the fact in the case of radium itself. One of the disintegration products is radon, the heaviest of the inert gases of the zero group of the periodic table; the other is the α particle, a doubly charged

---

in the upper left is our symbol for an α particle itself. This very stable nucleus of the helium atom contains two protons, $\oplus$, and two neutrons, $\bigcirc$ (see p. 315), which provide it with a $+$ charge of 2, and a mass of 4. We recall that only the protons are charged ($+$) and the *uncharged* neutrons have a mass approximately the same as that of the protons.

helium ion, which, on picking up its two attendant electrons from the multitudes available everywhere in the submicroscopic world, is just an ordinary, everyday neutral helium atom.

IF WE trace the history of the other disintegration product of the radium, that of radon, we find that it, too, is radioactive. It disintegrates in turn into a substance that is solid at ordinary temperatures, radium "A," a member of the sulphur group of elements. Simultaneously, as this change occurs the radon emits an α particle. Radium "A" soon becomes radium "B," and a third α particle is shot off; radium "B," which lies in the same group of elements as lead, next disintegrates into radium "C," and this time no helium ion (alpha particle) is the other product of the transmutation. However, as radium C was formed, a swift electron or β particle came out of the radium B. Only an inappreciable part of the mass (214) of radium B was lost, hence that of radium C is the same.* Of course, you ask then why is not

---

* Radium itself is the product of the disintegration of another element, ionium, which in turn is the result of the fourth disintegration of the parent-substance uranium.

The series is in part diagramed below. Alpha and beta ray changes are indicated. Below the name of the element are numbers giving its relative atomic weight, and its "half period," i.e., the time in years, months, days, or seconds, before any given amount of the substance will have been reduced to half by changing to something else. Its group numeral in the periodic table is above it.

VI		IV		V		VI	
Uranium I	$\xrightarrow{\alpha}$	Uranium X_1	$\xrightarrow{\beta}$	Uranium X_2	$\xrightarrow{\beta}$	Uranium II	$\xrightarrow{\alpha}$
238		234		234		234	
4.4×10^9 years		24.5 days		1.14 min.		3×10^5 years	

IV		II		O		VI		IV	
Ionium	$\xrightarrow{\alpha}$	Radium	$\xrightarrow{\alpha}$	Radon	$\xrightarrow{\alpha}$	Radium A	$\xrightarrow{\alpha}$	Radium B	$\xrightarrow{\beta}$
230		226		222		218		214	
8×10^4 years		1,590 years		3.82 days		3.05 min.		26.8 min.	

V		VI		IV		V		VI	
Radium C	$\xrightarrow{\beta}$	Radium C'	$\xrightarrow{\alpha}$	Radium D	$\xrightarrow{\beta}$	Radium E	$\xrightarrow{\beta}$	Polonium	$\xrightarrow{\alpha}$
214		214		210		210		210	
19.7 min.		10^{-6} sec.		22 years		4.9 days		140 days	

IV
Radium Lead
206
not radioactive

Note especially: (a) the great variation in half period, four thousand million years for uranium I and a millionth of a second for radium C'; (b) the loss of 4 in atomic weight for each α ray change; no loss for β ray change; and (c) the shift of chemical properties to the second group to the left for each α ray change and to the next group to the right for β ray change.

RADIOACTIVITY AND TRANSMUTATION

radium C regarded simply as a positive radium B ion. The answer is that chemically it has quite different properties from radium C, being a member of Group V instead of Group IV. Furthermore, an atom of radium C is an electrically neutral substance, not charged with either sign. The electron that radium B lost must have been lost in such a manner that it contributed

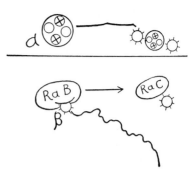

in no way to make up a net change of charge. There would seem to be no possible way to account for this without having radium C pick up another electron immediately upon formation.

Indeed, this is just what happens to α particles, is it not? As formed they are ions, doubly charged positively; but by the time they have finished their meteoric flight from the parent-atom they have become neutral helium by picking up not one but two electrons. The same thing you might suppose would be true of radium C. The important fact about the matter is that the electron (or α particle) that is lost in radioactivity signalizes a change in a *different part* of the atom than that in which electrons are subsequently acquired in the process of recombination or neutralization of the ions. The process of ionization and recombination in no way changes the chemical character of a substance. Here we have a different process that involves a profound change of character which is related to the atom's structure in some fundamental manner.

HAS an atom, then, so definite a structure that we may refer to different parts of it? Of this we are quite certain. The experiments that go to prove it quite fundamentally were first performed by Rutherford. Before we describe them let us answer one very common question.

When α particles were shot through a glass tube and helium was subsequently found therein, possibly you wondered about the holes left in the glass as the atomic bullets passed through. There were none left. But glass, you will say, is surely not porous, that these entities might find their way through holes already there? No. We have no evidence of porosity in glass. But glass, and everything else for that matter, is really not very solid. Indeed, most of the space that we consider filled with solid substance is quite empty, containing nothing but fields of force. This seems to you unbelievable, doubtlessly, and so it was to physicists until first the experiments of Rutherford, and subsequently a host of others in substantial agreement, have convinced us that the fact is really as we have just stated it.

The experiments referred to consist of firing swift a particles from radioactive substances at other atoms and finding out what happens. It is not necessary, we think, to go into details of experimental arrangements here which are rather complicated. We can photograph the tracks of a particles through a gas by condensing water drops on the débris that the a particle leaves in its path. This débris is a wake of ions, plus and minus, Solid matter which under proper conditions that we understand quite well and can control serve as nuclei for this fog that comes down on nothing else if we are careful.

That ions are formed at all shows that atoms must have been passed through and must have had some of their constituent electrons knocked out

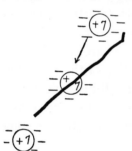

of them. These electrons picked up by uninjured atoms form negative ions with them; and the broken atom, without some of its constituent electrons, is the positive ion. In time, of course, the two kinds drift together; the detached electron (or another just like it) rides back to its former home; the breach is healed. The little waif, to change the figure, is picked up and adopted at once by some uninjured bystander atom and then soon returned to some one of the other atoms recently bereaved. Since all these children look just alike, no one ever knows or cares whether they get back the same one that they lost.

Now, as the atom's happy family is thus momentarily shattered, the ruthless a particle is slowed down only very little. Its acceleration in the negative sense, its retardation, is almost negligible in the early stages of its mad career. Change of direction counts in acceleration, you remember, just as well as change of speed. The speed of these particles (about ten thousand miles per second) is a bit too swift for us to study in detail, searching for changes in it that occur several thousands of times per second when they occur at all. That there is a dual test for acceleration is fortunate, since the direction of the particles is what we can clearly watch. Their paths persist before our eyes for several seconds and can be seen and photographed by a marvelous technique developed first by Professor C. T. R. Wilson, of the Cavendish Laboratory.

EXCEPT at the very end of the range THE PATHS OF α PARTICLES ARE VERY NEARLY STRAIGHT. (Note path-symbol in our sketches of them.)

In a flight of about 3 in. through the air the α particle passes through about 150,000 molecules or atoms. It is resisted by them almost not at all at first, each one slowing it up by but a trifling amount. However, it *is* slowed down a little, and the more it is slowed down the more do the atoms resist it. So it finally stops suddenly with a dull and sickening thud, as it were, at the end of its journey.

When a projectile passes through what is to it solidly filled space, the resistance it encounters is much greater when its velocity is high, and gets

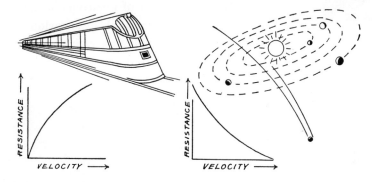

less and less as the speed decreases. Above we saw just the opposite effect. But if a projectile passes through space that is filled only with those fields of force we talked so much about in chapter 24, it exhibits behavior just like that of our α particle. A comet or a meteor or another small planet or a moon, if it could be fired through our solar system with sufficient speed, would stand a most excellent chance of getting through without actually colliding with anything except possibly the enormously massive sun at the center. The gravitational fields of force of every member of the solar system would be tugging at it as it passed through, but the faster it went the less time would it remain in a region where those forces had any appreciable strength, and the less would be their effect upon it. Most of the wanderers that our system has picked up from out of space are with us solely because they were not going fast enough. Delay due to lack of speed is fatal. They are caught and forever entangled in the web of gravitation that is woven around us.

Hydrogen

Helium

Lithium

Carbon

It is for these reasons that we regard material atoms as things analogous to our solar system in miniature. The electrons that enter into the composition of the various atoms, no two kinds of which have the same number, are often referred to as PLANETARY ELECTRONS. We have seen that hydrogen, the lightest atom, has one, and helium, two. Lithium has three; beryllium, four; boron, five; carbon, six; nitrogen, seven; oxygen, eight; fluorine, nine; neon, ten; and so on down the entire list of the chemical elements, ending with uranium, ninety-two.* The number of planetary electrons is often called the ATOMIC NUMBER, and of course it is equal to the number of positive charges that must be present *somewhere* if all these various numbers of negative electrons are to be held together and neutralized.† The ordinary garden variety of atoms, it will be remembered, gives no evidence of being electrically charged at all. What about this controlling positive charge—where is it and what is it? How can we find out about it?

Rutherford again has had much to do with supplying the answer. He not only has bombarded atoms of substances in the gaseous state with his α particle cannon balls but has shot them through solids as well. Of course, they will not travel very far through a solid without being stopped and lost to view. Human methods of detecting them avail only when they are in motion. One of these methods indeed you may at this very moment be wearing on your wrist.

The luminous dial of your watch

DID you ever happen to look at the luminous dial of a watch or clock (the type of dial that shines of itself in the dark) with a small magnifying glass? If not, you have missed one of the most interesting of spectacles, and you should not let another night go by without examining it. Under a

* There are ninety-two *naturally occurring* elements on earth. Recently several additional new elements heavier than uranium have been produced. Cf. pages 330 f.

† One of the isotopes of uranium, the most abundant one, has a mass of 238, a nuclear charge of 92. This nucleus is built up of 92 protons and 146 neutrons.

RADIOACTIVITY AND TRANSMUTATION

ten-to fifteen-power lens, after your eyes have become well adjusted to darkness—complete darkness, if possible—look at the luminous hands and figures of your wrist watch for about ten minutes. You will think that you are looking at a lot of shooting-stars. Under even this modest magnification you will see a field of sparkling points of light, no two ever in quite the same spot, each one only a flash, but hundreds of them visible at once. A year from now the phenomenon will be still going on; it never ceases day or night, whether you look at it or not. The luminosity in these dials, when properly made, is due to the fact that traces of radioactive materials have been mixed up with fluorescent ones. The scintillations you see are the splashes of light that α particles shooting off from disintegrating atoms make as they collide with, and are stopped by, the molecules of fluorescent substances such as zinc sulphide or barium platino cyanide.

RUTHERFORD'S experiment consisted of shooting the α particles through a very thin gold foil, hundreds of thousands of atomic diameters but only a few thousandths of a centimeter in thickness, and watching their splashes after they had passed through the gold and were finally caught and made visible under a lens by a bit of zinc sulphide. Every now and then one would be observed that had been deflected through a very large angle, sometimes as great as 150°, from its original direction by passage through 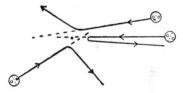 the gold. A study of the dynamics of these particles, whose masses and velocities we have already measured, is far beyond our scope, of course. The methods are modifications of those used in connection with the solar system, involving the law of gravitation and Kepler's laws of motion of a planet in its orbit, or of a comet wandering in toward the sun from a region far away. Applied to Rutherford's experiments these equations of momentum and energy change show that the central force which repels and deflects the positively charged particle, originates from a positive charge $n\epsilon$, where n is the atomic number (1 for H, 2 for He, etc.). The analysis shows, moreover, that this force emanates from a space extremely small within the center of this tiniest of systems.

This central region of an atom, where the positive charge $n\epsilon$ is located, is called the ATOMIC NUCLEUS. Around it as planets around a central sun are the attendant electrons n in number for the neutral atom. The region of the nucleus and that occupied by the attendant electrons are to be sharply differentiated. Processes of ionization—indeed, all chemical properties—have to do with planetary electrons and their arrangements cited previously (p. 310). Radioactive phenomena occur *within the nucleus,* and it is here the mass and positive charge resides. Any change of charge upon it, of course, demands a change in number of electrons outside; and their rearrangement alters the chemical character of the atom.

The analogy to a solar system, however, must not be pushed too far or be taken in too literal a sense. The mechanical laws and dynamical principles underlying these systems are quite different from their simpler forms with respect to larger bodies. We suspect that we still know rather too little about details of the submicroscopic world to be justified in giving you too definite a picture of it. We should not like to emphasize as important something that you might have to unlearn subsequently.

OUR ideas about such matters, for example, as the "orbits" of the electrons around the central nucleus have undergone great modification with the last few years. In some current texts you will still see pictures of the orbits as closed circles and ellipses, even approximately dimensioned. These pictures follow the earlier ideas of one of the greatest scholars in this field, Professor Neils Bohr, of Copenhagen. These Bohr orbits were a most suggestive and remarkable working hypothesis which for a number of years was our only guide to these matters. Then, as we began to apply crucial tests and as further experimental facts came to light, the picture was seen to be quite inadequate in some ways. Heisenberg, of Leipsig, showed that there was a very definite uncertainty with respect to experimental knowledge of both the position and the velocity of an electron. The more sharply we focus on one of these elements, the less certain the value of the other becomes.

RECENT work with X-rays has enabled us to look at atomic structure with sharper eyes. We find that the orbits are in reality more vaguely defined regions, nebulous in outline but agreeing, nevertheless, with very definite calculations of probability that were made earlier in these connections. The probability involved was that of being able to find the location of a particular electron. The greater the chance, the lighter was the ink, in the drawing representative of our ideas.* So different are these subatomic worlds

---

* Cf. Plate 99, p. 433.

RADIOACTIVITY AND TRANSMUTATION

from ours that many of our common ways of thinking seem to be meaningless as applied to them. It is a field where, in the main, as yet only a formal logic of great abstraction avails. Such a tool we have in mathematics, and only by the use of its language can we hope to keep ourselves free from error and ambiguity in our study. Even with the use of mathematics these things creep into the works of the ablest investigators.

SUCH elements as would, in picture language, correspond to spin or rotation both of electrons and nuclei undoubtedly exist. This idea led to the discovery that hydrogen gas was of two kinds—kinds that differ in no respect save that of orientation of their component rotating parts. These two kinds of hydrogen had certain predictable and different properties. Subsequently these two kinds of hydrogen actually were separated experimentally, and a pure concentration of one of them alone was made and found to have the predicted properties. Consequently, we stick to these working hypotheses, hard and abstract as they are to visualize and understand in more common terms. They answer their full purpose and perform distinguished services when they inspire and guide experimental research to definite and repeatable results.

NO BETTER example of this can be cited than recent progress in ARTIFICIAL TRANSMUTATION, now an accomplished fact, as we have said, in the case of *all* of the elements in the periodic table. To understand these results we must ask ourselves a little more definitely about what the α particle is and whether we cannot make some hypotheses with respect to its construction.

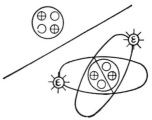

The α particle is first of all and quite obviously the NUCLEUS of a helium atom. It possesses at first a double positive charge and the mass of helium. Later, when it regains its two planetary electrons and is ordinary helium, it is no longer an α particle. Helium is never found with a positive charge of three, but it may have a charge of one. Thus two planetary electrons are its full complement; without them it becomes an unattended nucleus, an atom stripped completely of its electron "atmosphere." It is the most effective of all projectiles of less than atomic dimensions. It travels with a speed twenty thousand times that of a rifle bullet and, gram per gram, has three hundred and fifty million times the kinetic energy of the latter. Its characteristic shattering effect upon the planetary electron systems of other atoms has been recounted. Its occasional sharp divergence as it approaches the heavy nucleus of an atom like gold has given us the clue to the relatively small dimensions of the nucleus of even so massive a system.

MARSDEN first, without fully interpreting the meaning, and later Rutherford, Chadwick, and many others have observed that, when some of the lighter atoms, like lithium, beryllium, and nitrogen, are shattered by a particle bombardment, the destruction is not confined to the rather easily repaired outer electron structure of the system. On fluorescent screens set 60–75 cm. distant from the source of alphas and the target of other atoms under bombardment, scintillations have been observed in abundance which

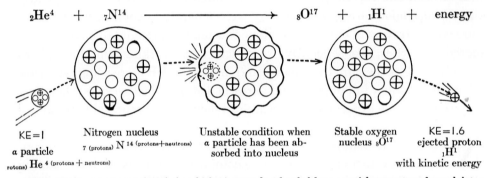

Rutherford's experiment (1919) in which nitrogen bombarded by a particles was transformed into oxygen and a high-velocity proton shot off. Protons, ⊕, and neutrons, ○, here drawn for clarity somewhat separated within the nucleus are to be thought of as in contact with one another—the "nucleonic fluid" of the water-drop-like structure of nuclei due to Bohr and Wheeler.

have been proved by deflection in magnetic fields to be due to particles of mass 1 and charge +1, i.e., to be protons, hydrogen nuclei.* So great a range for a proton means that it must possess a kinetic energy at the outset 1.6 TIMES AS GREAT AS THAT OF THE a PARTICLE. These results obtain in experiments where hydrogen gas, as such, *has never been introduced*. Whence can come these hydrogen nuclei with an energy greater than that of our bombarding projectile? There seems to be no place of origin conceivable unless it be the NUCLEUS of the beryllium or the nitrogen or other atom, that has been struck, and thereby ARTIFICIALLY DISINTEGRATED. The excess energy of the fragment that has been thrown out, that of this proton, was acquired by the violence of the explosion that was set off in the nucleus of the shattered atom. One of the component parts of this nucleus is, therefore, a proton.

* Within the last few years a particles have been produced by bombarding lithium atoms with swift-moving *protons*.

RADIOACTIVITY AND TRANSMUTATION

ALL THIS resembles radioactivity except for the fact that these atoms of beryllium, nitrogen, etc., are not radioactive, and that the phenomenon has been produced by ourselves and does not go on automatically and unamenable to any control.

Furthermore, since the discovery of neutrons by Chadwick (p. 315) a host of other transmutations have been effected, using not only α particle but also protons, deuterons (heavy hydrogen nuclei), neutrons, and photons (high-energy γ-rays) as projectiles and every element under the sun as targets. If the projectile energy is great enough, changes invariably take place in target nuclei squarely hit. New nuclei of *slightly* different proton-neutron content are formed. Any of the particles used as projectiles may be ejected from the final product—either simultaneously with its formation, or subsequently after a short, or even a very long, time. In the latter case the nucleus formed from the initial target nucleus is an ARTIFICIALLY MADE RADIOACTIVE NUCLEUS—often called a radioisotope of the element whose atomic number it has acquired.

This great amount of experimental data, now abundantly confirmed, has led to the conviction that NUCLEI OF ATOMS are COMPLEX STRUCTURES capable, under very special conditions, of having their structure modified—hence ATOMS can and have been transmuted from one element to another.

IN GENERAL, these transmutations shift the element only one, rarely two, places from its original position in the periodic table of elements.

It is very important also to know that another result of this modern work has been to deny to the electron any existence as such in the nucleus itself. The nucleus contains PROTONS and NEUTRONS—protons to the amount of the atomic number of the element (which, by definition, measures the positive nuclear charge) and neutrons to make up the total mass number of the element in question, thus,

$$A = Z + N,$$

where A = mass number, Z = atomic number (number of protons), and N = the number of neutrons.

We have mentioned (on p. 320) the β rays (electrons) that are ejected when uranium X_1 and X_2 disintegrate in the natural series of radioactive changes that begin with uranium I and end with radium lead. The decay of radium B, C, D, and E likewise is accompanied by electron emission (see note, p. 320). Whence come these electrons if electrons as such are no longer regarded as constituents of nuclei? Furthermore, how does a nucleus—consisting only of protons and neutrons—acquire the additional positive charge necessary to increase its atomic number by one, as chemical tests show that it does when the β ray transformations occur?

THESE are questions still not too clear, even after much recent work. The assumption made is that the two nuclear particles, protons and neutrons, can transform themselves—either into the other. Beta ray changes are then accounted for by the transformation of a *nuclear neutron into a proton;* this gain of a positive charge, a new proton, signalizes itself by the loss of a negative electron externally. Such an *ad hoc* assumption in itself might seem highly artificial, were it not for the fact that, in cosmic ray phenomena and in many artificial transmutations yet to be discussed in more detail, the emission of POSITIVE ELECTRONS* (positrons) is well attested. Where matter is abundant, positrons are very short-lived particles. They have the same mass as the electron and an equal electric charge in magnitude but one that is opposite in sign, i.e., the same charge as a proton. Under the circumstances of their ejection from a radioactive nucleus, the chemical properties of the daughter-element are those of an element of one atomic number *less* than the parent—one less nuclear charge, one less nuclear proton. Since the mass number is unaltered in positron (as well as in electron) emission, this means a neutron has taken the place of a proton in the nucleus or, in terms of the assumption formerly made, *a proton has transformed itself into a neutron.*

As examples of "β ray changes"—the phrase now broadened to include both negative electron (sometimes called "negatron") and positron emission —let us summarize a few of the better-known artificial transmutations without going into the details of techniques—a great variety of which are used to accomplish them.†

* For reference to the discovery of positrons by Anderson in connection with cosmic ray phenomena see below, p. 438.

† Of instruments, the cyclotron, invented by Lawrence, of the University of California, has been the most versatile so far. In it ions of hydrogen, deuterium, and helium are repeatedly accelerated by an electric field while being confined in a widening spiral path by a magnetic field at right angles. On release, the protons, deuterons, or α particles, nuclei, respectively, of the atoms of these three gases, are released with hundreds of thousands, or millions, of electron volts of energy as projectiles to bombard the nuclei of various target materials. More recently the betatron, invented by Kerst, of the University of Illinois, accelerates electrons whirling within a circular tube to prodigious velocities—comparable and exceeding those with which electrons are released from natural radioactive substances. The high-energy

RADIOACTIVITY AND TRANSMUTATION 330a

In the following table (p. 330b) you will find in Column 1 the name and symbol of the element; isotopes must be distinguished, so the name is followed by the mass number. In the symbol the atomic number is the lower prefix, the mass number the upper suffix.

Column 2 gives the half life of the element; Column 3 the type of radiation emitted as it decays—negative electrons, $\beta-$, or positive electrons, $\beta+$; Columns 4–7 give the process of manufacture of the isotopes in question; sometimes there are several processes for the same isotope. Column 4 contains the projectile used, Column 5 the target, Column 6 the ejected particle, Column 7 the "nuclear equation"—symbolically, on the left side the projectile and the target; on the right, the resulting isotope (same as Col. 1) and the ejected particle (same as Col. 6).

A GRAPH of the first eight elements of the periodic table (p. 330c) helps one to see the interrelations of their various isotopes. Suppose we arrange the elements in a vertical list, atomic numbers increasing downward, and spot the isotopes of each element horizontally alongside the element, the mass numbers increasing to the right. We will include all isotopes, the natural ones, with the percentages of them that make up the composition of the element as found on earth set down below each one, and the artificial ones. The former are indicated by circles, the latter by crosses. In the radioactive (artificial) ones an arrow indicates the direction of decay—downward to the next higher atomic number for $\beta-$ (negatron) decay; upward to the next smaller, for $\beta+$ (positron) decay. The half life is placed beside the corresponding arrow. This is the time it takes for half of any amount of the isotope present to be transmuted into its decay product.

In this graph special attention is called to several things.

1. In the upper left-hand corner are the two building-blocks of all other nuclei, the proton in the (1, 1) position and the neutron in the (0, 1) position. The electrons (− and +), if the *signs* of their charges are disregarded, have charges of 1 unit and masses negligible compared with the lightest atomic masses. Thus, without too great a stretch of one's conscience, these may be

---

γ rays associated with the impact of these electrons on targets are also effective in disrupting other atomic nuclei.

Quite different and requiring no huge mechanisms but in many ways a most effective technique is the use of artificial neutron sources. Of these, one important kind uses α particles from some natural source (like RaC), which impinge on a beryllium target. The beryllium becomes carbon, ejecting neutrons. High-speed deuterons from cyclotrons also produce neutrons quite effectively.

TABLE OF A FEW ARTIFICIALLY MADE RADIOACTIVE ISOTOPES OF THE LIGHTER ELEMENTS

(1) Name (Symbol)	(2) Half Life	(3) Type of Radiation	(4) Produced by	(5) Striking	(6) Which Eject from the Combined Nucleus a	(7)
Hydrogen three ($_1H^3$)	31 yrs.	$\beta-$	Deuterons ($_1H^2$)	Deuterons ($_1H^2$)	Proton	$_1H^2 + _1H^2 \rightarrow _1H^3 + _1H^1$
Helium six ($_2He^6$)	0.8 sec.	$\beta-$	Neutrons	Beryllium nine	Alpha particle	$_0n^1 + _4Be^9 \rightarrow _2He^6 + _2He^4$
Lithium eight ($_3Li^8$)	0.88 sec.	$\beta-, \alpha^*$	Deuterons	Lithium seven	Proton	$_1H^2 + _3Li^7 \rightarrow _3Li^8 + _1H^1$
			Neutrons	Boron eleven	Alpha particle	$_0n^1 + _5B^{11} \rightarrow _3Li^8 + _2He^4$
			Neutrons	Lithium seven	γ rays (photons)	$_0n^1 + _3Li^7 \rightarrow _3Li^8 + \gamma$
Beryllium ten $_4Be^{10}$	10^3 yrs.	$\beta-$	Deuterons	Beryllium nine	Protons	$_1H^2 + _4Be^9 \rightarrow _4Be^{10} + _1H^1$
Boron twelve $_5B^{12}$	.022 sec.	$\beta-$	Deuterons	Boron eleven	Protons	$_1H^2 + _5B^{11} \rightarrow _5B^{12} + _1H^1$
Carbon ten	8.8 sec.	$\beta+$	Protons	Boron ten	Neutrons	$_1H^1 + _5B^{10} \rightarrow _6C^{10} + _0n^1$
Carbon eleven	20.5 min.	$\beta+$	Deuterons	Boron ten	Neutrons	$_1H^2 + _5B^{10} \rightarrow _6C^{11} + _0n^1$
Carbon fourteen	10^3 yrs.	$\beta-$	Deuterons	Carbon thirteen	Protons	$_1H^2 + _6C^{13} \rightarrow _6C^{14} + _1H^1$
			Neutrons	Nitrogen fourteen	Protons	$_0n^1 + _7N^{14} \rightarrow _6C^{14} + _1H^1$

* When $_3Li^8$ radioactively emits a negative electron its nucleus acquires an additional + charge without mass change (np) and becomes $_4Be^8$, which also is unstable (radioactive) and splits into two α particles, $2(_2He^4)$.

Isotopes, natural and artificial of the lighter elements

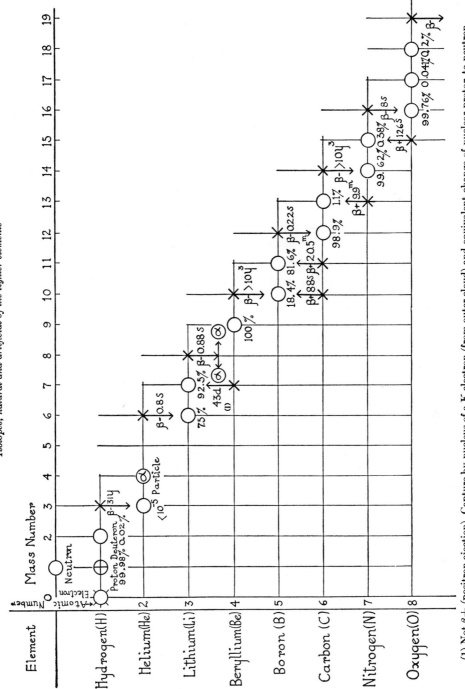

(1) Not β+ (positron ejection). Capture by nucleus of a K electron (from outer cloud) and equivalent change of nuclear proton to neutron.

placed in the (1, 0) position. The (0, 0) position is left free for a particle as yet hypothetical, of no charge and negligible mass, called a *neutrino*.

2. In general, the artificially made and unstable radioactive isotopes flank the stable ones on one side or on both, e.g., in Be, C, H, O, and in many heavier elements not shown in the graph.

3. Artificial isotopes of less mass, i.e., less neutrons than the stable forms, in general show positron ($\beta+$) decay. This scuttling, as it were, of some of their excess + charge, meaning (inside the nucleus) exchanging protons for neutrons, takes them to stable natural forms, in which numbers of neutrons and protons are more nearly alike. On the other hand, if unstable forms have an excess mass (neutrons), the decay is usually negatron ($\beta-$) in type. This loss of negative charge implies gain of positive protons over neutrons in the nucleus, and again closer neutron-proton balance is achieved.

4. The great range and haphazard character of the half lives of the radioisotopes from a few tenths of a second in lithium eight and boron twelve to more than a thousand years in carbon fourteen is a characteristic that shows up in the natural radio elements (see note on p. 320) likewise.

5. No isotope of any element having mass number 5 has yet been found or artificially produced. All others are present; indeed, there are some mass numbers—10, for example—that are associated with *several different elements*—in this case beryllium, boron, and carbon. Such isotopes of different atomic numbers and identical mass numbers are called *"isobars"* (equal weights) of one another.

Finally, this brief sketch of some of the outstanding features of this list of artificial isotopes of the lighter elements should not be concluded without a diagrammatic picture of at least two fairly representative examples of the manufacture of a radioisotope and its subsequent decay. We are choosing helium six and carbon ten (p. 330*e*).

AS ONE proceeds from the lighter to the heavier elements, processes of transmutation similar to those we have already described remain effective. Radioisotopes of every element have been produced in considerable numbers.* At least *nine* artificial forms are listed for iodine, element 53, its isotopic mass numbers running from 124 to 137. *Natural* iodine exists in but *one* isotopic form, $_{53}I^{127}$. Another extreme case is tin, which, in its natural

---

* Details of isotopes of all the elements as of this date will be found summarized by Glenn T. Seaborg, *Reviews of Modern Physics*, Vol. XVI, No. 1 (January, 1944); also in *The Periodic Table of the Elements* (Chicago: Museum of Science and Industry, 1946).

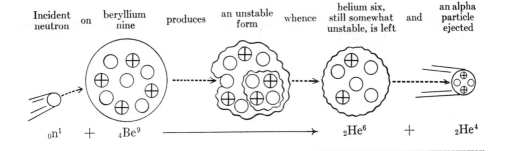

And subsequently the radioactive helium six decays thus:

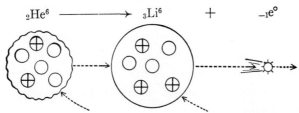

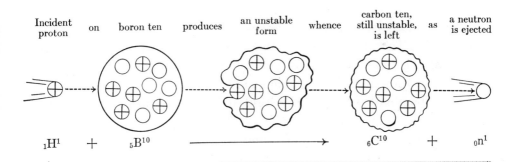

Subsequently, carbon ten, with more protons than neutrons, decays thus:

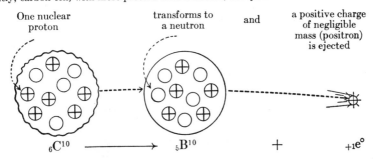

state, has *ten* isotopes of masses 112–24 and *twelve* additional artificial ones on record on January, 1934.

In bismuth two hundred and nine, we reach the *heaviest stable isotope in nature under* TERRESTRIAL *conditions*, $_{83}Bi^{209}$. From there, on up to uranium two hundred and thirty-eight, all atomic nuclei are more or less unstable and sooner or later disintegrate or decay in the processes long recognized as those of natural radioactivity. Indeed, the decay products of some of the heavier elements are found as far up in the periodic table as element 81, tellurium, whose isotopes 207, 208, and 209, are members of the actinium, thorium, and uranium series of natural radioactive elements.

A CURIOUS person, familiar with all this work on disintegration of all the elements up to and including bismuth, will naturally inquire: "What about trying similar experiments on some of the *less unstable* forms of these heavy elements?" Natural uranium, for example, exists in nature in three isotopic forms:

$\quad$ 0.006 per cent as $_{92}U^{234}$ with half life of 2.7×10^5 years,
$\quad$ 0.71 $\;$ per cent as $_{92}U^{235}$ with half life of 7.07×10^8 years,
$\quad$ 99.28 per cent as $_{92}U^{238}$ with half life of 4.51×10^9 years.

These isotopes can hardly be called unstable in the ordinary sense. True, they are radioactive, all three isotopes decaying by ejection of a particles into $_{90}Th^{230}$, $_{90}Th^{231}$, and $_{90}Th^{234}$,* respectively. This goes on so slowly, however, that the time required for half of them to disappear is reckoned in hundreds of thousands (10^5) to billions (10^9) of years. Suppose, then, this element is bombarded by the same kinds of projectiles, using the same techniques that have been so effective on all the other elements. May it not be possible thus to create new isotopes and perhaps directly or, from the ($\beta-$) decay of these, new elements of *higher atomic number than 92*—i.e., transuranic elements?

SUCH experiments were tried as far back as 1904. Intense and complicated new radioactivities resulted; and during the next five years the physicists, especially in Europe, were busy as bees trying to find out exactly what elements of atomic numbers 93 and upward were thus being manufactured. No agreement was reached, although about 150 technical papers were published. Not until January 6, 1939, did a paper appear in the German *Naturwissenschaften* by Hahn and Strassmann that offered, very hesitantly, a new, almost unbelievable, hypothesis (based on chemical evi-

---

* Th = thorium, of which these are all isotopes, is sometimes used as the symbol for all of them instead of their earlier individual symbols, Io, UY, and UX_1, respectively.

dence, however) to the effect that a uranium nucleus had SPLIT INTO TWO *not greatly unequal parts*. Mass numbers 138 and 101 (whose sum is 239) were suggested as of the appropriate orders of magnitude.

After five years of amazing secrecy, the dramatic events that followed in swift succession upon the release of the resulting ATOMIC BOMB have been more than fully exploited by the press and many newspaper writers.

Space here permits but the briefest summary of the physical phenomena involved, leaving study of the profound effects that these devastating experiments may have both for ill and, if that can be avoided, for good to future and more mature reading and reflection of the reader.*

FIRST of all, it was shown by Nier, Bohr, Wheeler, Dunning, and others in this country that the splitting of a nucleus, called FISSION, in uranium was confined to the rare isotope, U^{235}. Barium and a whole series of decay products, including elements 51–59 and mass numbers 116–44, resulted from one fragment. Another lighter series of elements from 34 to 49, mass numbers 83–116, resulted from the other fragment. In addition, and essential to the maintenance of the so-called "chain reactions" in these processes, there was an EXCESS of FREE NEUTRONS left over from the fission process—which in U^{235} is initiated by the *capture by that nucleus of a slow-moving (thermal velocity) neutron*.

Second and equally important was the discovery at the University of California of elements 93 and 94 and the development, largely by a University of Chicago group under the camouflaged name of the "Metallurgical Laboratory," of techniques by which these two new *transuranic elements*, 93 and 94, could be made in effective quantities. These have been christened neptunium (Np) and plutonium (Pu), respectively, and each exists in several isotopic forms, $_{93}Np^{237,238,239}$ and $_{94}Pu^{238,239}$.

Third, and of the greatest importance in providing additional fissionable material, was the discovery that the 239 isotope of plutonium, like the 235 isotope of uranium, was *fissionable by slow neutrons*. Thus this new element, made in tangible quantities from the common form of uranium (238), provided an alternate to the rare U 235, separable only with great difficulty from the chemically identical common uranium. Thus there was furnished the means for an additional supply of material for military purposes—material from which an equally prodigious supply of energy could be drawn.

* One of the best of these accounts is entitled *Atomic Energy in Cosmic and Human Life*, by G. Gamow (New York: Macmillan, 1946).

BY NOW you are doubtless wondering about the source of this enormous amount of energy so devastatingly revealed in the results of the two bombs dropped on Hiroshima and Nagasaki and with respect to the future wise control of which the future of mankind so largely depends.

To understand this matter we must consider some other aspects of radioactivity. In the first place, it has been known for years that the α particles that continuously come out from radium and other natural radioactive sources have prodigious velocities and, since their mass is by no means negligible, enormous energies. One gram of radium with its decay products emits about 140 calories per hour—more than enough to raise its own mass from the freezing- to the boiling-point of water every hour. Of this the α articles alone (and the atoms recoiling from them) supply the major amount of 125 calories, electrons (β rays) about 6, and γ rays 9.

The velocities of the α's range from $1\frac{1}{2}$ to 2 billion cm. per second, i.e., are of the order of 10,000 miles per second. This, combined even with a mass of only 6.6×10^{-24} gm., gives a kinetic energy ($\frac{1}{2} mv^2$) of $\frac{1}{2} \times (6.6 \times 10^{-24}) \times (1 \times 10^9)^2 = 3.3 \times 10^{-6}$ ergs per particle, or 2.06 *million* electron-volts* per particle. Now this is not only a very respectable but, indeed, a prodigous amount of energy for any atomic particle to have. Energies of chemical combination such as are needed to separate hydrogen from oxygen in water amount to about a millionth of this, i.e., 2.6 electron-volts. Even such violent explosives as T.N.T. possess energies of only about 35 electron-volts per molecule.

ALTHOUGH no chemical change is taking place inside the radium, we have had here for many years at least an intimation that there is a huge amount of energy of some kind or other locked up in atomic nuclei themselves (at least in the radioactive ones). No better testimony is needed than the great speeds and energies exhibited by these α particles as they get kicked out of radioactive nuclei when these decay into nuclei of near-by heavy elements. However, as we have said before, and now repeat for emphasis, during the first forty years of the half-century just elapsed since the discovery of radioactivity, no physical or chemical means was ever discovered in any way *to alter the course of radioactive changes*. The possibility of ever

---

* The electron-volt is the energy an electron acquires in falling through a potential of 1 volt. A million electron-volts (mev) is an energy unit commonly used in atomic physics and equals 1.6×10^{-6} ergs. 1 erg = 6.24×10^5 mev.

RADIOACTIVITY AND TRANSMUTATION

laying our hands to the slightest degree on the vast store of energy locked up in this Pandora box of trouble was entertained by few but the most daring.

WE HAVE already described how one of those most daring was Rutherford. What is far more important, he was most CAREFUL, PAINSTAKING, and PATIENT. He and Chadwick in 1919 used one of *nature's own* swift α particles from radium C' ($_{84}Po^{214}$), against nitrogen nuclei and disintegrated the latter into oxygen, a proton being ejected in the process. We have already mentioned that it was this same Chadwick who discovered the neutron in 1932. And in 1932 Cockroft and Walton in Rutherford's laboratory shot protons into lithium seven and produced helium, thus

$$_1H^1 + _3Li^7 \to 2(_2He^4) .$$

The two helium nuclei, *artificial* α particles so produced, were observed to leave the scene of their birth with speeds equivalent to *17 million electron-volts*. Whence came the energy for this swift flight? No natural radioactivity here! The light-weight incident proton had a negligible energy compared to this, and the target nitrogen only fractions of 1 electron-volt from its thermal velocity at the laboratory temperature. Physically, of course, we can say the energy came from work done by intense forces presumably existing between the like charged protons held so close together in all atomic nuclei.* Electrostatic forces, we recall, increase as the square of the distance between the charges diminishes. Other very intense forces are also recognized. There is, however, another, much more general point of view—purely theoretical—which in this particular experiment can be put to precise numerical test.

This comes about from the fact that the masses of all the characters in this particular little drama are very well known. The proton plus the lithium seven on the left side of the equation above weigh precisely

$$8.0241 \text{ atomic units.} \dagger$$

---

* How an assemblage of mutually repulsive, positively charged particles like protons have any stability at all even when associated with an equal or greater number of neutrons in nuclei is still a mystery. To say that the neutrons form some kind of a cement is, of course, to say nothing. Physicists *assume* the existence of attractive forces of very short range between neutrons and protons. In the close packing of neutrons and protons in nuclei (the only part of an atom that is not largely empty space) these short-range forces of unknown origin exceed the mutually repulsive electrostatic forces of the protons and prevent the assemblage from blowing apart.

† One atomic unit of mass is, by definition, one-sixteenth of the mass of the common isotope of oxygen, $_8O^{16}$, and equals 1.66×10^{-24} gm.

Two a particles together weigh

$$8.0056 \text{ atomic units}.$$

In the transformation of hydrogen one (proton), and lithium seven into helium four there seems to have been A LOSS OF MASS equal to

$$0.0185 \text{ atomic units, or } 3.07 \times 10^{-26} \text{ gm.}$$

This brings us to Einstein and something he suggested long ago.

ALBERT EINSTEIN, as early as 1905, in connection with his formulation of the theory of relativity had made the suggestion that the *mass* of an object depends on its velocity in this fashion

$$\underset{\text{Total Mass}}{M} = \underset{\text{Mass at Rest, } M_0.}{\frac{M_0}{\sqrt{1 - v^2/c^2}}},$$

where v is the velocity of the body and c is the velocity of light. It is seen that, as v becomes nearer and nearer to c, the denominator of this fraction approaches zero and the total mass of the moving object becomes indefinitely larger. Since ordinary terrestrial motions are extremely small compared to the velocity of light (3×10^{10} cm/sec), this increase of mass with velocity is not to be observed in them. *High-speed electrons, however, have been observed to increase their mass.* At a velocity imparted to electrons by half-a-million volts potential, their mass is twice their "rest mass" of 9×10^{-28} gm. Other conclusions of the special theory of relativity—largely astronomical in character—have also been observed as predicted.

Now if we take the above relation, rewriting it thus

$$M = M_0(1 - v^2/c^2)^{-\frac{1}{2}},$$

and, using the binomial theorem of algebra, expand it into a series, discarding all higher terms, we have

$$M = M_0 + \tfrac{1}{2}M_0 \frac{v^2}{c^2},$$

and then solve this for the change in mass, $M - M_0$, due to velocity we find that this change in mass (ΔM) is

$$M - M_0 = (\Delta M) = \frac{T}{c^2},$$

RADIOACTIVITY AND TRANSMUTATION

where T stands for our usual expression for kinetic energy of motion $\frac{1}{2}M_0v^2$, in which we commonly use for the mass M_0 the mass we get by weighing the body when at rest.

IN ADDITION to showing how mass depends on velocity and, if velocities are high enough, to an *observable amount* in accordance with the theoretical prediction, the theory of relativity goes further and, by analogy, *identifies* mass with energy in such a way that the ratio of the latter, measured in ergs, to the former, measured in grams, is the square of the velocity of light (c) measured in cm/sec, i.e.,

$$E_{\text{ergs}} = M_{\text{grm}} \, (c_{\text{cm/sec}})^2.$$

Because of the large numerical value of c, 3×10^{10} cm/sec, the number of ergs of energy represented by a gram of mass is very great—9×10^{20} ergs, or 25 million kilowatt hours (see table below).

EQUIVALENTS IN MASS AND ENERGY UNITS

Atomic Mass Units	Grams	Mev's	Ergs	Calories	Kw-hr.
1	$=1.66\times10^{-24}$	$=9.31\times10^2$	$=1.49\times10^{-3}$	$=3.57\times10^{-11}$	$=4.15\times10^{-17}$
$6.06\times10^{23} =$	1	$=5.62\times10^{26}$	$=9\ \times10^{20}$	$=2.15\times10^{13}$	$=2.50\times10^7$
2.81×10^{10}	$=4.66\times10^{-14}$	$=2.62\times10^{13}$	$=4.18\times10^7$	1	$=1.16\times10^{-6}$
1.07×10^{-3}	$=1.77\times10^{-27}$	1	$=1.60\times10^{-6}$	$=3.83\times10^{-14}$	$=4.45\times10^{-20}$
2.41×10^{16}	$=4.00\times10^{-8}$	$=2.25\times10^{19}$	$=3.6\ \times10^{13}$	$=8.6\ \times10^5$	1
6.71×10^2	$=1.11\times10^{-20}$	$=6.24\times10^5$	1	$=2.39\times10^{-8}$	$=2.78\times10^{-14}$

NOW IT is again to be emphasized that this generalization as to energy manifesting itself in three mutually interconvertible forms as mass, motion, and radiation is purely theoretical and, at least with respect to the mass-energy relation above, cannot be said to have been experimentally established except in a limited number of cases so far. One of these cases occurs in the experiment of transmuting lithium into helium, described above on page 330*i*, and to which we now revert. There was a measurable *loss of mass* in this experiment. The mass of the original proton and lithium seven nucleus with which the experiment began, was in excess of that of the two helium nuclei (a particles) with which it concluded by an amount of 3.07×10^{-26} gms.

The a particles, however, we noted, hastened from the place of their production with such high velocities that they possessed, together, kinetic energies T, of 17 million electron-volts, i.e., 27.2×10^{-6} ergs. Now *these observed facts, within the limits of experimental error, fit very closely to Einstein's mass-energy relation,**

$$\underset{3.07\times 10^{-26}\text{ gm.}}{M} \times \underbrace{\underset{(3\times 10^{10}\text{ cm/sec})^2}{c^2}}_{\substack{9\times 10^{20} \\ 27.6\times 10^{-6}}} = \underset{27.2\times 10^{-6}\text{ ergs}}{E}$$

$$= 27.2 \times 10^{-6}.$$

Thus in this particular experiment we have convincing evidence of the applicability of Einstein's mass-energy relation, at least to this nuclear transformation.

IN THE equation $E = Mc^2$, E stands for energy of any kind, kinetic energy of motion, thermal or radiant energy (hc/λ)† and M is a change (either loss or gain) of mass, the "mass energy" of which is transformed into energy (either gain or loss) of motion or of radiation.

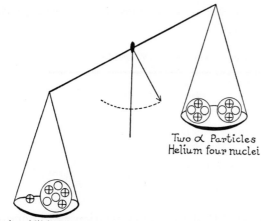

Two α Particles
Helium four nuclei

Proton Lithium seven

It is important, before we proceed further, to emphasize the fact that this interchangeability of energy between its three aspects— motion, radiation, and mass —is not a unique situation with respect to recent developments in nuclear physics. In chemical changes, for example, those associated with combustion, let us say, of carbon with oxygen to make carbon dioxide, the carbon dioxide molecule represents a more stable form of carbon and oxygen than that of these elements before they are combined. In accordance with the mass-energy relationship, the sum of the masses (C+2O) ought to be *greater* than the mass of the molecule of CO_2, but *so little greater,* as to be beyond any present possibility of measurement. Similarly, when T.N.T. explodes into less complex molecules, which are *more stable* than the T.N.T., this increase of stability means that the total mass of the products produced by the explosion should be less by a very slight amount than the mass of

---

* The figures given in this computation are actually those of Bainbridge, who repeated the Cockroft and Walton experiment later with higher precision.

† See chap. 35, where $E = h\nu$, and $c = \nu\lambda$, hence $\nu = c/\lambda$, and so $E = hc/\lambda$.

the original T.N.T. The mass lost is represented by the combined mechanical, thermal, and radiant energy exhibited by the explosion itself. In these two cases of chemical change only a few electron-volts of energy per atomic or molecular particle involved are released. These few electron-volts correspond to only about 10^{-27} gm. of mass loss. In these two cases, furthermore, only the *exterior* electron structures of the atoms suffer change—not their nuclei. In the so-called atomic bombs, however, *which would be better named* "NUCLEONIC" BOMBS, the atomic nuclei, as well as the outer electron structures, are altered, and from the nuclei energies of an altogether different order of magnitude (*millions* of electron-volts per particle) become available under certain circumstances—circumstances which the number of particles in the nuclei (protons and neutrons, often referred to conjointly as *nucleons*) control and to which the NEUTRON provides the trigger key.

Briefly, then, let us look at atomic nuclei with respect to the energies with which these combinations of protons and neutrons are held together.

IN OUR preceding discussion of isotopes, both natural and artificial, we have given them MASS NUMBERS that are integers, exactly whole numbers. For the relatively few isotopes whose masses have been precisely measured, these largely confined to the many of the lighter and a few of the heavier elements, the masses are *not exactly whole numbers*, nor are their differences exactly whole numbers.

The exact information that we have we owe largely to years of patient work chiefly by Aston, of Cambridge, England, Dempster, of Chicago, Bainbridge, of Harvard, and Nier, of Minnesota. Recent summaries give the following figures of isotopic weights on the arbitrary basis of choosing oxygen sixteen as being exactly integral. In the following table (p. 330n) we have divided the isotopic weight by the number of particles (neutrons and protons) in the nucleus and obtained a sort of average weight for these nucleons as they are packed together. We note that these figures *steadily decline in value*. This DECREASE OF AVERAGE MASS PER NUCLEAR PARTICLE continues until mass numbers around 65 are reached, *then slowly rises* to above 1.00 at mass number 200 (Hg).

These results, including a few heavier elements, are shown graphically in the diagram on page 330o. It is important to emphasize that the decline in the *average mass per particle*, which is the vertical value plotted against the mass number horizontally, represents energy that has gone into the nucleus or work that has been done by the forces that hold the nucleus together.

Elements in the vicinity of iron, cobalt, nickel, and copper (mass numbers near 60) have the most stable nuclei, since the binding forces are greatest here. Hence the largest amounts of energy (relative to other elements) are necessary to disrupt or tear such nuclei apart. At the left end of the graph, on the contrary, if neutrons and protons could be assembled *to make* helium

SOME ACCURATE ISOTOPIC WEIGHTS. AVERAGE MASS PER NUCLEON*

neutron	$_0n^1$	$1.00897 \div 1 = 1.00897$	
proton	$_1H^1$	$1.00813 \div 1 = 1.00813$	
deuteron	$_1H^2$	$2.01473 \div 2 = 1.00736$	
α particle	$_2He^4$	$4.00389 \div 4 = 1.00097$	note this exceptionally large drop
	$_3Li^6$	$6.01686 \div 6 = 1.00281$	
	$_3Li^7$	$7.0182 \div 7 = 1.0026$	
	$_3Li^8$	$8.0195 \div 8 = 1.0024$	
	$_4Be^9$	$9.01504 \div 9 = 1.00175$	
	$_4Be^{10}$	$10.01631 \div 10 = 1.00163$	
	$_5B^{10}$	$10.01671 \div 10 = 1.00167$	
	$_5B^{11}$	$11.01292 \div 11 = 1.00117$	
	$_6C^{12}$	$12.0040 \div 12 = 1.0003$	
	$_6C^{13}$	$13.0076 \div 13 = 1.0006$	
	$_6C^{14}$	$14.0077 \div 14 = 1.0005$	
	$_7N^{14}$	$14.0075 \div 14 = 1.0005$	
	$_7N^{15}$	$15.0049 \div 15 = 1.0003$	
	$_7N^{16}$	$16.0066 \div 16 = 1.0004$	
	$_8O^{15}$	$15.0079 \div 15 = 1.0005$	
arbitrary choice	$_8O^{16}$	$16. \div 16 = 1$	
	$_{10}Ne^{20}$	$19.99881 \div 20 = 0.999940$	
	$_{10}Ne^{21}$	$20.99968 \div 21 = 0.99985$	
	Cl^{35}	$34.9813 \div 35 = 0.9994$	
	Cl^{37}	$36.9788 \div 37 = 0.9990$	
	A^{40}	$39.97504 \div 40 = 0.99937$	

* Figures are from Bethe and Livingstone, *RMP*, IX (1937), 373.

nuclei (also very stable), the large drop in mass per particle indicates that a large amount of energy would be released in the process. This is precisely what was done indirectly in the Cockroft and Walton experiment described above (pp. 330*i* and *j* and diagram on p. 330*c*), in which the element lithium seven (mass 1.0026 per nuclear particle) had neutrons (mass 1.0089) added to their nuclei, which then ultimately disintegrated into 2 helium nuclei

RADIOACTIVITY AND TRANSMUTATION

(mass 1.00097) per nuclear particle, with the energy left over of 17,000,000 electron-volts in the kinetic energy* of the two products.

THESE LOSSES of mass-energy, equalized by energy of other forms, differ in no respect in *principle*, as we have said, from chemical changes which presumably would show the same phenomena, could they be measured. The difference between the two situations is *simply one of degree*, nuclear energy so released is about a *million times greater* than that of chemical change. Now with respect to the atomic bomb, one of the elements used in it

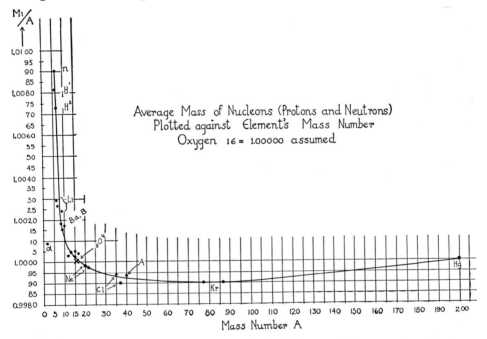

lies at the extreme right of the graph—indeed, beyond its range in the above drawing. Actual isotopic masses in this region have not yet been very accurately measured. However, the curve here is at least one division (0.001 atomic unit) above its minimum point at iron. This means these heavy nuclei are at least 0.001 mass unit per particle less stable than iron. Furthermore, there are *many nuclear particles* in uranium, 234–39 in different isotopes, so that the mass loss per nucleus is of the order of 0.23 (234×0.001) of an atomic unit. Now when *the uranium nucleus is broken into two approximately equal fragments*, these lie not so far as iron toward the curve's mini-

---

* In the expression $\frac{1}{2}mv^2$ for kinetic energy at high velocities v, the mass, m, also increases (relativistically) as indicated on p. 330j.

mum, but as far as the elements iodine, xenon, barium, tellurium, etc.—mass numbers 130–40. Hence energy of at least 0.20 mass units per atom is transformed into mechanical, radiant, or thermal form. Now

$$0.2 \text{ mass unit} = 196. \text{ mev} = 8.3 \times 10^{-18} \text{ kw-hr} .$$
$$(= 1.66 \times 10^{-25} \text{ gm.})$$

Since there are atoms to a number of the order of 6×10^{23} in a number of grams equal to the atomic weight, the energy released by 238 gm. of uranium so tranformed into elements of mass numbers 130–40 is about

$$6 \times 10^{23} \times 8.3 \times 10^{-18} = 5 \text{ million kw-hr} .$$

This figure times the number of seconds in an hour (3,600) gives $5 \times 10^6 \times 3600 = 1.8 \times 10^{10}$ kw-sec. Furthermore, in a nuclear change such as that produced in an atomic bomb, this energy is released in a *fraction of a micro second* (10^{-7} sec. or less), that is, *at a rate* of the order of 10^{18} horse-power.

LITTLE wonder, then, that it is said that the flash of light from an atomic bomb explosion had a brightness estimated to correspond to a temperature *far exceeding* that of the sun's interior, 20,000,000° (sun's surface, 6,000°), or that other effects of the violence of a release of this amount of energy in so short a time are proportionately great. Our simple calculations were based on a mass loss due to the transmutation of but 238 gm. of uranium. What amounts of what materials were actually used have not been disclosed.

From the official Smyth report* we learn, moreover, that, in addition to the fissionable isotope of uranium, U^{235}, separable by magnetic and other means from its hundred and forty times more abundant U^{238} only with great difficulty, an isotope of the new element, $_{94}Pu^{239}$, is also fissionable. Pu^{239} is formed as follows:

$$(_{92}U^{238}) + _0n^1 \rightarrow {}_{92}U^{239}) . \qquad (_{92}U^{239} \xrightarrow{\beta-} {}_{93}Np^{239})$$

the abundant uranium isotope + neutrons → a new unstable isotope of uranium. This decays ($\beta-$)-wise into a new element → neptunium.

$$_{93}Np^{239} \xrightarrow{\beta-} {}_{94}Pu^{239} \xrightarrow{\alpha} {}_{92}U^{235}$$

This isotope of neptunium, element 93, likewise decays into the *fissionable* isotope of plutonium; this *very slowly* decays α-wise into U^{235}

---

* H. D. Smythe, *Atomic Energy for Military Purposes* (Princeton, N.J.: Princeton University Press, 1945).

This fissionable plutonium, like uranium, is so slowly radioactive (half life, 24,000 years) that it "stays around." Being chemically different from its uranium grandparent, it can be separated therefrom by chemical procedures.

In summary, then, we have two fissionable substances—U 235 and Pu 239—from the fission of which large amounts of energy are released. For the fission of both of these materials a *supply of neutrons* is necessary. Also for the production of Pu 239 in quantity from U 238 a *continuous supply of neutrons is necessary.*

THIS IS precisely where Nature, impersonal and inscrutable, has provided men and women with the key to the door of a new age of technology, which may either be their undoing or lead to vast enrichment of their opportunities. When fission occurs in these heavy nuclei, not only are there fragments of much lighter elements formed, but AN EXCESS of from ONE to FOUR NEUTRONS is LEFT OVER. Thus a supply of neutrons is left in the material after the first fission occurs to provide other fissions, and the possibility of a self-perpetuating chain reaction is offered. To effect this chain reaction has been by no means a simple task; indeed, it was a major problem of the great group brought together for war research in this field.

IN A MIXTURE of different isotopes of uranium there are all sorts of competition for the stray neutrons—impurities of some kinds swallow them up with such avidity that these must be eliminated to the order of one part in a million or less. Whether neutrons produce fission in some of the uranium isotopes or are simply absorbed in some, forming other isotopes and other elements, depends on the speed of the neutrons. No small-scale experiments can be undertaken, since, unless a certain critical amount of material is assembled, the excess neutrons initially produced at very high speeds escape from the material altogether.

To produce both uncontrolled chain reactions for bombs and controlled chain reactions for plutonium production and, incidentally so far, continuous sources of large power was a tremendous and complex enterprise which space limitations prevent discussing here. Many aspects of the work are still under the censorship of military security. For such as have been disclosed the official Smyth report, now published in book form, should be consulted (*loc. cit.*, p. 330*p*).

IN CONCLUDING this part of our exposition, which has taken us from the dim beginnings of nineteenth-century evidence that suggested a close relationship between electricity and material substances to the astonishing pictures of the details of this relationship that form the working hypotheses of the scientific world as the close of the first half of the twentieth century ushers in the atomic age, we have to call attention to one of the most remarkable discoveries of very recent times. This discovery, which originated independently in a great industrial research laboratory in the United States and in the cloistered halls of a great British university, has to do with the wavelike characteristics exhibited by things that do not even remotely resemble anything undulatory except under the most searching line of attack. Electrons, protons, α particles—indeed, everything atomic or corpuscular—*have* wavelike characteristics. Nor is this all.

In the great field of physics that began to be studied in the dawn of civilization, in Egypt, Greece, and Arabia, that field that had to do with the phenomena of light, the nineteenth century had every reason to be complacent. The true wavelike character of light had been established beyond the possibility of challenge. The twentieth century saw the wavelike characteristics of X-rays also proved beyond reasonable doubt. But the last twenty years have shown that waves of radiant energy, X-rays, ultra-violet, visible, and infra-red light, and the longer electromagnetic waves, presumably have in addition to their wavelike characters another side, and one which is atomic and corpuscular in every sense. Before such statements as these can have any meaning to you at all we must embark upon a story that has to do with waves and wavelike characters entirely.

PART V
WAVES AND RADIATION

Photo, Florence Hendershot, Chicago

PLATE 72 ALBERT A. MICHELSON

CHAPTER 35

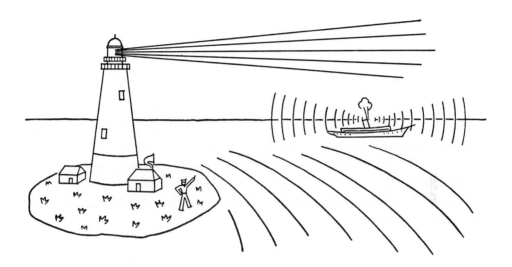

ABOUT WAVE PHENOMENA IN GENERAL

QUITE true to our prediction at the beginning of Part II, and as you have observed for yourself throughout Parts III and IV, nearly everything in physics begins with mechanics. This is true still even as we approach a subject that deals with what might well be thought of as the most intangible phenomenon in the physical universe, the phenomenon of light. Yet even light has mechanical properties; it carries momentum in its silent passage. Energy, too, is there, instantly converted into heat—molecular kinetic energy—in any object that it falls upon. Light is still to be classed among the most mysterious of nature's marvels; it is a subject that has provoked more discussion and controversy than anything else in the realm of physics. Nevertheless, light has been not only the greatest revealer of the more obvious and superficial aspects of the physical world, but its study has rewarded the scientific observer with a flood of illumination in what had hitherto been some of the darkest recesses in which nature's secrets are concealed.

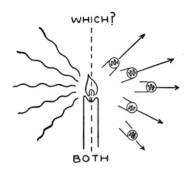

The last ten years have seen complete confirmation of suspicions that were taking form at the dawn of this century—suspicions that light was even more strange and difficult to comprehend than we had in our earlier ignorance supposed. It has two different aspects—one wavelike or undulatory, the other corpuscular—that were formerly regarded as mutually exclusive. As we have begun to see this, we have also begun to see that matter, too, possesses the same dual attributes. Our investigations into these things have but begun. What we do not ourselves understand any too well we cannot, naturally, explain fully to others. Possibly we may be able to suggest, in a few paragraphs, however, something of the interesting character of this situation, the solution of which is yet to be discovered.

IN PRECEDING chapters we have traced the development of our modern convictions that the world is essentially atomistic. Matter has been discovered to be of discrete character; not only to a first degree of subdivision, into molecules, but to the second, into atoms of the elements, ninety-two of which by their manifold combinations unite to produce all things inorganic or organic, lifeless or living. A third degree of subdivision has been revealed by our finding that electricity likewise is of atomic character, and that there are at least three, and possibly several more, combinations of these electric entities that combine as building-blocks to compose the larger systems of the atoms.

To this general characterization of atomicity, light also conforms. Like everything else it has corpuscular, atomic characteristics. Its individual packets are called PHOTONS; and the name, as representing a class of atomic characters, is accepted quite as universally and with as little dissent today as are any of the other words that have appeared in small capital letters throughout our pages. Their symbol denotes wave and corpuscular characters.

The atomic character of light, however, has been one of the most recent of our discoveries. It is not an obvious aspect of ordinary light. Perhaps you may feel that atomicity is not, after all, a very obvious aspect of ordinary matter; possibly you are not yet entirely convinced that we really know as

WAVE PHENOMENA

much about these things as we think we do, that you have been carried along by arguments that you only dimly understand and are, of course, quite powerless to refute, lacking scientific tools and training in their use.

But the case for the atomism of matter and electricity is, after all, far from being weak or unconvincing, even to the layman—at least on second reading.

The case for atomicity of light is, however, of a different order of difficulty altogether. Indeed, in these pages we are not going to be able to present this case at all; only to suggest it, assume it, or in exception to our chosen policy to ask you to take it entirely on faith. There is a corresponding situation with respect to matter likewise. MATTER—atoms, molecules, electricity, electrons, protons, alpha particles and all the rest—exhibits a group of properties that until the most recent times we never even suspected. Indeed, when first we had the suspicion rammed down our reluctant throats, the idea almost made us sick, because the properties were those of *continuous* and not discontinuous entities; they were ones that seemed *incompatible with atomicity*. The properties in question were those of WAVES.

Wave properties are as obviously properties of light and of radiation in general as atomic or corpuscular attributes are seen to be inevitably those that belong to matter and to electricity. But that light had atomic characteristics seemed nonsense until it appeared that matter and electricity had wave characters. Two wrongs never made a right, perhaps, but two reciprocal dilemmas sometimes result in clarification from the light they shed on one another.

WHAT, then, are these attributes that we call wave characters? Are they to be comprehended in language other than that of differential equations or at least of algebra, which is the common speech of physicists? We think they are.

We will begin as directly as possible with objects having a simple periodic motion; but although the motion is named SIMPLE HARMONIC, it is not quite as simple as those other motions we discussed in Part I. The harmonic cognomen really is well chosen, for the phenomena and principles involved are the fundamental physical ones behind sound and music as well as being fundamental to the study of light. The word "vibratory" carries the idea.

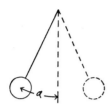

Consider the swinging of a pendulum. The problem was one of the first that attracted Galileo. He was the first to notice that the period of swing or the time of one single (or double) swing did not depend, as one might have supposed on the extent of the excursion made by the object swinging to and fro. He experimented and found that the time of one oscillation decreased only as the *length* of the pendulum was shortened.

Later more precise work confirmed this discovery; and the presence of another factor, the strength of the earth's attraction, g, was likewise noted. The ubiquitous "π" also put in an appearance, but with the exact formula* we have no present concern; yet some terms that we shall be frequently using in reference to the swinging pendulum must be defined.

The PERIOD, T, is the time of one complete vibration to and fro.

The AMPLITUDE, a, is the distance from the mid-point, lowest in the arc of travel, to the farthest limit of its excursion on either side.

The FREQUENCY, n, is the number of complete vibrations made in 1 sec.†

BUT what, you ask, have pendulums to do with waves? Suppose we have a row of pendulums all alike, and all at rest, and that we start them all swinging by running a light rod with a constant touch and uniform speed across them as if we were running it along a picket fence. In Plate 73 you can see a snapshot of the result a short time afterward. There is no motion to be seen in a snapshot, of course; but you can see that the bobs of the pendulums (or better, the white paper circles at the ends of the short projecting arms, one of which is fastened near the top and perpendicular to each pendulum arm) are momentarily in the form of a sort of "wavy" line of the kind which you yourself would naturally draw on the blackboard if some one asked you to draw the picture of a wave. Your imagination must supply the motion, which in this case is from right to left, since this was the

$$* \; T = 2\pi \sqrt{\frac{l}{g}},$$

where T = time in seconds of one complete swing, l = length of pendulum from point of support to center of bob, and g = acceleration of gravity.

† The pendulum of a grandfather clock marks seconds; in long pendulum clocks, a *half*-swing takes place in exactly 1 sec. if the clock is perfectly regulated. This period, as defined above, is 2 sec., and its frequency is one-half vibration per second. Obviously, period and frequency are reciprocal relations, $T = 1/n$, $n = 1/T$.

WAVE PHENOMENA

direction along which we moved the disturbing stick that set the pendulums all into motion.

We have shown you here a model only of a wave in order that we may analyze it a little further and define two or three other words. As a mechanical device this contraption carries no energy along with it, as do all real waves. This is because it lacks one important element which later we will supply to make a real wave out of it.

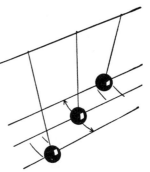

Before we do this, however, let us ask ourselves what it is that makes the appearance of a wave. We have nothing but a lot of pendulums. First of all they are all alike and swing with the same period. From the manner in which we set them going (by drawing the starting-rod across them with a uniform speed), they do not swing in unison, however. During their motion as they continue swinging each one has a definite relation to its neighbors on either side in being a little behind one of them, the one on the right, and an equal amount ahead of the other on the left with respect to the moment of passing through the lowest spot in its arc of swing. They are equally and regularly "out of step," as it were. We say they are out of PHASE with each other, and that there is a constant PHASE DIFFERENCE between adjacent individuals. These two features together, *their common periods* and *their constant*

← *Motion to the left*

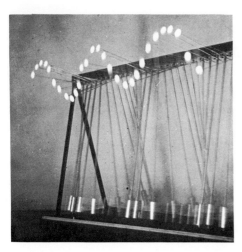 

PLATE 73 A Transverse Wave Machine (Stereophotograph). Undamped wave-form. No energy transmitted here, since pendulums are entirely unconnected.

phase differences, are what give the wavelike character to this motion of a group of pendulums.*

We could equally well have obtained a wavelike result if we had so manipulated our stick that those at one end were given a large amplitude of swing and those at the other end a small one. If this change had been done with a certain type of regularity, we would have produced a wave form that gradually got less and less in magnitude in the direction in which the wave form was moving. This we would call a DAMPED wave. One is illustrated later, Plate 75. It is the type that results whenever friction is present in a mechanical system that produces waves, or whenever in the case of any wave the energy it carries is gradually lost en route so that the wave becomes progressively enfeebled and finally disappears entirely.

ANOTHER way of describing a wave is to say that it is the motion *of a form* or *of an arrangement* which is assumed by successive portions of the medium through which it travels. In the case of water waves, at least in deep water, it is not the water itself that moves forward. If it were, it would all pile up on the lee shore. It is a form of disturbance on the surface that moves forward under the action of the wind or the commotion caused by a ship moving across the surface. The water moves, of course. A chip bobs around, up and down, and forward and back, as the waves pass under it. It may be blown forward by the pressure of the wind upon it, but this is quite a separate matter.

THE motions of our set of pendulums or of the markers attached to them is quite obviously at right angles to the direction in which the wave-form, that they have been made to assume, travels. If we shake one end of a taut rope, the other end of which is fastened to some solid object, the same thing is true. The rope moves up and down following the motion of our hand, but the resulting waves in it travel along the rope away from us. Waves of this sort are called TRANSVERSE. The motion of the individual particles participating in the wave motion is in this case at right angles to the direction of motion of the wave. Light waves are of this character.†

* As we started them in motion, they all were given also the same amplitude of swing; but this is not essential.

† The evidence for this statement is found in the study of the polarization of light, which is described in any extensive elementary textbook on the subject.

WAVE PHENOMENA

But the same is not true of that other class of waves that stimulate our other major sense, the sense of hearing.

Suppose, for example, we had a row of balls, all alike, each hung on double threads all of the same length so that they could not swing

PLATE 74 Longitudinal Wave. Photograph of a model of a wave in which particles oscillate to and fro horizontally in the same direction that the wave travels. Compare the clustering (condensations) of particles in some locations and their separation (rarefactions) in others with similar perspective effects of pendulum *rods* in Plates 73, 75.

transversely but only to and fro in the direction of the row. We might, furthermore, imagine them all to be connected by small coil springs. If, now, we should pull aside the one at the end of the row, the spring would be stretched and it would in turn pull aside the next one, and then the next, and so on down the line. We have at once satisfied the two conditions for the wave: (1) the periods of all the balls as they swing are the same, since the lengths of the supporting threads are all alike; and (2) there is a constant phase difference in their motions. This latter is true because of the inertia of each ball. The second one begins to move, not the moment the first one is disturbed, but an instant later than the first; also, the second moves an instant before the third, and the constant phase difference of the motion of any one with respect to the two on either side of it has been provided. A wave will travel, then, along the row. The displacements of the

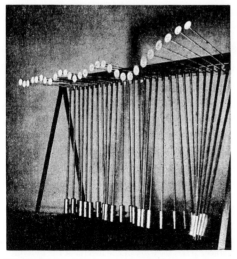 

PLATE 75 A Damped Wave (Stereophotograph). Transmitted from right to left. Pendulums connected by a rubber strip (not seen).

individual balls in this case are in the *same direction* as that in which the wave progresses. Such a wave is called LONGITUDINAL and is of the type of which sound waves are another example.

In the illustration Plate 74 we show a photograph of such a type of wave-form, but no true wave. In the case we have been describing, however, we have a true wave traveling along the row of balls. We disturbed only the first one with our hand. The elasticity of the spring transmitted the motion of the first on to the second, and the inertia of the second caused the delay that resulted in the constant difference in phase of the motion of these two. The same is true all along the line. Once something is started at one end, what takes place farther down the line is a result of the mechanical construction of the apparatus. The stiffer the elastic character of the springs and the less the inertia of the balls, the smaller the phase difference between any two and the greater the VELOCITY of the wave. This is in general true of waves of any kind.*

WE SHALL now proceed to add the elastic element that was lacking in the first model we employed, the transverse model, so that it will cease to give us just a semblance of a wave but will instead show us the propagation of a real wave along it (Plate 75). On the back of each pendulum, near the top is a short horizontally projecting pin. We have provided a long strip of rubber which has holes punched in it like a belt, so that it can be slipped over these projecting pins and thus provide an elastic connection from one pendulum to the other.

Having put the rubber strip in position, we now take hold of and disturb *only one* of the pendulums, perhaps the end one on the right. As we displace this and let it go, it disturbs the second through the agency of the rubber strip, and the second the third, and so on. A wave, transverse in this case, now travels down the line. When we introduced the rubber strip we introduced not only an elastic connection, but we greatly increased the friction in the whole apparatus. That there is friction within the rubber can be ob-

---

* The VELOCITY OF PROPAGATION of any wave is given by the relation

$$V = \sqrt{\frac{E}{\rho}},$$

where the medium through which the wave travels can be identified as a material substance. E is a measure of the elastic property responsible for the propagation of the wave, i.e., is the elastic restoring force against which one works in starting the original disturbance. ρ is a measure of the inertia of the individual particles involved; or, if these are too small to be apparent, it is the inertia (mass) of equal small volumes of the medium that transmits the wave.

WAVE PHENOMENA

PLATE 76 A Wave Going in Two Directions (Stereophotograph). Damped wave progressing outward in each direction from the center where it was started. Pendulums connected by a rubber strip.

served, because if it is twisted and untwisted vigorously it becomes sensibly hot.* Furthermore, the friction within the pivot supporting the pendulums is greatly increased by the side forces brought to play upon them because of this rather crude method of providing an elastic connection. As a result of friction, the wave that travels down the row of pendulums is strongly damped (see p. 338), the amplitudes of swing rapidly decline, and very little wave is to be seen at the other end of the apparatus.

Furthermore, if we disturb the middle pendulum of the row instead of one of the end ones, now that we have the elastic connection and thus a true wave-machine, we observe that a wave travels out IN BOTH DIRECTIONS. This is true for any medium that has continuity of character. If the medium is linear in character, like this apparatus or like a taut rope, a wave travels in opposite directions each way from the point of disturbance. If the medium were a stretched surface of rubber, the wave expanding from the point of disturbance would be of circular form. Ripples proceeding from where a stone has disturbed the surface of water are a well-known example of this.

* The heating of automobile tires, especially if they are underinflated, is an excellent example of this. It is usually supposed that they get hot from contact with the pavement. As a matter of fact, they get considerably hotter than the pavement because of their internal friction (viscosity), against which they are continually distorted from their original shape by the rolling of the car upon them.

If the medium is continuous in three dimensions of space, like a body of water with reference to some point in the interior of it, or like the atmosphere, or if it is a solid object of considerable extent, the waves that proceed from a point of disturbance are spherical in character if the body is homogeneous. If the medium is not homogeneous, these simpler geometrical forms may be very greatly distorted.

IN THE case of our model, disturbed at the middle, we notice that the wave that moves out in each direction is likewise strongly damped because of friction.

The energy we put into our wave-machine at the point of disturbance has not been carried forward very far in this case. It has all been dissipated by heat, thanks to the enterprise of the little imp from which we had some amusement at the beginning of chapter 11.

If it were not for friction, or if the friction were very small, it is quite easy to see that energy could be carried long distances by a wave from one place to another. Kinetic energy of motion put into the medium at one place is handed along to regions more and more remote. If the wave spreads out in all directions like the widening circles from a stone thrown into a pond, this energy becomes enfeebled also by being distributed continually over a larger and larger circle.

In all mechanical waves, in sound waves, and in waves on the surface of water, the progressive enfeeblement is due both to the spreading of the wave and to friction. In the still rather mysterious phenomenon of light the enfeeblement of intensity due to the spreading of the wave to greater and greater distances is also well known and easily to be observed. But there is another aspect, a corpuscular aspect, to light, as we have told you in our

introductory paragraphs. This corpuscular aspect is recognized partly through the fact that certain energy characteristics of the light are *not* diminished in the process of spreading out. These characteristics of the energy of light are just as strong and powerful in the light that is received on the earth from atoms in a star in the great nebula of Androm-

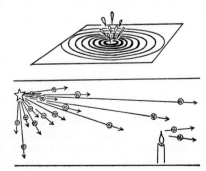

eda* as they are in the immediate vicinity of those atoms in the atmosphere of that star whence the light originated. You probably do not understand what we have been saying very well—the energy represented by the intensity of the light gets less and less with distance but some other aspect of radiant energy does not. How confusing this all is! You are quite right. We ourselves and all other physicists have been much confused about it too. The aspect of light that is unchanged by distance is related to the *color* of the light, of which more anon.

IN ADDITION to transverse and longitudinal waves there is another very common kind of wave about which we shall have quite a bit to say—a kind of wave that is neither longitudinal nor transverse; indeed, it is *both*. We called your attention, a few pages earlier, to the fact, with respect to waves on the surface of water, that the motion of the water, as indicated by that of a chip upon its surface, was up and down and to and fro. If you move your hand at one and the same time up and down and to and fro you would perforce be moving it in a circle or an ellipse or back and forth in a straight line inclined at an angle to the horizontal. The motion of the particles (any given small volume) of water is either in circles or ellipses. In only one very special case, that of the water at the very bottom of a shallow shore on which waves are rolling in, is the water's motion longitudinal. In all other cases the motion is circular or elliptical with both a transverse (vertical) component and a longitudinal (horizontal) one.

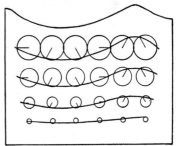

Form of disturbance in deep water waves at surface and below it

Every strong swimmer who loves to swim out and in through a breaking surf knows this. As one swims in through a surf he is both lifted up and carried forward by the wave that overtakes him. As the wave passes, he is lowered and retarded in his forward progress. The momentum imparted by the forward-falling, breaking crests may ultimately cast him up upon the shore, but otherwise the wave contributes no

* From this spiral nebula, light, traveling at the rate of 186,000 mi. per sec., takes 950,000 years to make its passage.

forward progress. What the crest (if it does not break) contributes, the trough that follows takes away.

Some further description, a little more detailed, of the appearance of wave forms is needed for future use and for present better visualization.

A SIMPLE transverse wave in section is a regular "wavy" line as we have seen. A fairly simple trigonometric formula describes it,* and for this reason these curves are often called SINE CURVES, because of the name of the trigonometric function used in the formula.

WITH respect to longitudinal waves the *appearance* is very different. Instead of crests and troughs, since the motion of the particles is not across but along the direction of wave travel, there is a bunching of particles together, instead of a crest, and a spreading of them apart, instead of a trough (look at Plate 74, p. 339, again). These groupings are called in a purely descriptive sense, for want of any better terms, CONDENSATIONS and RAREFACTIONS, respectively. If the particles are too small to see, and the medium from our gross-scale point of view seems to be continuous, or is, as in the case of air, entirely invisible, there are, nevertheless, pressure differences to be found by direct observation within the wave. Corresponding to the crest on a transverse wave, or to the bunching of particles in the condensation in a mechanical longitudinal wave, there is in the condensation of a sound wave in air an increase of pressure above the normal average atmospheric pressure at the moment. In the rarefaction of the sound wave, there is a corresponding reduction of pressure below the average. Indeed, by methods of shadow photography by means of electric sparks it is possible to record directly on a plate or film the actual forms of sound waves in motion.†

* $y = a \sin 2\pi \left(\dfrac{t}{T} - \dfrac{x}{\lambda} \right)$,

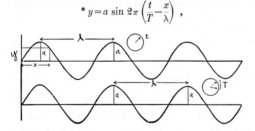

where y is the displacement of any particle at a distance x from some chosen origin, and at a moment t, by the clock; a is the amplitude of the disturbance; T is the full period of the wave, i.e., time between the passage of the crest of one wave past a given point and that of the next wave crest; and λ is the wave-length, the distance from one wave crest (or trough) to the next one.

The velocity V is given by λ/T. A $+$ sign in the ( ) means the wave is going in the opposite direction.

† See illustrations in Knowlton, *Physics for College Students* (McGraw-Hill, 1928), pp. 249, 250.

WAVE PHENOMENA

THE waves on the surface of water are, of course, visible; but they are usually very complicated by wind and tide and "ground swell" (waves that have come from a great distance). As observed in nature they are usually mixtures of waves of all sizes and periods, also of waves that may be going in several different directions, as they may have their origins in several different places. We can not make out much about their essential form unless we go to sea and, on a calm day after quite a blow, observe them at some distance from the commotion caused by the ship. Then we discover the characteristic form is this.

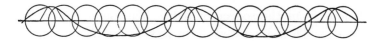

The shape betrays a character of the disturbance that is at once both longitudinal and transverse. Combination of condensation *and* crest makes for a narrow, lifted crest. The combination of rarefaction and trough makes a broad, gently sloping valley several times wider than the crest. Many people are sufficiently familiar with this form* to have a drawing of it say "deep-water waves" to them.

With these pictures before us, the two or three remaining definitions of the language we shall have to use in future pages become easy.

By WAVE-LENGTH, λ,† we mean the distance from crest to crest or from trough to trough in the "sine curve" that represents simple transverse waves. In longitudinal disturbances it is the distance between two condensations or between two rarefactions, if these can be directly observed. If not, it is the distance between two regions of maximum pressure or of minimum pressure. In water waves it is the distance from wave crest to wave crest, or from trough to trough at mid-points, the latter being rather more difficult to see and determine because of their broad, flat character.

WHEN we throw a stone into quiet water, we see that the disturbance reaches in equal times to regions equidistant from where the stone fell in. We infer this really, for what we see is an ever widening *circle* which indicates that the water wave has the same speed in all directions over the surface. We often call that assemblage of points in space which the wave disturbance reaches at the same moment by the name WAVE-FRONT.

---

* Which approximates to the mathematical curve called a "cycloid."

† Cf., also, drawing at bottom of opposite page (p. 344).

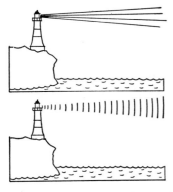

The wave-fronts of waves caused by commotion at a point on the surface of water are expanding circles. This, at least, is clear. In still air, uniform in temperature, sound wave-fronts are expanding spheres.

When we draw conventional pictures of light proceeding from a source, we would probably never think of drawing the wave-fronts unless we were quite sophisticated in these matters. We draw, instead, a few radial lines out in all directions from the source of light. This says "source of light" in picture-language. These lines we call the RAYS. Any phenomena of transmission, reflection, refraction of light or sound, or any other kind of wave disturbance may be geometrically described by a SYSTEM OF RAYS or equally well by a SYSTEM OF WAVE-FRONTS. These two systems of lines (or lines and surfaces, in space) are always mutually perpendicular to each other. Detailed discussion and illustrations we will not enter into here. We shall recur to these ideas quite naturally in connection with the discussion of certain properties common to all waves in the next chapter.

CHAPTER 36

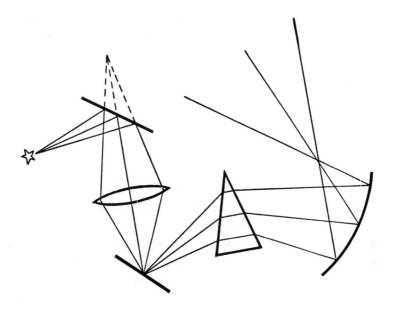

THE SPECIAL ATTRIBUTES OF WAVES

PERHAPS the most obvious property of any kind of wave disturbance is the fact that in a homogeneous medium it travels in straight lines. We have already met this fact in connection with wave-fronts and rays. The rays are usually straight lines; the wave-fronts, concentric circles on a plane surface or, in space, concentric spheres.

To observe water waves at their best, we suggested in the preceding chapter that you go to sea. This was not really necessary because great deep-water waves are not the only kind of wave disturbance on the water. Perfect water waves of the *ripple* variety may easily be produced on the surface of a liquid, and with a suitable means of illumination they may be observed in great detail. Such observations will tell us nearly all that ocean waves can tell and some things that the latter cannot. Special illumination, however, is necessary to observe these ripples, because they travel so swiftly across the surface of water that the eye cannot readily follow them.

The velocity assumed by waves on the surface of a liquid is rather curious. Very tiny waves travel very swiftly, and very large waves likewise move with express-train speed. Out in the Pacific Ocean, waves are some-

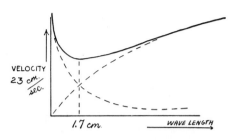

times encountered that run about five to the mile*—thousand-foot wavelengths. Such waves travel at a rate of about fifty miles an hour; and when they roll in, onto a gently sloping beach, surf-board riding of them is a thrilling sport, not to be practiced by the unwary. Natives of the Hawaiian Islands, and not a few Americans and Europeans, perform this feat with enviable skill. No motor boats are needed. A long flat board, rounded underneath, suffices. If one can get it properly placed in front of an advancing comber and then quickly raise himself, first to a kneeling, then to a standing position on it, keeping the board meanwhile at just the proper tilt, one will travel forward at a furious pace just ahead of the crest of the wave as the wave rolls in.

Smaller waves on the surface of water travel more and more slowly, until a minimum speed of 23 cm. per sec. is reached by waves about 1.7 cm. in length.† Waves smaller than this move more rapidly again, and the very small ripples we wish now to exhibit are altogether too swift in their motion to be seen at all, except by very special methods.

A METHOD most easily employed is to view the water surface by intermittent illumination so timed that it has about the same frequency as that of the waves themselves. By "frequency" you recall we mean the number of waves that pass a given point per second. Now, if we illuminate the water with a flash of light that has *exactly* the same frequency as that of the waves, any single flash will show the wave pattern almost motionless; but it will not endure long enough for the eye to take in any amount of detail. If the *next flash* comes as the *next wave* comes by the point at which we are looking, it will show this second wave right where the first one was before; and so the succession of flashes will catch a succession of waves each at the same point as its predecessor, and the whole pattern will appear motionless. If the flashes come faster than about 16 per sec., the eye will not be able to see even the

---

* Another definition, important in some connections is the WAVE NUMBER, or number of waves in some unit of distance as here, 5 per mi. The wave number times the wave velocity is the frequency, i.e., number per unit *time*. For example, with 5 waves per mile moving at 50 mi. per hr., 250 waves pass any fixed point per hour.

† This, indeed, is not a new idea to you if you stop a moment and try to recall whether you ever saw a wave crawling at a snail's or worm's pace along the surface of water.

THE SPECIAL ATTRIBUTES OF WAVES

flicker of the light.* It is important, also, to use simple and uniform waves, so that each one is just like every other one. By this means we can have before us the entire phenomenon, frozen, so to speak, and absolutely motionless in appearance (Plate 77).

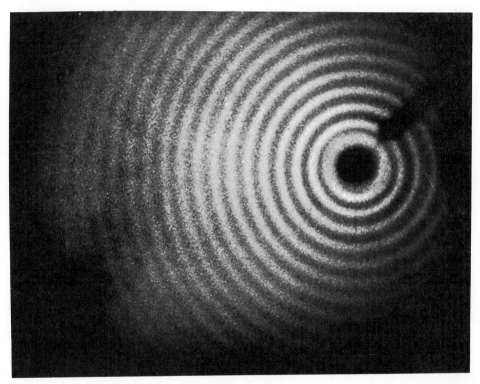

PLATE 77 Ripples on the Surface of Water Photographed by Intermittent Illumination of the Same Frequency as the Waves (Stroboscopic Illumination). Note the perfectly circular wave-fronts indicating that the **velocity of** the waves is the same in all directions.

If the sequence of flashes is a trifle faster than the wave frequency, the wave next succeeding any particular one will not have had quite time to move into its predecessor's position, and the whole pattern will appear to be moving slowly backward. On the other hand, if the flashes come a trifle too slowly, the pattern will appear to move slowly in a forward direction.†

* It is well known that in the projection of motion pictures the light is on the screen intermittently. No flicker is perceived, however, if the flashes are sufficiently rapid (over 16 per sec.).

† A similar result is sometimes seen in movies where the wheels of a car or wagon moving forward appear stationary or going backward.

This trick of illumination is very common in scientific and engineering practice. It is called by a difficult name, STROBOSCOPIC ILLUMINATION. "Stroboscopic" must not be confused with "stereoscopic," which applies to many of the binocular photographs in this book; nor should either of these two words be confused with "stereopticon," which is a sort of "magic lantern" for the projection of transparent slides upon the screen, the forerunner of the motion picture. It is possible to have stereoscopic pictures, projected stroboscopically by a stereopticon. If you can understand that sentence, you will have no further difficulty with these three words!

TO RETURN now to our ripples which we produced electrically on the surface of very clean water, then projected stroboscopically in a stereopticon, and finally have photographed to show you—you will note first of all the very perfect circular wave-fronts (Plate 77). The waves travel in straight lines across the surface and with equal speeds in all directions.

If we place a piece of glass, coated with paraffin, in the water and projecting above the surface, the waves will be REFLECTED. You can see the reflected wave-fronts traveling back in the opposite direction just as if they had emerged from a point as far behind this "mirror" as the source of waves is in front of it (Plate 78). Of course there is no source behind the mirror any more than there is another person very much like yourself behind a

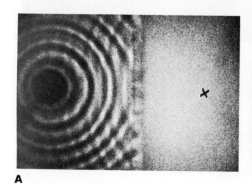 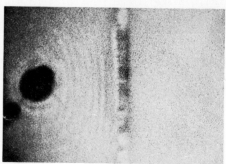

A   B

PLATE 78 (A) REFLECTION OF RIPPLES. Ripples on the surface of water proceeding from a source at the left are reflected from a plane mirror. Note reflected wave-fronts proceeding as if from the virtual image at **X**.
(B) STATIONARY WAVES ON WATER PRODUCED BY REFLECTION. (See p. 365.) Same experiment as the foregoing except as seen by continuous, instead of interrupted (stroboscopic), illumination. The progressing wave-trains are invisible, but the stationary pattern appears.

THE SPECIAL ATTRIBUTES OF WAVES

mirror when you stand in front of it. It is only that the reflected wave-fronts proceed as if from such a source. We call this apparent source a VIRTUAL IMAGE of the object in front.

OUR experiment is exhibiting wave-fronts rather than rays. But the latter may be found by drawing the series of lines that are everywhere perpendicular to the latter. Here these are, of course, the radii of the concentric circles that form the wave-fronts. If we take any one of these lines and follow it from the source of disturbance to the mirror surface and then away again along the other set of reflected wave-fronts, we will find that the angle it makes with the perpendicular to the mirror, as it approaches the latter, is exactly equal to the angle it makes with this same perpendicular as it recedes. On recession, of course, it lies on the opposite side of the perpendicular.

Our ripples have here illustrated a very important principle of optics. Whenever a ray of light strikes a mirror, the angle of INCIDENCE equals the angle of REFLECTION. This fact, true of any kind of waves and of any kind of mirrors that reflect them, is thus a perfectly general property.

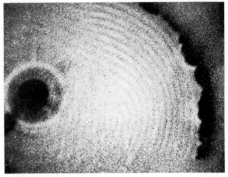

A　　　　　　　　　　　　　　　　　　　　　　　　　　　　B

PLATE 79　(A) REAL IMAGE (**X**) FORMED BY REFLECTION OF RIPPLES IN A CONCAVE MIRROR. Reflected waves converge upon (**X**). Stroboscopic illumination.
(B) STATIONARY WAVES IN FRONT OF A CONCAVE MIRROR. Interference pattern of stationary waves produced by two wave-trains shown above (see p. 365).
The set-up is unchanged in these two photographs; only the illumination has been altered. (A), intermittent (stroboscopic); (B), continuous.

If the mirror is hollowed out ("concave" is the word, caved toward the source of the waves), the principle we have just outlined results in the possibility, under proper conditions, of the rays being brought to a "focus," i.e., of the wave-fronts being caused to converge upon some point. A concave mirror is used in the next illustration (Plate 79) for the ripples, and it is seen that they converge upon the region marked X. It is further seen that, having converged, they diverge again beyond this point just exactly as if the latter were another real source of disturbance like the original one.

The real and intense disturbance here at this point of focus is called a REAL IMAGE of the original disturbance. In another plate (Plate 80) we have photographed an electric-lamp filament and alongside of it its real image made by a concave mirror some distance away. You could probably have recognized the difference between the lamp filament and its image even if their relations to the mirror and lamp bulb had not been so obvious, because the image is a little fainter than the original. No mirror reflects back quite all of the light that falls upon it. You will also note that the image is inverted.

NOW we must be careful of our language for a moment. As a matter of fact, what we photographed was not the lamp filament and its image but only a real image of the lamp bulb and another real image of the real image of it made by the mirror. All cameras, usually by means of lenses (discussed later, p. 355) rather than by mirrors, focus light received from objects into real images of those objects, which are then recorded by the light photoelectrically on the sensitive film (see p. 287) and subsequently developed into visible impressions. Strictly, our eyes likewise see, not the objects we say we see, but only real images of those objects formed by the lens of the eye on the sensitive end organ of the optic nerve, the retina. With these comments we shall revert to the less careful language of everyday speech.

IF WE reflect our ripples from a convex mirror (one rounded out toward the source, instead of hollow), the rays, already divergent, are diverged more widely and appear to come from an object behind the mirror—a virtual image in this case again. This type of reflection in this kind of mirror produces a reduction in apparent size and a corresponding widening of the field of view seen in it as compared with that of a plane mirror of equal area (Plate 81). As rear-vision mirrors in motor cars, these find a very common field of usefulness.

PLATE 80 INVERTED REAL IMAGE OF A LAMP FILAMENT IN FRONT OF A CONCAVE MIRROR.

THE SPECIAL ATTRIBUTES OF WAVES

WE HAVE hitherto been emphasizing the fact that rays travel in straight lines and with a constant speed in media that are homogeneous. If the medium is *not* homogeneous, then the speed of the waves is not constant; in general the rays are *not* straight lines, *nor* are the wave-fronts circles or, in three dimensions, spheres.

A common example of this is to be seen in the case of light passing from water out into the air, or vice versa. Light travels more slowly in water than in air.*

The simplest illustration we can think of that is analogous to the behavior of wave-fronts in these connections is that of a column of men marching abreast, or a series of such parallel columns. On a smooth parade ground the columns will maintain an even front, all men maintaining equal steps. If the column as a whole suddenly comes to an area of soft loose sand, its speed will be retarded and the entire column will be slowed up. After the series of columns has all come out on the loose sand, they will be a little closer together because of this retardation, since their steps are now shorter and we assume that they are not taking any more steps per minute than before.

PLATE 81 Two Erect Virtual Images of a Lamp Filament behind a Plane and Convex Mirror Combination (a Plane Piece of Glass over a Convex Mirror).

The distance between columns is roughly analogous to wave-length; thus the wave-length is smaller when a wave passes into a medium where the going is slower.

If the columns pass off hard ground into loose sand obliquely, then an additional result is produced. The end that comes off first into the loose sand is slowed up while the other end continues at its original rate; the straight front of the column is broken where the two different media are in contact; and when the column is all out in the sand, the front will have been swung out of its original direction.

* This may be determined by direct measurement of the velocity of light (see p. 402) in the two different media.

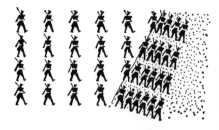

The line of march, at right angles to the column, corresponds to the ray; and the rays will thus have been bent toward the sand, i.e., will make a smaller angle than they did before with the line perpendicular to the edge between sand and hard ground. This bending of ray or of wave-front when it encounters a medium of different density in which a different wave velocity obtains is called REFRACTION. We will now look at some optical examples.

A straight stick placed obliquely in water and viewed from above appears bent at the point it enters the water (Plate 82). When you stand waist deep in clear water and look downward, your legs seem curiously foreshortened—the bottom of the lake or tank seems nearer than it really is; and if you look farther away, you get the impression that you are standing at the bottom of a saucer-shaped depression. The more obliquely the line of your vision meets the surface the more elevated does the bottom of the lake appear.

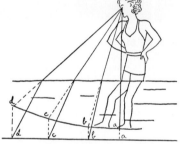

The drawing may help to make these things clear. It must be remembered that the eye is not conscious of any previous kinks that may have been in the ray of light that proceeds into it. It assumes that the ray has traveled straight from its source. If the ray has not, then the eye sees the source of the ray not where it really is but as if it were located on the backward projected portion of the ray as it enters the eye.

PLATE 82 A Straight Stick Appears Bent when Immersed at an Angle in Water.

The wave-fronts that we were suggesting in our illustration of the marching columns of men, above, were parallel wave-fronts, and the rays were therefore parallel also. Such wave-fronts and rays correspond to those that proceed to us from very distant objects. By the use of the phenomena of refraction, as well as in the case of reflection, the direction of rays may, as we have

THE SPECIAL ATTRIBUTES OF WAVES

just seen, be altered. The curvature of wave-fronts thus may be changed at will. Suppose our parallel-marching columns encounter a circular patch of sand in diameter somewhat less than the total width of the column. The center of the column will be retarded most, having to travel the maximum  width of the patch. Toward each end the men will encounter less and less sand and at the extreme ends none at all. Thus the line of men that was originally straight will be bent into an arc concave on the forward side. Their direction of travel will thus be so changed that the men in the central portion of the column will converge toward a common point—become focused, as it were.

The insertion of a spherical ball of glass into the path of a parallel beam of light, such as comes to us from the sun, will produce a similar result. The glass indeed need not be spherical. It is better for several reasons to have a disk of glass with outbowed, convex surfaces, a DOUBLE CONVEX LENS.* If this is 3 or 4 in. in transverse diameter, it will serve as a burning-glass and ignite readily inflammable objects by the focusing of the sun's light and the attendant concentration of this radiant energy into one very small region. With moderately large lenses rather high temperatures of a thousand degrees or more may be thus obtained.

At the focus of such a lens we have formed a real image of the object whence the light proceeded. All camera lenses are of this sort, although they are not simple lenses, for reasons you may see later. A convex lens† and a concave mirror, one by refraction, the other by reflection, produce, in the main, quite the same result.

Virtual images may be produced by means of lenses of the concave variety. This type of lens is illustrated most commonly in the view-finders often attached to cameras. These show you on a miniature scale the entire compass of the field of view in which you are interested.

---

* A convex lens is one that is *thicker* at its center than at its edges; a concave lens is one that is *thinner* at its center than at its edges.

† A lens with one convex surface and the other plane or even concave (if its concavity be less than the convexity of the other side) will do the same. Such lenses are called "positive" lenses. The measure of the degree of curvature either of concavity or convexity is *one* divided by the radius of the curve, in the case of a curve in a plane. For a curved surface the sum of the reciprocals of the radii of curvature measured in two directions at right angles to each other on the surface is taken.

THE human eye is constructed like a tiny camera. It has a liquid double convex lens, equipped like a camera with an automatic-working iris-diaphragm that reduces the lens aperture if the light is so intense as to threaten damage to the sensitive RETINA which takes the place of the sensitive film in the camera. Unlike the common camera, all the region between retina and lens in the eye, corresponding to the body of the camera, is filled with a transparent fluid also.

Defective vision due to faulty optics alone and not involving damage or injury to the delicate sensory mechanism of the retina is very common. Practically none of us escapes having some degree of it at some time during the course of one's life.

If the eyeball is a little too long, the image will be formed in front of, rather than upon, the surface of the retina, and it will be seen somewhat blurred. The addition of a concave lens in front of the eye, as spectacles, will reduce the convergence, and throw the point of focus back on the retina where it belongs. If, on the other hand, the eyeball is too short and the retina too close to the lens, so that the focus lies behind it, the addition of a convex lens will increase the convergence of the rays, and the image will thus be brought up to where it can be sharply seen.

Astigmatism is another common optical defect of human eyes. This is the result of a defect in the cornea or in the lens of the eye or in both. Pressure from the eye socket makes the lens surfaces slightly cylindrical instead of truly spherical. This defect either alone or in conjunction with one of the other difficulties cited above may be corrected by placing lenses that are correspondingly cylindrical, or cylindrical and convex or concave as well, in proper orientation in the spectacles.

By suitable models, one of which is illustrated (Plate 83), these various optical defects of the eye may be more carefully studied.

THE case of refraction, and the use of lenses to modify wave-fronts by virtue of it, is more complicated than the corresponding cases of wave-front change by mirrors. All *colors* of light are reflected in the same direction by a mirror set at a given angle to them. On the other hand, light of one

1

A tank of water with a lens (not seen) inserted in the water behind a glass window, and having a white screen (retina) upon which images may be projected, makes an excellent model of the eye. The retina may be moved toward or away from the lens, illustrating near- and far-sightedness; and suitable correcting lenses may be placed in front of the "eye."

2

An eyeball too short, far-sighted, does not have objects sharply focused on its retina unless—

3

provided with a convex spectacle lens. (Note improvement of star-shaped image on the retina as compared with that in illustration above.)

PLATE 83 MODEL OF AN EYE

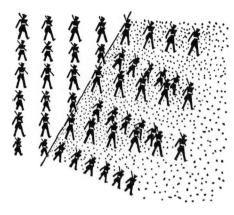

color is refracted somewhat differently in amount than is the light of another color, blue and violet suffering larger deviations than do red and yellow. Green is intermediate between them.

In our illustration of marching men, it is as if there were both long-legged and short-legged men mixed up in any given column. On hard clay or on concrete by suitable individual adjustments they can all manage to keep in line. But on striking sand, which retards and deviates all of them, the short-legged fellows are interfered with most, and they are bent farther off their course than are their longer-limbed companions. Thus there would be a sorting-out of men by "leg-lengths" in the process of marching through an extended sandy area, the shorter ones lagging farther and farther behind, as well as being deflected the more out of their original line of march. The analogy of course is very crude.

In the case of light, violet rays travel appreciably more slowly than do red, in water or in glass or in other dense transparent media; whereas all the colors that go to make up white light travel with exactly the same speed in empty space.* Violet light is shorter than red not in leg-length but in wave-length. The other colors are intermediate between these two. Later we shall see how we can determine this fact quite independently.

ISAAC NEWTON, who has so much to his credit already in the discovery of fundamental matters about this physical world, did pioneer work in these connections also. It was he who first passed a beam of light through a triangular-shaped piece of glass that we call a PRISM and noted the *difference* in the *refrangibility* (amount of bending by refraction) of the different colors. We call the phenomenon DISPERSION. As we have said, it is a sort of differential aspect of refraction. Newton showed that white light by this means could be broken up into a colored band which began with a deep red, the least deviated color, and which then merged by infinitesimal gradations into orange, yellow, green, blue, indigo, and violet. The division into seven distinct colors is very arbitrary. Every combination of any adjacent two is to

---

* Their speeds in ordinary air are all very closely identical and very nearly that *in vacuo*.

THE SPECIAL ATTRIBUTES OF WAVES

be found between them, as, for example, all shades of blue-greens and yellow-greens. Newton further showed that if any very small section of this colored band is separated out and passed through

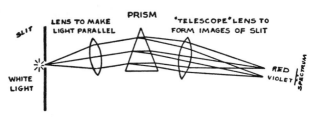

Dispersion of a parallel beam of white light by a prism into a spectrum. A prismatic spectroscope

another prism, no further breaking-up resulted. Furthermore, if the entire band of colors was passed through an oppositely placed prism and so reunited, by this or by any other means, pure white light was the result. Newton called the colors PRIMARY for this reason, and the band of color we call a SPECTRUM.*

Although in connection with refraction and dispersion we have been speaking of light and using illustrations with respect to this particular kind of wave phenomenon, these properties are common to all kinds of waves, in general.† Waves on the surface of water exhibit change of velocity as they encounter a drop of oil upon the surface. Hence the phenomenon of refraction is present. Furthermore, since, as we have seen, the velocity of surface waves depends on the wave-length, the effect of dispersion accompanies refraction always in waves on the surface of water.

SOUND waves travel with different speeds depending on the pressure and density of the air through which they pass. Changes of temperature change the density of the air even though the pressure may remain the same (Charles's Law, p. 161). Hence sound waves are refracted if they encounter either sudden or regular changes in temperature of the air along their path.

The well-known clarity with which sounds sometimes carry over calm water to distances where they would normally be quite unheard is an example of refraction of sound waves.

If it is quite calm, the air cools more rapidly at lower levels. Often a layer of cooler air flows down and over the water from surrounding wooded shores, and this accentuates the temperature difference near the water.

---

* It is the same band of color that we also see in a rainbow where the dispersion of sunlight by a water drop becomes visible when myriads of drops are involved.

† There are a few noteworthy exceptions. Sound waves, for example, do not show dispersion.

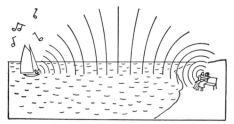

Thus it will be found that warmer layers of air often lie above cooler ones below them and nearer to the water surface. Of course, for this to result there must be no wind; otherwise the air is stirred up and becomes more homogeneous.

Spherical expanding wave-fronts from some source of sound near the water thus will have their lower portions retarded, since the speed of sound diminishes with the temperature and the wave thus becomes flatter and may even be reversed in curvature, becoming concave in a forward direction. This tends to focus the sound again at some distance away, where it is heard with much greater intensity as a result.

The phenomenon of dispersion is not present in the case of sound, since the velocity of sound does not to any appreciable extent depend on the frequency or wave-length of the sound waves.

Sound waves may, and often are, focused by reflection from mirrors also. By means of large parabolic ones made of plaster, sounds such as the ticking of a watch, normally inaudible at a few feet, may be directed into parallel beams and sent with very little loss for considerable distances. Then if refocused by another mirror, the ear can hear the ticking very distinctly. Whispering galleries are due to accidental combinations of surfaces in architecture which act as mirrors refocusing otherwise divergent rays at some distant point where the sound is greatly intensified.

ANOTHER very common and extremely important aspect of wave phenomena appears when there is relative motion between the source of the wave train and the person observing it. Take the case of our ripples on the surface of water. If you are motionless and observing them from a distance, you can count a certain number passing by each second and learn thereby what the frequency of the wave train is and consequently how many disturbances in the form of waves are sent out per second by the emitting source. However, if you advance toward the source, you meet the waves more frequently and count a larger number per second. This larger number is the normal number of waves plus those contained in the distance that you have moved up in one second's time.

Motion of ear

THE SPECIAL ATTRIBUTES OF WAVES

If you move away from the source, the frequency of the waves passing by you decreases. Obviously, if you move away with the speed of the waves, none will go by you. The frequency will have dropped to zero. If, on the other hand, you had moved toward the source with the speed of the waves, their frequency as observed by you would have been doubled.

Motion of the observer thus produces a change of frequency. In the case of sound waves the frequency of their reception by the ear determines what we call the "pitch" of the sound. As you ride swiftly on a train past a grade crossing where a bell is sounding, you can easily observe the falling pitch of the note as you pass by.*

A VERY similar situation, but one nevertheless of fundamentally different character, is produced when the source of sound, or of any wave train for that matter, is in motion. Then an actual *change in the length of the waves* results. The source, in motion in any given direction, by following up the waves it has sent out while sending out new ones all the time, makes the lengths of those in front of it shorter. The reverse action on the waves behind it makes them longer. An observer toward whom the source is moving notes the shorter waves which, by being shorter, consequently pass by him more frequently. If they are sound waves, he hears a higher note if the source moves toward him, just as he hears a higher note if he moves toward the source. The change in frequency in the case of the moving source, however, is due to a real change in the wave-length of the sound. There was no change in wave-length of the sound when the observer moved.

Such changes in wave-length may be readily observed in the ripples that we produce electrically on a clean water surface if we "freeze" them stationary by suitable stroboscopic illumination and then move the source of ripples across the surface of the water. We have succeeded in getting two different photographs of this effect to show you (Plate 84).

In general the equation for the change of frequency observed when both the source and the observer are in motion is a rather complicated expression, and we will have no need to put it down. You probably ought to know, however, that such changes of frequency or of wave-length or of both are charac-

* In detail the effect is as follows: While you were approaching the bell, your ear encountered the wave-fronts from the bell at a rate greater than their normal frequency, and the pitch you heard was higher than the true pitch. As you came opposite the bell, you heard the normal pitch. After you passed the bell, you were receding from the source, fewer waves per second reached you than normally, and you heard a pitch lower than the true one. Thus, during the interval as you approached, passed, and then receded from the bell, its note may have fallen in pitch by several tones. The closer the bell is to the track and the greater your speed of travel, the greater and more sudden is the fall in pitch.

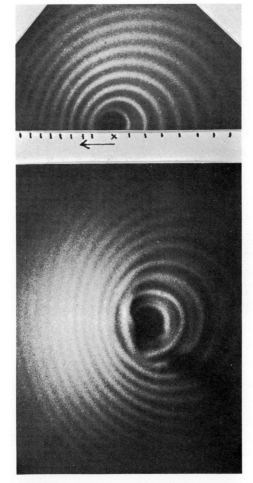

PLATE 84 DOPPLER EFFECT IN RIPPLES. Motion of source to the left in both photographs has produced a charge of wave-lengths in different directions.

teristic of all types of wave phenomena, and go by the name of the DOPPLER EFFECT. The Doppler effect has important contributions to make to our knowledge in the case of light waves, so that we shall recur to it again.*

One final very significant and useful property of waves is the INTERFERENCE that is produced when several systems of identical wave trains pass across the surface between two media or within the body of a homogeneous one. Indeed, the phenomenon takes on its simplest form when the medium is of one dimension only, as in the case of a stretched rope.

We shall look at this case first. Suppose that along a stretched rope we have a train of waves traveling in one direction, say toward the right, and that simultaneously we have an exactly similar train of waves traveling oppositely, toward the left. We can find what the result will be by taking the equation that represents each wave, adding the two together, and simplifying the result. Since you would probably skip the entire matter if we followed this procedure, we suggest an alternative in picture-language.

We will take an instantaneous picture of each wave. Then, supplied with these two wave-forms, we will put them both down one above the other and draw a line and add the graphs instead. If, as we set them down, the

* In the case of light, however, we do not need to try to distinguish between motion of source and motion of observer, because it is only the *relative* motion between source and observer that is of any significance.

THE SPECIAL ATTRIBUTES OF WAVES

crests of one line up with the crests of the other, the sum will be a wave-form of twice the amplitude of either. Next we will move one of the wave-forms a bit to the right and the other an exactly equal amount to the left, and add again. Now we will obtain another curve that resembles the first sum, but its amplitude will be smaller. We repeat the process until we have moved the crest of the right-traveling wave exactly over the trough of the other one; and the sum is everywhere zero, i.e., a straight horizontal line. Following through the rest of the additions, as many as we wish to make in our imagination, we visualize the result as a vibration of the string in segments up and down but WITHOUT there being any FORWARD MOTION whatsoever of the changing resultant form. This resultant of the sum of two equal and oppositely traveling waves is called, therefore, a STATIONARY WAVE. It contains, in essence, the basic phenomena of interference.

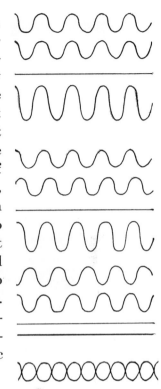

EXPERIMENTALLY we can demonstrate this also by shaking one end of a rope or string which is fastened at the other end. The wave that travels down the rope away from our hand is reflected back from the fixed end of the string and gives us experimentally the result that we desire. There is one limitation, however, in dealing with a very limited extent of medium in this manner, the velocity or frequency of the disturbance must be properly adjusted so that by the time the reflected wave has returned to our hand, where it is again reflected, it is in step with the waves which our hand is continuing to send down the string.

This, fortunately, is easy to accomplish, since we can change the speed of the waves by altering the tension of the string. The tighter it is drawn, the faster the waves will travel along it.*

* The velocity of a wave along a stretched string is given by

$$S = \sqrt{\frac{T}{\rho_1}},$$

where T is the tension under which the string is stretched, and ρ_1 is its linear density, i.e., the inertia (mass) per centimeter of length.

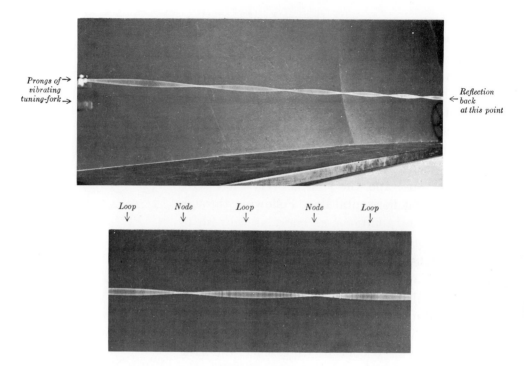

PLATE 85 Stationary Waves in a Stretched String. Close-up, showing nodes and loops

Since, perhaps, our hand cannot be controlled to shake the string with exact precision, we use a tuning-fork electrically vibrated which maintains the shaking at an accurate frequency. In the illustrations (Plate 85) which are time exposures of the apparatus in operation, you will be able clearly to discern the blurred vibrating LOOPS and between them the stationary NODES, as these two aspects of a stationary wave phenomenon have been christened.*

You might ask why did we not take a snapshot of the result so that the entire length of the string was sharp. Of course we might have done this; but the results might have been a moving wave, for all you could tell, since snapshots have the property of making moving objects appear stationary. That the string is vibrating you can see quite well by the blurred aspects of portions of it. That these vibrations are *not traveling* along but are stationary

---

* From the construction by which we derived this result, it should be obvious to you that the distance between two adjacent nodes (or loops) is half the wave-length of the waves in the moving wave trains.

THE SPECIAL ATTRIBUTES OF WAVES 365

you can see by the fact that certain portions of the string, the nodes, have not moved appreciably during the time of the exposure which was that of many score vibrations.

THE production of stationary waves in strings, in vibrating air columns, in metal or wooden rods, and in flat surfaces like the diaphragms of a telephone receiver, or even in curved metal shapes like those of bells, is of the utmost importance in music. All musical sounds, and many that do not appeal to us as musical, have their origin in the vibration of rigid objects that are either rigid by the fact of their being very solidly constructed or are made to be rigid by their being put under considerable tension. Often the sounds are strengthened and reinforced by having suitably chosen inclosed columns of air in their immediate vicinity. These may be put into the same period of vibration as that of the vibrating solid, sympathetically, or, as we say, through RESONANCE. We shall discuss this at greater length in the next chapter.

NOT alone in connection with vibrating mechanical systems such as those that produce sound waves, may stationary waves be demonstrated. Suppose, for example, we reflect our ripples as we did before from the surface of a mirror, and look at them not with the special illumination that is of the same period as the waves but with ordinary continuous light instead.

The medium in front of the mirror along the line of the perpendicular from the source to it is quite analogous to our vibrating string. A wave train is passing along it from the source directly to the mirror, and from the mirror the reflected wave is passing back in the opposite direction. There should be, therefore, stationary waves on the surface of the water between the source and the mirror. If stationary, they should be visible in ordinary continuous illumination, and it should be possible to photograph them with a time exposure under such illumination. The illustrations on preceding pages (Plates 78B, p. 350, and 79B, p. 351) give adequate testimony of the existence of the stationary wave pattern not only along the line we have been considering but over a very considerable area between the source and mirror. The lines of crests and troughs of these stationary waves are not straight. Detailed consideration of the geometry of the situation would show them to be hyperbolic curves, in the case of the plane mirror.

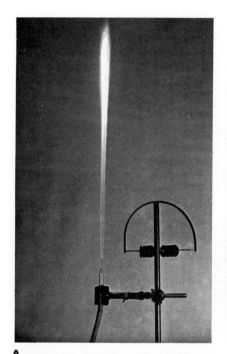 

A
B

PLATE 86 A Flame—Sensitive to Sound. (A) Flame burns quietly if there are no high-pitched sounds, such as might be made by shaking a bunch of keys or the string of metal washers shown. (B) To the left in this illustration a hand is seen blurred as it shakes the string of washers. The flame flares and drops to half its former height.

SOUND waves in air likewise can be discovered in the stationary condition when they are reflected from a mirror. One way to do this is to use very high-pitched sounds which show more obvious reflections and which can more readily be projected in straight lines than those of lower pitch and much longer wave-length.

Although very high-pitched sounds of over twenty thousand vibrations per second are inaudible to the ear,* certain types of flames are sensitive to them. A long flame from gas under considerable pressure and issuing from a tip of the appropriate size, if turned up just to the point of flaring, will flare violently (Plate 86) if a bunch of keys is jangled near by, in response to the very high-pitched sounds that we do not hear but that accompany the audible jangle. More regular sounds of this sort can be produced by an air blast through a specially constructed whistle.

* Cf. p. 376, n

THE SPECIAL ATTRIBUTES OF WAVES

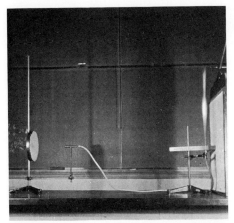
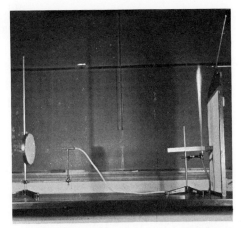

Flame at loop   *Flame at node*

PLATE 87 Stationary Sound Waves Detected by the Sensitive Flame

Such waves may easily be rendered into a parallel beam and focused again by small mirrors of plaster. They can be reflected from smooth surfaces, and the direction of the reflected beam can be located by our SENSITIVE FLAME. If such a sound beam is sent perpendicularly toward a smooth wall, in the region in front of the wall stationary waves of sound may be found, the air vibrating with rapid pressure changes in the loops and being entirely quiescent in the nodes between loops. (Compare with illustration of vibrating string, p. 364.) As the flame is moved slowly out from the wall, it will alternately flare strongly and then become perfectly quiet: it flares when it is in the "loops" and is quiet at the "nodes" (Plate 87).

Since the distance between any two nodes or any two loops is just half a wave-length of the sound (see p. 364, footnote), the full wave-length is twice this distance; and if we know the velocity of sound in air (about 1,100 ft. per sec.), the frequency of the sound may be determined. The relation between wave-length, frequency, and velocity is very simple. If the wave is λ cm. long and n waves pass by per second, then the sound must be traveling with a speed of $V = n\lambda$, cm./sec.* If you cannot visualize this in terms of algebra, try arithmetic. Imagine the waves to be 6 in. long and that ten waves travel past per second, then any one of them must have moved 60 inches in a second and the velocity is 60 in., or 5 ft., per sec.

*In the footnote on p. 344 we gave the alternative expression for velocity of a wave, $v = \frac{\lambda}{T}$. Of course as there defined, $n = \frac{1}{T}$.

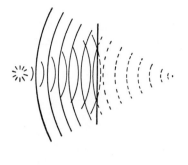

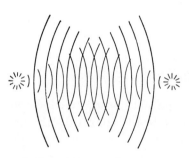

There are still other ways of producing stationary waves than by the use of mirrors. A few pages back (p. 350) with respect to waves reflected from a plane mirror we remarked that their direction was such that they seemed to come from a virtual image of the source that was as far behind the mirror as the source was in front of it. Of course, in the case of reflection, there is nothing like the source behind the mirror. Suppose, however, that we place another *real source* of waves exactly in the position of this virtual image and then remove the mirror. In the region that was formerly in front of the mirror, everything about the waves is now just the same as it was before the second source was set in operation and the mirror removed—the same two wave trains moving in opposite directions are there and the same stationary waves that make up their sum. In the region that was formerly behind the mirror, now everything is just the same as it is in the region that was formerly in front of the mirror. The mirror might have been facing the other direction, and then our new source would become the original source and the old one would be taking the place of the virtual image of the other. Thus, between two different but otherwise identical sources (same periods and wave-lengths) there will be found a system of stationary waves. We show you this phenomenon by means of photographs—one stroboscopic (Plate 88), showing the two expanding wave systems; the other in continuous (Plate 89) light, showing their stationary aspect.

THESE manifold aspects of stationary waves all are cases of what we have called in general by the name of "interference." The word is especially used with respect to light, but this is conventional rather than logical. Indeed, the phenomena of interference in general comprise what might be called a crucial test of undulatory or wavelike character. From the point of view of modern problems as to the nature of radiation and of matter, as suggested at the beginning of chapter 35, this is the reason for the great importance of this subject and is why we have treated it in such detail.

PLATE 88

INTERFERENCE

Two identical wave-trains of ripples on the surface of water under intermittent illumination. Interference is found throughout the region between the sources and when the illumination is continuous the stationary wave pattern may be seen. (See Plate 89, also 78B.)

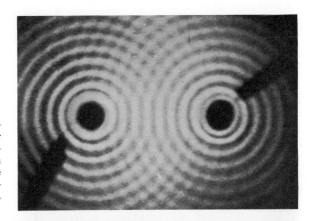

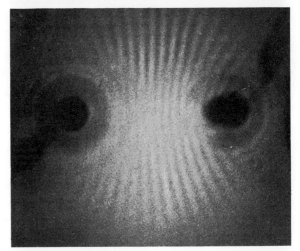

PLATE 89

STATIONARY RIPPLES

Surface of water over which two identical wave-trains are proceeding (photographed under continuous illumination). The interference of these two sets of waves produces stationary waves on the surface of the water. (See Plate 78B and 79B.) The difference in the number of waves in the two illustrations is due to the difference in distances between the two sources.

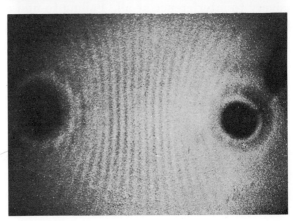

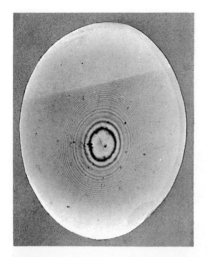

PLATE 90   NEWTON'S "RINGS." Concentric circles of brilliant colors seen between convex and plane surfaces of glass in contact at the center. An example of interference of light not interpreted as such by Newton.

The early workers in the field of light and in studies of optics for many generations were divided into two camps with respect to their hypotheses as to the nature of light. One group maintained it was a wave phenomenon, and the other that it was corpuscular in character. Waves would show phenomena like reflection and refraction, but so would solid elastic particles, if you postulated the proper attributes for them and for their relations to ordinary matter. Corpuscular entities, however, could not be imagined to exhibit wave characteristics. The great Newton recognized this and searched for evidence of interference in light. He discovered it. The famous Newton's "rings," circular bands of beautiful colors seen when a slightly convex piece of glass is in contact with a flat one, are true examples of optical interference.* The distances between the flat and concave plate corresponding to the various zones of color around the point of contact were measured by Newton. These distances have definite relations to the wave-lengths of the colors seen at those zones, but the distances were so small that Newton could not conceive that any waves could be so tiny, i.e., of the order of a few ten-thousandths of a millimeter (0.000,05 cm.) and he did not relate his discovery to interference at all. Failing to find anything he could interpret as interference, he quite logically then embraced the corpuscular theory of light as against the undulatory, since in his judgment the crucial evidence for waves—interference—was wanting.

Because of the great weight of Newton's opinion, those who followed him with other evidence of interference in light and a clear recognition of the excessively minute character of light waves encountered great indifference, and indeed hostility, from colleagues less well informed in these matters but otherwise able scientific men. Some of the details of this story of interference in light are important in so many connections that we shall resume some further discussion of them in a later chapter (chap. 39).

* The colors of thin films of oil on water are due to the same cause.

THE SPECIAL ATTRIBUTES OF WAVES

WITH this broad and general survey of the properties of waves we are now finished. We have discussed the properties primarily, and have taken illustrations from mechanical systems such as vibrating strings, or from easily demonstrated wave systems such as we can set up by ripples on the surface of water. We have taken as examples certain of the simpler phenomena of sound and light also. We have not, however, discussed with any completeness, or even in outline, those two great aspects of natural phenomena that bring us messages from the world about us, through our two chief senses, hearing and sight.

For the space of a brief chapter now, we shall discuss the subject of sound on its own account, not merely as furnishing convenient illustrations in another and much larger field.

CHAPTER 37

ON SOUND

IT IS with a sense of satisfaction that we take up a short exposition of the subject of sound. There is no subject in the entire field of physics that is as well understood or as completely worked out in detail as this one. The fluttering wave impulses that originate in the faintest whisperings of the wind through grasses by the water's edge, or that come to us from the call of the night hawk, or by electrical mechanism from the voice of someone we know very well a thousand miles away, enable us instantly to recognize the source.

This ability of the ear to differentiate between the various thousand and one origins that sounds may have is one of nature's greatest wonders. As to the details of this receptor mechanism it is beyond our purpose to inquire, except very superficially. Both the ear and, to a much lesser extent, the human voice are extremely complex biological mechanisms. Their mechanical aspects, however, are primitive and simple if contrasted with the nervous mechanism that controls and translates.

From a purely physical point of view, however, the sound waves themselves must carry, coded as it were in some way, all the information that the ear can pick up and translate from them. If you think for a moment of what the trained, expert ears of the conductor of a great symphony orchestra can detect and interpret in the tremendously intricate flood of sound waves over which he presides and dominates, you naturally will wonder what are the aspects of these wave trains that make them capable of transmitting such great wealth of detail.

SOUND waves may be carried by any material medium—solid, liquid, or gas. That sound travels considerable distances under water is well known to anyone who swims below its surface. The whir of the propellers of different boats in the vicinity, the clicking together of stones beneath the surface by someone at a distance, are things that the majority of the present generation of swimmers, using modern strokes that keep the ears largely submerged, have often listened to.

Sound travels through solid substances as well; witness the distress of many dwellers in the most modern types of apartment houses where, alas, the sound-proofing and insulation of one domicile from another is far from being ideal because of the difficulties involved in preventing the free transmission of these waves through brick, concrete, and steel for considerable distances. However, in what follows we shall limit our discussion entirely to that type of sound waves to which we usually are listening, sound waves in air.

THE passage of sound through air or any other gas is due to a longitudinal wave in it. Superimposed upon the kinetic, thermal motions of the gas molecules, are oscillations back and forth *in the direction that the sound is proceeding.* These oscillations originate usually from the vibration of some rigid object or from some stretched string or membrane. A tuning-fork, upon release of its prongs after compression, vibrates elastically, and the air in its immediate vicinity has alternate condensations of pressure and rarefactions created by the swift "fanning" action of the vibrating prongs. The vocal cords in your throat that can be made to stretch across and partly close the air passage to the lungs are set into vibration by the flow of air past their edges. They then superimpose upon the outgoing air a vibratory motion, in the form of condensations and rarefactions, that follows the same sequence as their own vibrations. These waves are profoundly modified by the forms of the air passages through which they pass outward. These modifications supply the vowel characters in speech. All the consonant sounds are added

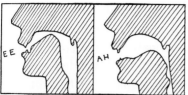

Redrawn from Saunders' "A Survey of Physics"

by motions of tongue, teeth, and lips. The character of the winged words, once they have escaped, is formed and fixed in the sound wave train. This wave train can never be recalled, but it may be recorded and preserved; and another one almost exactly like it may be sent out again, thanks to modern electrical methods of sound recording and reproduction.

What are the aspects of this wave train that give it its individuality or QUALITY? We have already learned that the amplitude of the vibrations determines loudness, and the frequency, pitch; but there is another characteristic quite as important as the other two in the case of sound,

Since it is very difficult to visualize compressional waves or even to make drawings of them in any detail, we shall have to resort to a trick, and make graphs of *transverse* form to represent these waves that have nothing but longitudinal characteristics.

THE vibration set up in the air by a simple tuning-fork or gently bowed violin string or softly blown organ pipe that makes a longitudinal wave form like this in reality:

would be represented like this graphically:

A louder sound would be thus shown:

and one of higher pitch but of the same intensity as the first, as follows:

A tuning-fork and a higher-pitched but fainter whistle sounded simultaneously would give this kind of a graph:

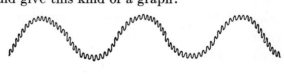

What is most important for our human purposes is that if the two (or, indeed, several) pitches are very different the ear is able to analyze this complex wave, and we hear two (or, indeed, perhaps many) distinctly different sounds in the one combined wave train that enters the ear.

NOT all vibrating objects execute such simple vibrations. A string, for example, may oscillate as a whole, or in two, three, four, or many parts depending upon how its length, tension, and the resulting mechanical wave velocity along it cause the stationary wave pattern to take form. Indeed, it is quite possible for it to vibrate in several of the modes at one and the same time. In this case the sounds it sends out, if graphed, might look like any of these curves.

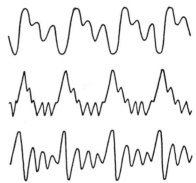

Now the physicist has instruments, called "harmonic analyzers," designed to represent mechanically certain theorems of the mathematicians about such complex types of vibration. By such means as mathematical analysis directly or by means of instruments, we can dissect such complex forms into their simple characters and find out their component parts. But in general THE EAR IS UNABLE TO HEAR AS SEPARATE these components that do not differ widely in frequency. It is this unanalyzed aspect of the incoming wave that gives the sound what we call its QUALITY or TIMBRE. Training and experience alone teach us that one complex character is our mother's voice, that another is the sound of a violin, and that a third comes from a bell. From a purely physical point of view, these various qualities may be analyzed into their component parts, although to do this for any one of them with any degree of completeness may be a very laborious task.

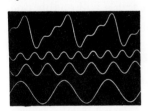

Analysis of a compound wave form. The complex wave form shown at the top can be shown to consist of the sum of the three simpler component waves below whose periods are as $3:2:1$ and wave-lengths as $\frac{1}{3}, \frac{1}{2},$ and 1. Machines known as "harmonic analyzers" have been designed to reproduce with fidelity the much more laborious mathematical calculations that are required to discover in the case of any particular complex what are its component parts, i.e., what simple waves added together produce it.

Thus we see that the human ear is a very marvelous instrument but not a very analytic one. Of course its range of perception is very distinctly limited, also, both as to the frequency and intensity of the sounds it can hear. To vibrations that take place less often than about 20 times a second, and to those that occur more than 20,000 times a second, it is insensitive. The sensitivity of the human ear to any frequency depends to a considerable extent upon the intensity with which that sound is produced.*

VERY important in connection with the phenomenon of sound is the manner in which a limited portion of the medium through which it travels (a column of air in a pipe, for example) may be set into vibration, sympathetically, as it were, by the waves from some other vibrating object and, as a result, produce a sound of far greater intensity than the original. It is unfortunate that sounds cannot originate from the printed pages of a book that deals with these matters. The experiments we wish to describe might then be greatly multiplied in number and could be presented with far greater force.

You will have to imagine the sounds in the following experiment. We have a vertical glass tube which can be filled to any height with water from an opening in the bottom which communicates to a larger tank or water faucet and to the sink as desired. We strike a tuning-fork and hold it above the pipe, then rapidly shift the water level up or down. A first trial will reveal a certain approximate location for the water level at which the sound of the tuning-fork seems to be reinforced. Subsequent trials will enable one to fix this position more exactly. If the pipe is long enough and the pitch of

* Acoustical engineers are in the habit of representing the audibility of the human ear graphically as an area inclosed by a curve. The X-axis of co-ordinates represents frequency, and the Y-axis intensity. Since the widest range of frequency is heard only at the higher intensities, this area has its greatest width nearer the top than the bottom. Indeed, for very low intensities sounds are audible only over the rather narrow frequency range which is not widely different from that used by the human voice—an interesting example of the adaptation of our physical mechanism to our needs.

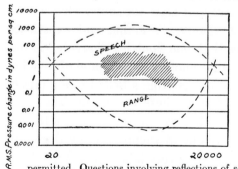

A great many interesting applications of curves of this sort to the study of modern electrical methods of sound reproduction in phonograph, telephone, and radio could be cited if our space permitted. Questions involving reflections of sounds from the walls of rooms, especially of auditoriums, and the ways and means of securing satisfactory acoustical results are likewise very interesting but rather too technical to find place in these pages.

For a brief graphical treatment of these matters through the medium of motion pictures we refer the reader to two of the University of Chicago series of sound pictures, entitled, *Sound Waves and Their Sources*, and *Fundamentals of Acoustics* (University of Chicago Press).

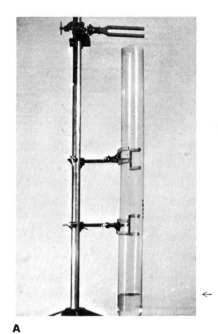
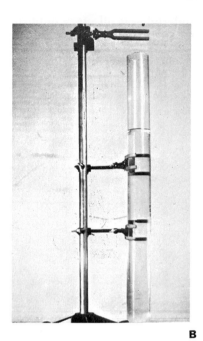

PLATE 91   Resonance of an Air Column to a Tuning-Fork. (A) Length of column is $\frac{3}{4}$ wavelength of sound wave emitted by fork; (B) length of column is $\frac{1}{4}$ wave-length of sound wave emitted by fork.

the fork not too low, i.e., if the wave-length of the fork is short compared to the total possible length of the air column, there may be found not only one, but several, such places where the air column within the pipe is just the right length to reinforce the sound of the fork. (Plate 91 shows two such lengths.) When these have been determined and the pipe length carefully adjusted to one of them, it can be shown that, even if the tuning fork vibrates so feebly that no sound can be heard from it even a few feet away, when it is placed over the air column, the sound, which must come largely from the latter, may be heard distinctly over a good-sized auditorium. The phenomenon goes by the name of RESONANCE.

Resonance is most important in the design and construction of musical instruments. The shape of the air cavity that forms the body of a violin or cello and of the air passages of flutes and all other wind instruments, has profound effects on their tone quality. While a great deal of the details and technique of construction is undoubtedly empirical (i.e., based on "cut-and-try" methods) at the present time, there is an unquestioned scientific basis underlying it all. It is only the complexity of many details that has stood in the way of the more accurate scientific basis of these practices having been completely worked out.

We have already commented on the fact that in the passage of sound waves from the vocal cords up through throat and mouth their character is further modified. The resonance of these air chambers is in part responsible.

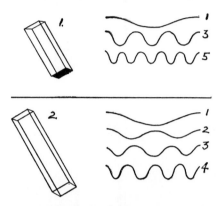

Of far more fundamental importance in scientific work with respect to resonance is the fact that by measurement of the lengths, which in our experiment we found were capable of making an air column resonate to a given tone, we can find the wave-length, or frequency, of the note itself.* Details of the analysis we will not include. They may be found in any technical elementary textbook on the subject. The result is that if one end of the pipe is closed, as ours was by water, and the other open, the pipe will reinforce the note when the length of the pipe is $\frac{1}{4}$, $\frac{3}{4}$, $\frac{5}{4}$, or any *odd* number of *fourths* of the wave-length.

If, on the other hand, we use a pipe open at both ends, it also will be found to produce resonance to a given note when the length of the pipe is $\frac{1}{2}$, $\frac{2}{2}$, $\frac{3}{2}$, $\frac{4}{2}$, or any *whole* number, odd or even, of *half* wave-lengths.

The shortest length of a closed pipe that will resonate to a *given note* is therefore half of the length of the shortest open pipe that will respond to the same note; $\frac{1}{4}\lambda = \frac{1}{2} \times \frac{1}{2}\lambda$.

CLOSELY related to this is the fact that if an irregular blast of air is sent upward through a pipe or blown across its open mouth, the pipe, by reinforcing only those pulses that come at just the proper distance apart to correspond to the conditions set forth above, will soon be found to be emitting a tone on its own account.† The lowest tone or fundamental tone of a closed pipe will have a wave-length FOUR TIMES the length of the pipe.

* Another method using a sensitive flame and stationary sound waves in a beam of reflected sound has been described on p. 366. As there stated, the method is not well suited to sounds in the audible range.

† Filling a jug with water causes sounds that enable us to tell when it is nearly full. This is another example of these effects.

ON SOUND

For an open pipe so blown, the wave-length of the lowest tone will be TWICE the length of the pipe. If the open and closed pipes used in these experiments have the same length, the wave-length of the fundamental tone of the open pipe, being only twice the pipe length and therefore only half the wave-length of the fundamental of the closed pipe, will be one OCTAVE higher in pitch than the latter. By our definition of the word "octave" we mean that the ratio of the frequencies (or wave-lengths) of the two notes related by this interval is as $2:1$ (or $1:2$).

When pipes are strongly blown, however, other tones than the fundamental we have been describing may be produced. We shall not detail these cases. The reader is usually invited to work such matters out for himself. Indeed, it might be said to be easy for him if he will but remember that, in all cases for either type of pipe, the HARMONICS above the fundamental in frequency that will be emitted are exactly those other notes that correspond to the pipe lengths being 3, 5, 7, etc., fourths of the wave-length, for closed pipes, and 2, 3, 4, etc., halves of the wave-length for open pipes.

A FLAME, a piece of red-hot wire gauze, or, best of all, a small electric heating coil that can be raised to incandescence may serve quite as well for a source of irregular disturbances (convection currents of heated air in these cases) as an air blast across the lip of the pipe. When such heating elements are inserted at different places within the inclosed vertical air column, a great variety of tones may be produced, by no means as simply related as those we have been describing. Such SINGING PIPES are usually regarded with great curiosity and interest by the uninitiated in these matters. To them, of course, the cause of the sound is quite mysterious.

ONE other item in sound should be discussed because of its very close relation to interference mentioned in the preceding discussion of the general properties of waves, and because of the great importance of your having as clear a picture of interference as possible since, as we shall see in several connections, its presence is a crucial test of wave characteristics.

Suppose we have two sources of sound of exactly the same pitch sending out their separate wave trains. If these are of a normal well-audible pitch, about middle "C" (261 vibrations per second) on the piano scale, their wave-lengths will be about 4 ft. We should expect interference to exist between these two wave trains and a stationary wave pattern to be present. With the many reflections of sound that take place if the experiment is tried indoors, this pattern is rather difficult to detect. This is because each reflection contributes a virtual image with which goes another system of primary waves, and much additional interference results besides that of the original two sets of waves.

If the two forks are sounding out of doors, however, in the absence of reflecting surfaces the stationary waves may be detected by walking around in the vicinity, especially if we use but one of our listening devices and plug our other ear with cotton. The intensity of the combined sounds will be heard to rise and fall as we change our position. When loudest, our ear is near a loop of the stationary wave pattern; when the sound dies away almost to silence, the ear is at a node. A rough map may thus be made of the interference pattern and, of course, will be found to be a greatly enlarged reproduction of what we photographed for you in the case of ripples in Plate 89, p. 369.

LET us think of what we know about the Doppler effect now in this connection. As we changed our position with respect to either of the sources, in so far as our motion had a component in the direction of either source, we altered to a slight extent the frequency of the note we heard. Furthermore, since the two sources were not at one and the same location, we could not move at random with respect to the waves from one of them without moving somewhat differently with respect to the waves from the other.

Consequently, we must have altered in different degree the frequencies of the two notes we were hearing; and then, to us, they were not of the same pitch. In other words, by moving across an interference pattern, the effect upon the observer is that of listening to two notes of *slightly different pitch*. As we so move around, the intensity of sound rises and falls, as we have seen.

LET us now try a somewhat different experiment. We will maintain a *fixed point of observation* and slightly *alter the pitch* of one-tuning fork so that it is a little different from that of the other. As we listen to them both, the sound swells and falls in intensity with perfect regularity. The phenomenon is now called by the name of BEATS of sound. In terms of two trains of waves of slightly different lengths, one can readily see (Plate 92) that at certain times the two sets of waves will be in phase, crests or troughs from the waves in each arriving at the ear simultaneously. An intensity represented by a wave of double amplitude will then be heard. As the trains

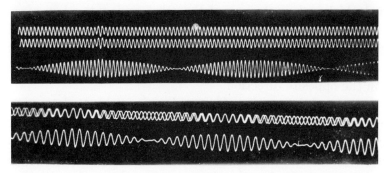

PLATE 92 BEATS. Two graphical illustrations of this phenomenon. *Above:* Thirty-nine waves in the upper curve occupy the space of thirty-eight in the lower one. The sum of the two waves is below them. *Below:* One curve of nineteen waves has another in which twenty occupy the same space drawn on top of it. The sum of the two is in the lower line.

progress, however, the one having the slightly shorter waves begins to get out of step, its crests arriving more and more ahead of those of the other series until finally they arrive simultaneously with the troughs of the other. One set of waves now in its effect on the ear quite exactly neutralizes the other; the ear hears nothing, or at most a very faint sound. The double, amplitude has now become zero amplitude. One wave train is one-half wave-length ahead of the other.

The process continues. Soon the shorter waves have caught up another half on the others, crests are again adding to crests, and the sound has swelled to its maximum again. In the time between one maximum of intensity and the next, one wave train has gained one entire wave-length upon the other. If the longer has 265 waves passing per second and the shorter 266, then in 1 sec. there will be one "beat" between the two. If, however, the two wave systems are closer in wave-length or frequency, so that, for example, the beats come about one every 5 sec., this means that it takes 5 sec. for one train to gain one wave-length on the other, i.e., that it gains but $\frac{1}{5}$ of a wave-length per second. Thus, if the longer has 265 as its frequency, that of the shorter is $265\frac{1}{5}$, or 265.2.

As the pitch of the two notes approaches identity, then the beats become less and less frequent until they can finally be no longer distinguished. *The* MOVING *interference pattern which their combined effects represent moves more and more slowly.* When it stops, the two sources are identical in frequency.

Of course, if a shift in the frequency of one of the sources is continuous in one direction, as when the higher frequency changes, decreasing through that period which is characteristic of the other, and on to a lower frequency, the beats will slow down, cease, and then begin again as the source, formerly of higher frequency, becomes the one of lower frequency. In terms of a *moving interference pattern*, this pattern slows down, stops, and then begins to move in the opposite direction.*

Observation of the *number* of beats alone is unable to determine WHICH is the lower frequency. This can be done by making a known shift in the frequency of either source and observing if the number of beats per minute increases or decreases. If a known increase in frequency produces fewer beats, then it is certain that the change was made on the source of longer waves and lower pitch.

THE tuning of musical instruments to a standard pitch and the tuning of various strings into their correct relationship with one another is done by means of beats exactly following the practice and principles we have just described.

AS TWO sources become more and more different in pitch, the beats increase in rapidity. If the two original sounds are in the audible range, a very dissonant and unpleasant effect is produced as the beats become too swift to count. At certain very definite differences between the two individual notes the beats, now so rapid that they produce an audible note themselves, blend with the two original tones and produce consonant and agreeable combinations of tone—complex tones, of course.†

The various intervals of musical scales—octaves, thirds, fourths, and fifths of both major

---

* You undoubtedly note the close parallel of what we have been describing to the phenomena that are seen to occur in stroboscopic illumination of periodic motions in which one observes essentially the beats between two frequencies—one, that of the series of moving objects (waves as they pass each into the position occupied by its predecessor); the other, that of the illumination.

† Because of the fact that the notes on a piano (an instrument with a fixed keyboard) are not quite adjusted in frequency to exactly those ratios of 4:5:6:8, which are among the most harmonious, a certain degree of dissonance and in some cases audible beats, as between the notes "C" and "E," may be detected. Compare any text or encyclopedia under "diatonic," and "equally tempered scales."

ON SOUND

and minor character—have very definite relations in these aspects of frequency ratios and differences that often are of interest to those whose main occupation is with musical matters. More often the artist, however, is quite unconscious of the mathematics and the physics on which his art is based. For the detailed relations of the various musical scales any standard reference book may be consulted.

In case the two sound wave trains are of frequencies that are too high to be audible, i.e., above twenty to twenty-five thousand vibrations per second or wave-lengths less than about 1.5 cm., the beats between them may be quite audible. Suppose we had two sources of 25,300 and 25,044, respectively. Sounding simultaneously, these two would beat at a rate of 256 times per second. This is a frequency often used for a standard in scientific work and is not far from that of middle "C" (international pitch) on the piano. Thus the beat between these two very high inaudible notes appears as a definite pitch in the audible range. By this device, sometimes called "heterodyning," an entire range of frequencies over the whole range of human audibility may be produced. The vibrating mechanism in these cases is usually electrical and consists of alternating currents of the frequencies desired produced by means of generators of one form or another. These currents through their magnetic effects set diaphragms into vibration and thus sound waves of corresponding periods in the surrounding medium, air or water, are produced.

The various weird whistlings, screeches, and whining noises that come out of radio sets may sometimes be attributed to the performers; but more often these sounds come from the simultaneous reception of two stations or from interference, in the literal sense, between electrical waves set up by your set and those from some other in the vicinity. The frequency difference between the two trains of waves becomes transformed by beats into an audible sound wave of most unpleasant character coming from the loud speaker.

Indeed, the electromagnetic waves of radio devices are as perfect examples of wave phenomena as can be found. It is true that the longer of these waves are not to be seen, heard, felt, tasted, or smelled. Of their reality and existence, however, this is no criterion. The energy of these waves is just as definite as the energy of anything that can be perceived directly by the senses. Momentum, likewise, these waves possess; and thus they have mass and inertia. Totally blind we are to all but one narrow band, an octave (see definition, p. 379) in width, that we call "visible light."

Although the connection between the electrical vibrations which we may artificially set up and produce in electrical circuits and the form and character of the radiated energy WHILE IT IS IN TRANSIT from source to receiver remains one of the greatest mysteries of nature, we know a tremendous amount about the details of what is going on at each end of the line. From a practical point of view this is about all that is necessary.

For the remainder of our story, then, we shall devote ourselves to the study of electromagnetic waves, long and short, ranging in distance, from crest to crest, from miles to billionths of a millimeter. It is a subject that was almost entirely unknown one hundred years ago.

CHAPTER 38

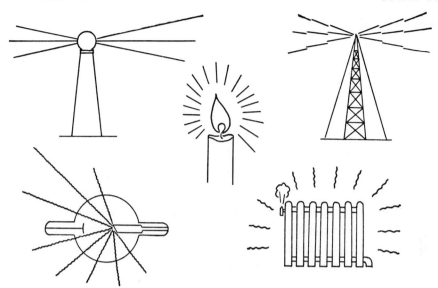

ELECTROMAGNETIC WAVES

RADIANT energy of all forms (that transmitted by wireless, radiant heat, light, X-rays, and some aspects at least of cosmic rays) is carried through space in the form of what we call "electromagnetic waves." The phenomena can be described adequately and succinctly only by means of mathematical equations. This is to say that the periodic changes that take place in the electric fields and in the magnetic fields that are associated with the changes in the electric ones may be so described. It does not mean that, even if you are thoroughly familiar with the language of mathematics and quite able to write the famous equations of Maxwell in this connection, you really understand the phenomena in any fundamental fashion.

To describe, be it ever so perfectly, a set of relationships does not necessarily imply an understanding of the causes behind them.

With respect to electromagnetic waves, a very fundamental type of difficulty appears at the outset—a difficulty recognized early in the nineteenth century and still unsolved in any direct and simple fashion. It has been described as the difficulty of thinking of waves without a "waver."

For descriptions of ripples on the surface of water there is the water to move in circles, large or small, and so carry the wave and the energy. If there is momentum in one region, this is momentum of water that has mass, or density, and velocity. That this momentum, and the corresponding aspects of force and energy may be handed on from one region to another seems perfectly clear. Momentum of water it was to begin with, momentum of water it remains, and momentum of water it is when received.

Quite similarly with sound waves. These consist of periodic variations of air pressure among other things. Invisible the medium may be to direct sight, except under special conditions; but it has a definite objective existence to which a host of experiments attest—mass, density, inertia, weight, and all the other attributes of any other kind or state of matter that are possessed by any gas. That periodic changes of pressure take place and are transmitted from one region to another by a wave phenomenon affords no stumbling-block, no gap across which one's sense of logic must leap blindly before it can proceed with the argument.

WITH respect to electromagnetic waves, however, and all of those phenomena called RADIATION—wireless impulses; radiant heat from an oven whence the fire has been drawn; infra-red by means of which we now can photograph by the "light" of a heated flatiron; also that less "infra" red by which we can "see" through cloud and mist; ordinary visible light; ultraviolet, so potent in stimulating certain kinds of chemical change, so valuable to manufacturers of "health" lamps and other gadgets for the ills to which flesh is supposed to be heir; X-rays, important in enabling us to look through solid and opaque substances, therapeutically of importance, too, as are their shorter sisters, gamma rays from radioactive materials; even the cosmic rays, if waves they really be—all fall at the outset in the unfortunate position of being waves *without* a "waver."

If there really is a medium along through which these waves take form, we have never detected the slightest evidence on any other grounds for its existence (see later, pp. 410 ff.). A luminiferous "ether" formerly was postulated; but since it never was verified, the physicists have ultimately come to lose conviction with respect to it. Therefore, we cannot explain to you what electromagnetic waves *are*. Energy, mass, momentum, and pressure travel across the vast reaches of interstellar, intergalactic space, measurable in quantity, in quality, in speed. This energy, upon its receipt, is found to possess wavelike attributes. It has other attributes, those of atomicity, also, as we have mentioned in an earlier connection. Both of these aspects—

ELECTROMAGNETIC WAVES

wavelike ones of amplitude or intensity, of frequency, and wave-length or color; and corpuscular ones like mass, momentum, or pressure—are predominant in importance whenever electromagnetic waves are received by material objects to which their energy is imparted, recognized as such, and measured. What aspects radiant energy possesses on its journey while in transit we cannot say from experience, since we cannot have experience of physical phenomena in the absence of material objects. We assume the energy is there in space and speeding through it, so many ergs of radiant energy for every cubic centimeter of volume. Although this is pure assumption, it is one we can hardly avoid making, since we can know that so much energy left the distant source, and that in a measurable time later was all or in part received by some other object. In terrestrial experiments, if all is not received, any dissipation that occurs may be accounted for to balance the account. Across interplanetary spaces there is no evidence of any loss at all except for very slight scattering and some absorption by diffuse, finely attenuated matter that we know to be there by this very loss itself. Whether in intergalactic space any energy is transformed or disappears in the radiant form is still a speculative matter.

THE VERY MOMENT THAT WE CAN LAY HANDS ON RADIANT ENERGY WE EXPERIENCE FROM IT ALL THE MANIFESTATIONS THAT MATERIAL BODIES HAVE. HENCE WE CALL IT MATERIAL, AS WE CALL EVERYTHING ELSE IN THE PHYSICAL WORLD MATERIAL.

IN AMOUNT the sum total of the energy sent forth by the sun in the radiant form is staggering to contemplate:

Heat is received from the sun at the surface of the earth at a rate of 1.98 cal. per min. on a square centimeter at right angles to the beam of sunlight. This is a power flow (see chap. 12) of a little less than 5,000,000 h.-p. per sq. mi. If we were to pay at the rate of 1 cent per kilowatt hour (a very low rate, as prices go, for energy in the electrical form) for *sunlight*, the amount the whole earth gets would cost about a half billion dollars per second. The earth, of course, is small and a long distance from the sun. The amount our tiny planet receives is but one part in more than two thousand millions of what the sun sends out in all directions.

Enormous as is this prodigious flow of energy, we do not know the manner of its coming. The mystery is related to the mystery of the action of one electric charge upon another at a distance from it, and to the corresponding effect of magnetic poles upon each other. How a moving charge can affect a magnet, or a moving magnetic field lay hold of charges and cause electric currents to travel in a closed circuit, is part of the same great mystery.

We dwelt at length upon these matters in a descriptive way in Part III. We are thoroughly conversant with the laws that govern the actions and reactions. But at no time have we given you an explanation in terms of concepts more fundamental, as we did when we were discussing the phenomena of heat. We have no explanation to make. We do not understand this "action at a distance" any better than did the men of the eighteenth or the nineteenth century.

THE practical school of thought in these matters, represented by the engineers and experimental physicists, has followed the lead of Faraday. Lines of force were his invention to "explain" the action between charges and between poles. Maxwell made noteworthy advances from Faraday's experimental data. Maxwell possessed mathematical talents of a high order. We know that Faraday's lines of force are unreal but convenient; to Maxwell they were unnecessary. The situation is quite paralleled in the consideration of electromagnetic waves. Since we have in mind continually the importance of telling this story in words as far as possible and not by means of equations, we shall of course adopt types of explanation here that are extensions of the imagery of Faraday. We shall speak of electromagnetic waves as if they *had* a "waver." The "waver" will be Faraday's lines of force.

Let us imagine we have a charge of q units of negative electricity quite a distance from all other charges. Its $4\pi q$ lines of force (see p. 217) will stretch out radially in all directions about it. Taken together with all the others, each line behaves as if it repelled all other lines and also as if it were under tension along its length. It is by the tension that we visualize the attraction between the unlike poles or charges at the opposite ends of the line, and by the property of mutual repulsion of lines that we "explain" the fact that they distribute themselves as they do in the characteristic patterns that describe the reality behind them. This reality is that, after all, they *are* the lines of direction which a + charge or a N. pole would take in the region represented.

To interpret electromagnetic waves in terms of lines of force we must endow the lines with a third attribute, inertia. Let us look at but one line. If

ELECTROMAGNETIC WAVES

it is "shaken," it will not all move as a whole. Parts of this line near the end that is shaken will change position before those parts that are more remote. It behaves, when shaken, just as a rope behaves; having inertia, it does not all get set in motion at once. WAVES may thus be set up in lines of force.

If you think that this is taking too great liberties with these lines of force and letting our imagination run rather wild, our justification once more is simply that we know no other way to describe the phenomena graphically.

SUPPOSE, then, that our line of force is attached to an electron or to $q/4\pi$ charges that may be moved together as a unit, and also imagine this line of force stretching away a half-million miles or so, and we are ready to proceed. Very quickly let us shift the position of the charge by some arbitrary distance—let us say a foot. The part of the line of force immediately adjacent to the charge will shift instantly with it. Farther and farther away along the line of force the new position will be assumed at later and later moments. Continuing to be explicit, 186,000 mi. away the line will shift to its new position just 1 sec. after we shifted the charge; 558,000 mi. away the impulse that results in the shifting of the line will occur just 3 sec. after we moved the charge. This is only a way of saying that an impulse due to a shift in position of one end of the line travels away from that place with a velocity of 186,000 mi. per sec. If the charge is given a vibratory motion, a whole series of these impulses travels down the line of force. If it were a rope, we should call these impulses waves. The faster we shook the rope, the shorter would be the wave.

Now, when electric charges oscillate, something actually does travel away from them with this velocity of 186,000 mi. per sec. This something is the radiant energy with its wavelike attribute of frequency and wave-length.

That our analogy is not yet complete you may have noticed. Suppose we move a charge in the vicinity of a magnet, i.e., needle. There is a magnetic field sent out by the motion of the charge and the compass needle *near by* is deflected at once. Suppose the compass needle were 93,000 mi. away; when would it be deflected, at the instant the charge was moved? No, your guess would be, to complete the analogy, $\frac{1}{2}$ sec. after the charge was moved.

When electric charges move or oscillate, magnetic, as well as electric, fields are in motion in their vicinity. Suppose, for example, we have a separation of charges at opposite ends of a long conductor. The charges we will also suppose are oscillating in a vertical line in the plane of the page. As they oscillate, the lines of electric force must oscillate and reverse their directions end for end. If the central portion of one of these lines had too great inertia, a loop would be formed, since it could not respond soon enough to the motion of its ends on the conductor. Imagine that this loop broke off and went traveling outward. A closed line of force like a smoke ring puffed out from a round aperture would result. What about the magnetic field? Use the left-hand rule. As the negative charge moved up its path, an N. pole to the right of it would have been moved out from the paper toward you. As it moved down, the pole would have moved back into the page through it and out behind an equal distance. Since the magnetic impulse travels with the same speed as the electric—indeed, the two are part and parcel of one and the same phenomenon—a closed line of magnetic force, a loop at right angles both to the

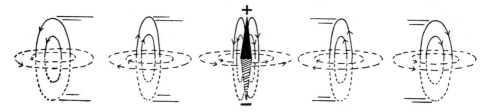

paper and to the electric loop, would represent the magnetic aspect of the radiation. The two loops, furthermore, would not be in phase but out of step. At the extreme positions of the charge its motion has ceased; it is about to swing the other way. At this moment, however, there being no motion, the magnetic force is zero. The electric displacement at this moment is a maximum.

Our two curves really represent, you see, (1) the position of an oscillating electric force (ELECTRIC VECTOR), and (2) the simultaneous position of the corresponding magnetic force (MAGNETIC VECTOR). Imagine these two oscillating arrows (one, one-fourth of a complete oscillation behind the other) shooting out in all directions in space around every oscillating charge, and you have as close to a "picture" of what radiation is as anyone can have. We said in "every direction." That was not right; there are two directions where the disturbance vanishes no matter how vigorously the charges move. Those two directions are the two represented by the line of motion in which the charge is oscillating, extended indefinitely in each direction.

Furthermore, the magnetic field also vanishes along the line of the motion of the charge. The fact is that these waves differ from almost all of the

ELECTROMAGNETIC WAVES

others, mechanical or otherwise, that we have spoken of. They are lacking in any longitudinal component. ELECTROMAGNETIC WAVES are purely TRANSVERSE.*

As in the case of all other wave phenomena, the wave-length (λ), frequency of vibration (n), and velocity of transmission (c), are related.

$$c = n \times \lambda.$$

Slow oscillations of charges, such as occur in ordinary alternating-current circuits of 60 cycles (complete oscillations per second), give, using metric units, waves of length:

$$\lambda = \frac{c}{n} = \frac{3 \times 10^{10} \text{ cm. per sec.}}{60 \text{ per sec.}} = 5 \times 10^8 \text{ cm.} = 5,000 \text{ km.}$$

BECAUSE of certain other aspects of radiant energy (those of corpuscular character, as we shall see later), such very long waves, thousands of miles in length, carry very little energy in the radiant form away from the circuits where the waves originate.

The longest electromagnetic waves of scientific or practical usefulness are not much longer than a hundred miles, corresponding to a frequency of oscillation, using British units, of

$$n = \frac{c}{\lambda} = \frac{186,000 \text{ mi. per sec.}}{100 \text{ mi.}} = 1,860 \text{ per sec.}$$

Wireless telegraphy on the longer ranges of wave-length employs waves of from 15 mi. to $\frac{1}{4}$ mi. in length, of frequency from 12 thousand to 700 thousand cycles per second (kilocycles).

Radio waves, used in telephony and broadcasting, are still shorter, and run from 1,700 ft. (566 m.) to 600 ft. (200 m.) in length, the shorter ones representing frequencies of as high as 1,500 kilocycles or 1.5 megacycles. The more recent "short-wave" radio sets can be attuned to frequencies as high as 25 megacycles, which correspond to 40-ft. waves.

Laboratory experiments have been and are being carried on with still shorter waves indeed, down to sizes of a few hundredths of an inch. Here the electric circuit is just a spark gap and a pair of plates to radiate the energy. The receiver is another tiny spark gap (or lamp) and a pair of rods to pick up the energy.

* It is phenomena that go by the name of polarization, which are rather too complex to be described in these pages, that attest to this fact.

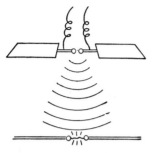

Very short-wave transmitter and detector

Waves as short as this are readily absorbed by solid objects, and the relatively much larger amounts of energy they carry are transformed into heat, the temperature of the absorbing material being raised thereby. Artificial fevers, so called, in animals may be produced by these waves.

Indeed, we usually class as heat (radiant heat) waves that lie between 0.01 and 0.000,028 in. To avoid long decimal fractions we have various subsidiary units of length we use in these connections. They are the μ (mu), a thousandth of a millimeter; the $\mu\mu$ (mu-mu), or micromillimeter, a millionth of a millimeter; and, most important in general scientific use, the Ångstrom unit (A.U.), equal to 10^{-8} cm. and equivalent therefore to 0.1 $\mu\mu$.

Even Ångstrom units are inconveniently long to measure the shorter electromagnetic waves. The X-ray unit of length, (X.U.), is a thousandth part of the Ångstrom unit.

Kind of Waves	Lengths
Long electromagnetic	Hundreds of miles to 15 mi.
Wireless telegraphy	15 mi. to $\frac{1}{4}$ mi.
Wireless telephony	$\frac{1}{4}$ mi. to 120 ft.
Short electromagnetic (experimental and RADAR)	120 ft. to $\frac{1}{100}$ in.

	In Scientific Units		In Decimal Fractions of an Inch	
Heat	300. to	0.7 μ*	0.01	to 0.000,028
Visible light	7,000. to 3,500.	A.U.*	0.000,028	to 0.000,014
Ultra-violet	3,500. to	130. A.U.	0.000,014	to 0.000,000,5
Far ultra-violet	130. to	45. A.U.	0.000,000,5	to 0.000,000,17
X-rays	45,000. to	100. X.U.*	0.000,000,17	to 0.000,000,000,38
γ-rays	100. to	5.6 X.U.	0.000,000,000,38	to 0.000,000,000,02
Secondary gamma rays from cosmic ray primary particles down to and below		0.001 X.U.		0.000,000,000,000,005

* $1\mu = 10^{-4}$ cm.; 1 A.U. (Angstrom unit) $= 10^{-8}$ cm.; 1 X.U. (X-ray unit) $= 10^{-11}$ cm.

The table herewith summarizes the entire range of electromagnetic waves that we have now explored. It will be noted that visible light comprises but one small category (one octave, as it were), from 7,000 A.U. to 3,500 A.U. out of a total range sixty octaves altogether.

By what means, you may quite naturally inquire, are we enabled to detect and study those ranges of the electromagnetic frequencies for which we have not been provided by nature with any sensory mechanism? The means are as varied as are the waves themselves in size.

For the detection of the longer waves we utilize principles of resonance very similar in character to what we described by this name in the chapter on sound (p. 377). Since the oscillations of electric charges produce these waves, it might be expected that the waves falling on circuits through which charges could be set in motion might cause the charges to respond to the grasp of the electromagnetic field combination sweeping over them. This indeed they do. Over short distances the waves sent out from oscillating circuits produce currents sufficient to produce visible sparks in nearby circuits of the same electrical period. However, by the time the waves have traveled great distances, they have become so enfeebled in intensity that the response is very small, and it is necessary in some way to build it up or to amplify it.

The first step in the process of receiving these faint impulses is to set up our receiving circuit so that its electrons have exactly the same period of oscillation as that of the incoming waves. Then each feeble impulse as it comes in will be hoarded and added to the next and by the time a few score of thousand have come in, the TUNED receiving circuit will have built up in it a detectable current. This, then, by combinations of vacuum tubes may be amplified many hundred thousand times by adding current from a local supply so that enough energy is brought in for diaphragms to be vibrated and the impulses thus communicated to the air. Of course the high frequencies that transmit these signals when impressed upon the air are still inaudible. But they may be made to beat with other frequencies locally generated by oscillating vacuum tubes,* so adjusted that the beats lie within the range of audition.

It is quite needless to dwell at any length on the remarkable development of modern engineering with respect to wireless telephony and the radio in these connections. Each month that passes sees new complexities—devices to enable sounds to be less distorted, more free from extraneous noises, apparatus more sensitive to the feeblest of impulses. The waves generated by oscillating circuits travel through the earth in the longer ranges and outside the earth almost entirely at higher frequencies. The reception of short waves over great distances and around the curved surface of the earth would not be possible were it not for the fact that the upper reaches of the atmosphere reflect those waves whose frequencies are measured usually in mega-

* This is the method used in so-called HETERODYNE receivers.

cycles. The reflection of electromagnetic waves occurs at any medium where there is a separation of electricity or where electric charge is mobile and free to circulate. Plates of metal act thus as reflectors.

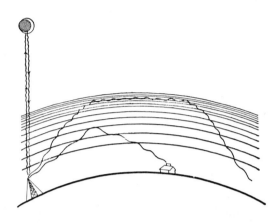

In modern ultra-short waves produced by ultra-high frequency used in RADAR—waves approach in properties those of far infrared light, which, indeed, they are. They are carried in hollow-tube metallic "wave-guides"; they are reflected back from solid objects, even from local atmospheric condensations, such as thunderstorms. Most amazing of all, perhaps, is the fact that the upper ionized layers of the atmosphere become transparent to them, and they have been generated with sufficient energy to be detected, after being sent to the moon, upon their reflection back to us from that companion satellite—a round-trip journey of something less than HALF A MILLION MILES.

As to the origin of this conducting and reflecting layer in our upper atmosphere that traps all short-wave radio waves, excepting only the ultra-short radar, the ultra-violet and shorter radiations in sunlight are its cause. Fortunately, for us these deadly short light waves cannot penetrate our atmosphere.

Of the very intense flood of the ultra-violet rays of the sunlight upon the atmosphere which absorbs them strongly, only a small fraction of the total ever reaches the earth's surface. Ionization of the molecules in the upper layers is thus continually maintained. At the low pressures 50–75 mi. aloft recombination of ions is infrequent, since the distances between individual molecules is very much greater than at the bottom of our atmospheric ocean. Thus, free charges and very mobile ones exist there in the form of gaseous ions of both signs. To the existence of these we owe the fact that we are surrounded by a spherical mirror, as it were. All but the shortest of radio waves that travel radially outward from a source in all directions when they encounter this layer of the air are reflected back, ofttimes to very distant portions of the earth's surface.

The mirror is far from regular, of course. It varies in height and degree

ELECTROMAGNETIC WAVES

of ionization from day to night and also seasonally. The radiation of the sun likewise is inconstant having a definite eleven-year period, betrayed superficially by the numbers of cyclonic storms or sun spots that mottle its surface. In addition to its radiant light and heat, the sun, like any other incandescent object, emits vast swarms of free electrons, many of which succeed in penetrating as far as our earth and producing ionization in our atmosphere. We see the evidence for this as swarms of them from time to time sweep in spiral fashion around the earth's magnetic poles, giving us the northern lights, the aurora. This light is due to the excitation of radiation as a result of the ionization there produced by these clouds of electrons. Accompanying these auroral displays it is very common to have strong magnetic disturbances, the so-called "magnetic storms" that have the same origin. These effects, of course, greatly modify the character of these reflecting layers upon which we rely for long-distance transmission for short-wave radio.

THOSE bands of frequency in the spectrum of radiant energy that are higher than the shortest wireless and radio waves are explored primarily through their heating effects when they are absorbed. Thermocouples, or strips of metal whose resistance is measured as a function of the temperature (bolometers), form the most important of these devices. In recent years we have begun to learn how to sensitize photographic plates quite far out into the longer waves of the infra-red. Although scientifically there is much information of no small importance with respect to the nature of many types of molecules that we can derive from these wave-lengths, there is little of current practical interest in this region save to the research chemist.

TO THE study of visible light and wave-lengths still shorter we shall devote the remaining chapters. Our means of detecting wave-lengths shorter than the deepest visible violet are chiefly photographic. Unfortunately glass, and then quartz and fluorite from which we make our lenses, become opaque successively in this order before we have proceeded even two octaves below the visible region. Mirrors by means of which we can form images as well as by lenses also lose much of their reflecting power for very short waves. To cap the climax of our difficulties, even the air, so limpid and clear, and of high transparency for longer waves, becomes murky and opaque. Two octaves below the visible, air is quite opaque, and what explorations we make must be made in apparatus where the radiation travels entirely *in vacuo*. This region has been consequently one of the most difficult to explore, but its study has reaped a rich harvest of information, largely technical, which has done much to clarify many of the clues as to the structure of atoms—clues that we are able to obtain from the study of the radiation that the atoms emit. A discussion of this sort of thing will also be deferred to the next chapter.

MANY hundreds of times shorter than even these deep ultra-violet radiations are X-rays. Discovered, as we have seen, as an accompaniment to the acceleration, positive or negative, of an electron, they are not unique among electromagnetic waves in this respect. All electromagnetic impulses arise from electron accelerations. Electron accelerations in turn arise from every motion of a vibratory character in which these tiny atoms of electricity can be induced to indulge.

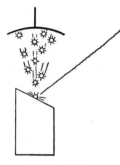

Origin of X-rays

When streams of electrons in the form of cathode rays of any kind are directed against a target, the sudden loss of velocity gives rise to radiation that is especially conspicuous. This is in part due to the fact that under experimental conditions easy to obtain this acceleration may be made so great that the resulting waves are exceptionally short—so short, indeed, that matter, even solid matter, is unusually transparent to them. The opaqueness exhibited for longer waves that we discussed in the last paragraph has quite disappeared by the time waves of the length of X-rays are reached.

Those corpuscular properties that we have frequently referred to, but at no time very definitely described, are found, by the time we reach the consideration of waves as short as these, to be *so much more conspicuous than the wave properties* that many years of experimentation on X-rays elapsed before it was suspected *that they had wave aspects at all*. X-rays travel in straight lines, it is true; but they show no reflection or deviation, when ordinary lenses or mirrors are used. Thus there is no refraction, they cannot be focused, and they form no images. Only by means of those crucial phenomena of interference that we can obtain from them do their wave characteristics make themselves known.

X-rays are powerful in producing ionization, a statement for which we have already given you ample evidence on page 278. This is one of the methods by which they may be detected, of course. They are absorbed, as we have said, much less than most other forms of radiation. However, they *are* absorbed somewhat, and the absorption power of different materials for them varies directly as the atomic weight of the material. There are many important aspects of this fact. By incorporating salts of a heavy element like lead in photographic films, these can be made to absorb enough energy to reduce the radiation-sensitive compounds into the developable condition; and so it turns out that photography is one of the most important and, by all odds, the most common method of detecting and recording X-ray phenomena. X-ray photographs, however, except for certain very special cases, are simply *shadow* photographs. The "pictures" taken by this means of broken bones, or of foreign objects lodged within the body, are solely records of the different degrees of density of shadow cast by the objects in the beam upon the plate or film that lies beyond them and away from the source of radiation. The metals form denser shadows in general than do the less dense materials that go to make up flesh and blood or even bone. Heavier metals cast much denser shadows than do the lighter ones. For these reasons X-ray pictures show no definition or perspective; for the detection of rather minor differences in density within an object such as may often exist, as for example, in the case of dental abscesses, their delineation is often uncertain and frequently ambiguous.

ASIDE from these practical uses, however, and certain therapeutic and still somewhat uncertain differential destructive actions on various types of living cells where they are being used more and more in combating malignant diseases, X-rays and gamma rays, their still shorter cousins, have afforded the physicist one of his most productive fields of investigation. Again the story becomes a story of the relationships between electricity and matter, with the radiation not only bringing to us some knowledge that would otherwise not be readily obtained but illuminating the entire picture and revealing a beauty of design, a harmonious balance, in a landscape of laws and relationships that are so far removed from everyday affairs that understanding on the part of the investigator and learning on the part of his students become matters of the utmost difficulty.

Into this field we shall proceed in our closing chapter. Prior to doing this, let us give a little more time and attention to a part of the important background for these studies that comes to us by means of the ordinary visible electromagnetic radiation that we call by the name of LIGHT.

CHAPTER 39

VISIBLE LIGHT

IT IS rather hard to believe that the common light of every day, and the artificial illumination by means of which we see at night, is of precisely the same character as the many octaves of invisible radiation we have been describing. We are used to seeing light of the visible range, reflected back to our eyes from the objects in our immediate environment, or streaming directly into our eyes from the source if it is farther away and not too intense. In these cases we interpret the sensation as book, friend, something to eat, approaching automobile, airplane beacon, stop light, or star; and we are usually totally unconscious of how we see, since we think and react almost automatically to what we see.

YOU probably never caught yourself wondering as you entered a dark room how long it would take after you turned on or struck a light before that radiance would stream forth and, by being returned from every nook and cranny, reveal the room to you. As far as any practical purpose is concerned, we may consider the passage of light under such circumstances as instantaneous. As you have watched a distant ship or railroad train or factory chimney, however, you have most likely seen a puff of vapor that said "a whistle" to you some seconds before you heard the sound. In thunder-

VISIBLE LIGHT

storms you always hope the crash of thunder will not follow too swiftly the flash of the lightning. The relatively slow speed of sound is familiar to all of us.

On the other hand, did you ever wonder whether perhaps, if the object you were watching were far enough away, possibly the light might shine forth before you saw it, just as in the case of the not-so-distant whistle the sound requires a definite time to cross the intervening space?

GALILEO asked this very question. He did more than that. Since it was his conviction that these matters were not best settled by debate but rather by experiment, he devised an experiment by means of which he thought the answer might be determined. The Florentine Academy subsequently tried it out. Men with lanterns equipped with shutters were stationed on two hilltops that faced each other across the valley in which the city of Florence lies. The distance between these hills was several miles. On taking their positions after dark with lanterns lit, but shutters closed, one group uncovered their light so that the other group could see it. Upon the very instant that the others saw the gleam they returned the signal by uncovering their lamp. The first group attempted to observe the interval between the uncovering of their own light and their seeing the signal from the opposite hill.

Little did Galileo or the other men of Florence dream that the speed of those flashes of light was such as to take them a distance of eight times around the entire earth in one second of time, 186 THOUSAND MILES per second. The irregular and unconvincing data that they obtained measured only the speed of their own physiological reactions and motor responses. They concluded, correctly from their limited experience, that the speed of light was infinite.

FROM the field of astronomy, thirty-four years after Galileo's death, came the first definite evidence that light had a finite velocity. Olaus Roemer, the Danish astronomer, observing with his telescope that wonderful spectacle of the planet Jupiter and its satellite moons, noticed that their periods of revolutions about the planet suffered a curious irregularity. Since our point of view from the earth lies almost in the same plane as that in which these moons circulate, they pass

out of sight behind the planet at frequent intervals and with a suddenness of extinction that makes the precise instant accurately observable to within a few seconds or less.

You may recall that Jupiter is over five times as far from the sun as we are and takes nearly twelve years to make one complete circuit of his vast orbit. During each year, therefore, as we swing around our circuit, we catch up with him and pass on ahead again. For almost all the year we can see Jupiter in the sky. When we are exactly on the same side of the sun as he is, we see the planet on the meridian at midnight. When we are exactly opposite, he lies on the meridian at noon and is then invisible because the sun is also in this position. But a little earlier or later than this, being a bright object and in spite of his then greater distance from us, he is to be seen shortly before sunrise or after sunset as a morning or an evening "star."

ROEMER observed that each year, during the season that the earth was swinging around in its orbit toward that side of the sun on which the great planet also lay, the eclipses of its moons were a little more frequent than later in our season when we were receding from Jupiter as we proceeded on

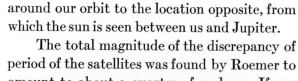

around our orbit to the location opposite, from which the sun is seen between us and Jupiter.

The total magnitude of the discrepancy of period of the satellites was found by Roemer to amount to about a quarter of an hour. If one began his observations on the moons as soon as the planet could first be seen emerging from behind the sun before sunrise, the moons appeared to be circulating a little more rapidly than they did when one watched the planet as the most conspicuous object in the southern skies at midnight. Then, as one followed the periods as Jupiter drifted into the west farther and farther as the year progressed, they were observed to lag more and more. The rate of gain or lag was greatest when our earth was moving on its orbit in a direction at right angles to Jupiter's motion.

ROEMER'S interpretation of this fact was that, although the periods of these moons, like that of any other gravitational system, was very constant, as we receded or approached, the light took a longer and longer time or a shorter and shorter one to reach us from the planet. The total

VISIBLE LIGHT

discrepancy of sixteen odd minutes represented the time taken by light to travel across the orbit of the earth a distance of $2 \times 93{,}000{,}000$ miles.*

Roemer's interpretation met with opposition by some of his contemporaries, notably Descartes, who asserted that the passage of light must be instantaneous. Newton and Huygens, however, supported Roemer, seeing in his work complete vindication of Galileo's plan for measuring the speed of light by the retardation in the arrival of a signal over a great distance.

NEARLY two centuries elapsed before scientific techniques were sufficiently developed to measure a speed so great as this by laboratory methods. Fizeau in 1849 measured the speed of light by a mechanical device that was in essence the old proposal of Galileo. A bright light from a lantern was shut off and turned on by a shutter which consisted of a series of teeth along the edge of a wheel.

The intermittent flashes sent out as the wheel was swiftly rotated were collected on a mirror several thousand meters distant and returned back and focused upon the wheel. When the wheel was at rest, the light sent out between two teeth was focused so that it could be seen through a telescope directed between two other teeth. Upon a sufficiently swift rotation that caused several thousand teeth to pass per second, it was found that the beam no longer could be seen. Each returning flash of light encountered an opaque tooth instead of an empty space in its path, this much displacement of the wheel having occurred while the light was in transit. At twice this speed of the wheel, the light could again be seen, since now during the time of travel of the flash back and forth the wheel had moved the entire distance between the openings between the teeth. This experiment gave 315,000 km. per sec. for the velocity, but subsequent refinements by Cornu and other workers using the same method reduced the value to approximately 300,000 km./sec. (3×10^{10} cm./sec.).

ANOTHER method using a rotating mirror was proposed in 1834 by Wheatstone, whose name is better known for his "bridge"† for the measurement of electrical resistance. Foucault was the first to employ the method which was so greatly improved by Michelson (cf. Plate 72, p. 332),

---

* There is a curious numerical coincidence in this connection. The $16\frac{2}{3}$ min. retardation of the eclipses is, of course, 1,000 sec. The diameter of the earth's orbit, 186,000,000 mi., divided by 1,000 gives the 186,000 mi. per sec. for the light speed. Roemer, not having quite as exact knowledge for the size of the earth's orbit, actually obtained 192,000 mi. per sec.

† Cf. p. 120, n.

the great American physicist, that his experiments, carried on over a period of nearly half a century, represent almost the ultimate of human achievement to date in a precision measurement of this most important of physical constants.

IN THE Foucault and Michelson experiments a flash of light is sent from a swiftly revolving mirror to a distant one. On being reflected back to the place of origin, the revolving mirror, having shifted in position meanwhile, reflects the spot of light to a position to one side of the place it would have come had the mirror been at rest. A most ingenious arrangement of lenses and mirrors by Michelson enabled the flash to be sent over twenty miles and back without a loss of light great enough to render it invisible. Michelson's arrangement makes it now possible to demonstrate and measure the prodigious speed of light within the confines or an ordinary room or laboratory. Such an arrangement is illustrated in Plate 93. As a result of his work the speed of light is known to approximately a mile per second. Michelson was the first American physicist to receive the Nobel Prize, and was given this award as a recognition of this work and of his other contributions in the field of interference, to which we shall recur later.

ONE of the first triumphs of the wave theories of light was achieved when the velocity of light through a transparent fluid medium was measured in the laboratory and found to be less than its velocity in air (or *in vacuo*). This was in accord with the wave interpretations of the manner in which refraction and dispersion took place in such media. To account for the same phenomena on the corpuscular theories that Newton upheld, the velocity of light should have been greater in such media, and not less.

The discovery of *interference*, however, was the crucial and convincing matter. We have noted that in Newton's rings the evidence for interference was complete, but it failed to be recognized as such by its discoverer. A little later Huygens, famed also for inventions in other fields of physics, proposed a theoretical construction for the representation of wave-fronts, based on wave conceptions, that is used in physics classes to this day.

His proposition was this: If you take the position of a wave-front of a beam of light at any instant, then any subsequent wave-front of this same beam may be found if you consider that each point of the first wave-front is the source of a tiny optical disturbance that progresses outward in all directions

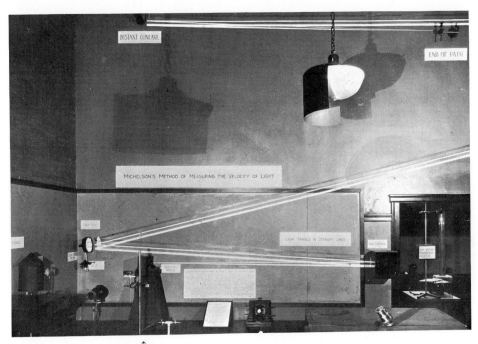

↑
Rotating mirror

DEMONSTRATION OF THE METHOD OF MICHELSON FOR THE MEASUREMENT OF THE VELOCITY OF LIGHT

AS SET UP IN THE PHYSICS MUSEUM IN THE UNIVERSITY OF CHICAGO

PLATE 93

Light from an incandescent lamp is focused on a rotating mirror (not seen) on top of the pyramidal base left of center. This beam is flashed out over the path indicated by the lines on the walls, being reflected by five mirrors. It returns over the same path to the rotating mirror, whence it is reflected into the observer's eye, and viewed through a microscope.

The path distance is 65 ft., and the time taken by the flash to traverse it is about one fifteen-millionth of a second. During this time the mirror, driven about three hundred revolutions per second by a turbine, shifts in position sufficiently to show in the microscope a deflection of the beam by a perfectly measurable amount, to one side of the position to which it returns when the mirror is at rest.

By suitable timing arrangements for the speed of rotation of the mirror, this apparatus is capable of measuring the speed of light to an accuracy of about 1 per cent.

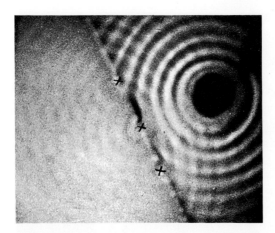

PLATE 94 DIFFRACTION OF RIPPLES. Waves on the surface of water emerging from source on right encounter an obstacle in which there are three holes (**X**). Proceeding through these holes, the individual semicircular wavelets may be seen. This illustrates the fact that wavefronts progress in the manner described in Huygens' geometrical construction.

from it. Any surface that can be found that then makes tangential contact with the surface of all of these tiny wavefronts at any given subsequent moment is the surface of the wave-front at that later time. Whether or not Huygens was dominated by the idea of interference and its crucial character for settling this controversy when he proposed his construction is for us unimportant. That his construction is entirely correct can be shown by the form that ripples on the surface of water assume as their wavefronts pass through small apertures in a screen (Plate 94), and has been attested by the manufacture of innumerable lenses the forms of whose surfaces have been calculated by methods that are extensions of those of Huygens. Furthermore, in terms of this construction it is very easily demonstrated how interference in the case of light may be made visible. The experiment we are about to describe was first performed by Thomas Young in 1801.

Suppose we take a source of light, focus its beam on a small aperture, and then let the expanding cone of rays on the other side illuminate two other apertures. The apertures may be either small holes or narrow slits perpendicular to the plane of the paper and thus appearing as small holes in cross-section.

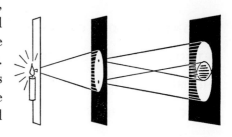

The cones of rays diverging from these second apertures will ultimately expand enough to overlap and still not become too faint to see if the distance between the holes is not too great. In the center of this overlapping region, on a screen to catch the light, there will be seen some faintly colored, fine bands of light and shade. A magnifying lens is a great help and may be

VISIBLE LIGHT

used so that you look through it facing the light at this exact region. Then the intensity of the phenomena is greatly enhanced, and the character of the bands may be studied under considerable magnification.

How can there be shadows in the center of two overlapping beams of light when no objects are interposed to cast them? Obviously these bands are not shadows. How, then, is it possible for light added to light to produce darkness? We saw that in the case of the ripples, in spite of the fact that there were two trains of waves passing over the surface, there were locations where the surface remained quiescent—the nodes of the stationary wave pattern. Waves reflected back and forth along a string showed the same effect even more clearly. In light a region of no disturbance means absence of light, which is darkness; but to argue this implies, of course, that light, at least in certain of its aspects, is a wave phenomenon.

ADHERENTS of the corpuscular theory in those early days died hard. Indeed, when a great authority stands on one side of an argument, his side is particularly hard to overcome even in scientific matters where the discussion is entirely objective. The great French scholar in optics, Fresnel, obtained the same effects by two other methods both entirely different in arrangement from the experiment of Young, and the conclusion became finally accepted. It was not, however, until the invention of the INTERFEROMETER by Michelson, that every possible, as well as every probable, objection was finally anticipated and answered.

The details of interference methods in optics are far too technical for us to discuss with you to any advantage. Only with respect to one device—important, above all others, for the analysis of light—whether of visible, infra-red, ultra-violet, or X-ray character, must we go a little more into the details of interference. We can develop the idea from the experiment of Young, using the principle of Huygens.

SUPPOSE we have spreading out through two apertures two identical wave trains alike in wave-length, frequency, amplitude, and also in exact phase relation with one another.* Let us consider what will happen at various points on a distant screen that receives the light, neglecting, for simplicity of drawing, any lenses that may be needed to form images so that we can focus our eyes on the result.

* This last condition is necessary if the resulting interference phenomena are to be visible to human eyes.

In the accompanying figure, A and B are the apertures, and the point P is midway between them. The line P_0P is at right angles to the line AB, and P_0 lies on the screen where we are to observe the light. Obviously, P_0 is equally distant from A and B. We have drawn it on a scale greatly distorted with respect to the sizes of the waves coming from the apertures which are represented diagrammatically as they first emerge from the holes A and B and whose lengths are marked off on both the lines AP_0 and BP_0. We will suppose there are just 7 waves exactly contained in each of these distances. In phase as they leave A and B, these two waves will be in phase after having

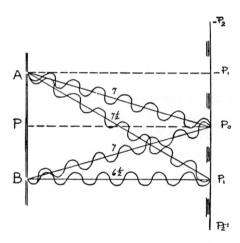

traveled equal distances and upon their arrival at P_0. Their effects will add, so that at the point P_0 there will be an amplitude of that disturbance we call light just twice as great as if we closed either one of the apertures.

The point P_1 next takes our attention. This is offset from the center, so that the distance AP_1 is greater than BP_1. We select P_1 so that the difference between these two distances will be just one entire wave-length. We have drawn $7\frac{1}{2}$ waves in one path and $6\frac{1}{2}$ in the other. Thus, at P_1 the wave trains from the two apertures will arrive in step, and their disturbances will add. P_1 is like P_0 a point of intense illumination.

Between these two points P_0 and P_1, however, it must be possible to find some intermediate position that is just one-HALF wave-length farther from A than it is from B—say $7\frac{1}{4}$ waves distant from the former and $6\frac{3}{4}$ from the latter. At such a point the waves will arrive *exactly out of step*, and the disturbance made by one train will always exactly cancel that made by the other. The sum of the disturbances will be zero, and the region will be devoid of any light.

Thus on the screen, where we have drawn dark rectangles on the right-hand side of the figure, will appear equally spaced bands that grade from intensity that corresponds to double amplitude of wave at points $P_{0,1,2}$, etc., below and $-P_{1,2}$, etc., above, to intensity zero at positions halfway between these points.

VISIBLE LIGHT

IF WE move our eye across this region we will see BEATS of light. If we look at the entire region from a larger viewpoint, we are looking at stationary waves of light. The aspect is precisely that of interference in any kind of wave phenomenon. The fact that such an effect may be produced in more than a score of different ways is proof positive that visible light possesses wave characteristics.

Long electromagnetic waves, infra-red, ultra-violet, and, by methods much more refined, X-rays and gamma rays, may be caused to produce similar patterns. The difference in all of these cases is but one of SCALE. The size of the wave-length determines the scale, of course, since P_1 was so chosen that the distance from it to A was one wave-length farther than it was to B. For longer waves P_1 would have had to be farther out, and for shorter ones farther in toward P_0. Indeed, this experiment is one of the very best ones that we have for the MEASUREMENT OF WAVE-LENGTHS.

FOR precise work, since it is difficult to make measurements upon the rather broad fringes that are given by only two apertures, and if many different waves not very different in length are present, it becomes impossible to discriminate between them. We can improve the pattern both for the measurement of any one wave-length and for the discrimination between two but slightly different, by using, instead of just two apertures, a very great number of them equally spaced and very close together. We accomplish this by ruling a great number of very fine lines—indeed, many hundred thousand—on the surface of a polished mirror and then reflecting therefrom the light that is to be analyzed. One single slit from which the light is emitted and a lens to focus it after reflection from the ruled surface are also required.

Such ruled surfaces are called DIFFRACTION GRATINGS. If white light is reflected from them, we find that it contains all wave-lengths, since we get as a result a continuous band of color—red at one end, and graded through all the intermediate hues of yellow, green, blue, and indigo, to violet at the other. Thus a ruled grating gives us a spectrum of all wave-lengths *sorted out* just as a prism does.

THERE are two or three important differences. First, the colors are diffracted to one side of the position of the direct beam in the *opposite* order to that in which they are spread out by a prism. Violet, the shortest wave-length in the visible light, is displaced the least, as it is shorter, and red; the longest wave is displaced the most. In the prism it was the shorter waves that were most slowed down in glass, and consequently bent aside the most. With grating spectra, it is possible so to arrange matters that the actual amount of the displacement is exactly proportional to the wave-length. Such spectra are called NORMAL. Prismatic spectra are not spread out in color arrangement in a way that is simply related to the wave-length. Indeed, no two types of glass give exactly the same degree of displacement, and in all the violet region this color is spread out to a much greater extent relatively than is the red. On the other hand, the grating is rather wasteful of the light. Quite a little of it goes into the central undeviated image; the rest is spread over several different regions on either side, $P_{1,2,3}, -P_{1,2,3}$, etc. In the prism, except for a little reflection and absorption in the glass, all the light goes into one, and only one, band of color.

For faint sources of light, such as stars, prisms are still widely used by astronomers to analyze the light. For precision work in the laboratory or with bright sources like the sun we use gratings, of course.

SO FUNDAMENTAL is the relation between the amount of deflection of the light and the wave-length, that we give you this relation in a footnote accompanied by a figure for a many-lined grating in a dimensioned drawing.*

---

* We will here make the construction of Huygens more explicit and draw the wavelets that pass each aperture. Light is produced at an angle θ_1 from the perpendicular, where $\sin \theta_1$ (defined, in a right triangle, as the ratio of the side opposite the angle to the hypotenuse) is here given by $n\lambda/nd = \lambda/d$.

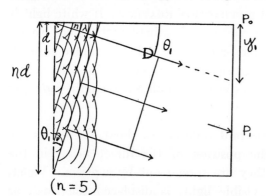

(n = 5)

There is also another direction, θ_2, where $\sin \theta_2 = 2\lambda/d$.

These directions are those where the first, second, etc., ORDERS of spectra are to be found. Of course, the same angles θ are to be found in the larger triangle where their sines are equal to y_1/D, so that there is really none of this mysterious "sine" business of trigonometry involved if you refuse to learn anything about it.

By similar triangles, and without any trigonometry at all, for the first order, $\lambda/d = y_1/D$,

$$\lambda = \frac{dy_1}{l}.$$

VISIBLE LIGHT

If we work out the details of an actual experiment of this sort, we will find that the dimension of a light wave, i.e., its wave-length (λ), is excessively minute.

Yellow-green light, to which the eye is most sensitive, consists of waves the distance of which from crest to crest amounts to but 0.000,055 cm., and the actual length for any particular color may be determined by good gratings with a precision of .000,000,000,001 cm., i.e., 10^{-12} cm. In this field of measurement of wave-lengths of light with great precision, as well as with respect to the speed of light, Michelson made many original contributions and also many refinements of older methods of measurement.

SUCH precision is possible through the use of the principles of interference of which he was a master. Stationary waves in any other type of a wave phenomenon may be produced, as we have seen (p. 364), by sending two identical wave trains in opposite directions past one another. The resulting nodes and loops, which in the case of light are alternate regions of light and of darkness, are of course separated, as we have also seen, by a distance of half the wave-length. With waves as small as these, such distances are impossible to observe in stationary waves, produced in this manner directly, with any degree of precision unless, by some means or other, these patterns can be greatly magnified. Passage of light through a few slits, as we have shown, makes it possible to magnify the stationary wave's pattern, as it were.

MICHELSON invented another device to magnify them which is called the "interferometer." This actually enables one to send two identical beams of *large cross-section*, not in opposite directions, but in the same direction or rather at any chosen excessively minute angle with one another. The stationary waves that then may be seen are tremendously brighter than in Young's experiment, and the full beauty of these interference fringes may be appreciated. They are gorgeously colored parallel bands (or fringes) in white light, or alternate bands of light and of darkness in monochromatic light of (a single color) just as in Young's experiment. The details of the apparatus are rather technical, and you might not understand them any too

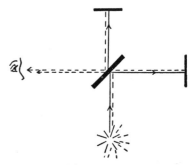

Path of light in interferometer

well even after reading many pages of description. The applications of these interference methods in optics are of signal importance in many other fields of investigation, and we will recount but a few of them.

One of Michelson's first applications was to use a highly accurately determined wave-length of a certain color of red light to serve as a standard of length to take the place of the international standards if for any reason these standards should accidentally be lost or destroyed. As a result, these could now be replaced by others with such precision that no significant difference could ever be detected. Consideration of the fact that upon these fundamental standards of measure depend all transactions involving the sale and exchange of commodities will at once show you a practical side to purely scientific studies that perhaps we have not stressed sufficiently in these pages.

IN ASTRONOMY, likewise, Michelson was the first to show how, by means of ingenious methods of producing and measuring the interference of light from distant stars, the actual diameters of those stars might be measured. By such devices attached to telescopes it has been found, for example, that Antares, the bright ruddy star rather low in the southern skies in midsummer evenings at latitudes of about 45°, has a diameter of 400 *million miles.* So vast is the size of this star that if our sun were placed at its center with its planets around it, Mercury, Venus, the earth, and Mars, as they circled round the sun, would all lie within the star. Many other of the brighter stars of the class known as "giants" have also thus been measured.

With respect to many problems of atomic structure that will be discussed in the next chapter in connection with the radiation that atoms and molecules emit when disturbed, the precision of interference methods has been of the utmost importance in enabling us to obtain entirely independent confirmation, by radiant means, of hypotheses we have built up as a result of electrical and other quite different types of experimentation.

FINALLY, Michelson's real purpose in inventing the interferometer was to discover something about the nature of this mysterious medium that we had to imagine was necessary to carry waves and transmit energy. Michelson's reasoning may be explained by an analogy.

Suppose we have an oarsman who can row 5 mi. per hr. across still water. To row to a point 7.5 mi. away and to return will thus require 3 hr. If, however, the man rows along a flowing river upstream at first and then back along with the current, he will never be able to make as good time over any distance to and fro as he would if the water were not in motion.*

The physicists of the nineteenth century reasoned that if there were a medium that carried the light, in the sense that the river carried the oarsman, then a beam of light which was carried on an object that itself was moving through this medium would require a longer time to go upstream and back again than if it were sent in a direction to and fro in a direction along which the motion of its vehicle had no component of velocity through the medium.

The difference in time involved in the case of a velocity as swift as that of light would be exceedingly minute. But the utilization of the principles of interference provides similarly a most delicate and sensitive detector. It is a situation similar to the putting of a Sherlock Holmes on the trail of a Raffles.

MICHELSON'S invention of the interferometer was made specifically for the testing of this effect. Two beams of light were split up out of a single one by a mirror set at 45° to the incident rays. In this way these two beams were started out at right angles to one another. They were reflected back and forth upon themselves several times and then reunited in such a manner that they would produce interference—visible interference taking the form of the bands of light and darkness that we previously have referred to as the "beats of light" (p. 407). These bands could be produced only if the paths of the two beams, after being split apart and before being reunited, were almost exactly alike—so nearly alike, as a matter of fact, that only interference methods like these could determine that they were at all different.

As the earth travels around the sun in its yearly journey, its rate of motion is of course considerable. It covers a distance of about 46 million miles

* To convince yourself of the truth of this statement, which is not obvious to everyone, invent some particular examples and then work them out for yourself; e.g., if the river flows 2 mi. per hr., he can make progress at the rate of only 3 mi. per hr. upstream and can return at the rate of 7 mi. per hr. To go to a point 6 mi. upstream will thus take him 2 hr.; and to return, $\frac{6}{7}$ hr., a total of 2.86 hr. up and back. On still water, however, rowing 5 mi. per hr., he can cover the round trip total of 12 mi. in $1\frac{2}{5}$ or 2.4 hr., i.e., in nearly half an hour less time. The limiting case is always obvious. If the river should flow with a speed equal to his rowing speed, he could row downstream to any given point in just half the time he could on still water; but since he could make no progress back, the *round trip* would require infinite time.

each month, which is better than a million and a half a day, or about 18.5 mi. per sec.

A person with a well-trained sense of spherical geometry and knowledge of the relative positions of earth and sun, and of his own position on the earth, by a little figuring is able to tell approximately how he should place himself so as to be facing in the direction in which he is moving or at right angles to it. Thus he can aim one of these beams of light so that it gets the maximum component of the orbital motion of the earth and then the other will get the minimum.

IF HIS interferometer has been previously adjusted and he slowly rotates the whole instrument about a vertical axis, twice in each revolution, each one of the two beams will be in the proper direction for that beam to be affected by the maximum component of the earth's motion. Ninety degrees from this position of the instrument the other beam will be in this orientation; hence the two beams are continually interchanging positions where the effect of the motion will be the most detectable.

The effect of the passage of one beam to and fro over its path in a longer time than the other took would be indicated by just the same effect as would occur if that path actually became a little longer than the other. The position of the bands of interference would shift a trifle, but nevertheless by an amount amply sufficient to be not only detected but measured.

THIS was the famous ETHER-DRIFT EXPERIMENT, known by the names of Michelson and Morley. From the known velocity of the earth and the known sensitivity of their instrument, a well-measurable effect was expected. To their astonishment and that of the entire scientific world, no effect whatsoever was detectable.

Most of us are often tempted to rationalize for ourselves experiences which we do not entirely comprehend. We must admit that the physicists, for quite a number of years after these results were announced, were guilty of a certain amount of rationalizing. However that may be, the negative result of this experiment became, many years later, the origin for the famous theory of relativity of Einstein.

VISIBLE LIGHT

D O NOT be now totally dismayed! We shall not attempt an exposition, in these pages, of the theory of relativity. You may read several famous and able expositions of it, alleged by their authors to be popular, simple, and easy to understand. We would not want to enter into competition with them. Our purpose has been mainly to point out how some of the simplest conceptions, such as rowing up and down a flowing stream, may lead to matters that are most utterly profound. Of course a large part of our difficulties in these connections is that we *assumed* that the passage of radiant energy was to be compared with the passage of the oarsman. Evidently this is not justified when we are dealing with these waves with respect to whose "waver" we know nothing at all. The light-bearing "ether" of the nineteenth century has been in serious difficulties ever since the experiments of Michelson and Morley.

OTHER important applications of interference, by-products of this brilliant effort to storm one of Nature's yet impregnable citadels, have been amazingly successful. We have mentioned several practical connections of this type of work in former pages. With respect to the more technical applications—as to how interferometers can be used with considerably greater precision than gratings to measure the light waves themselves, how they give us additional evidence in conjunction with other optical devices as to the nature of the origin of radiation from atoms and molecules, and of the rôle that the electric building-blocks of matter play in the process—we likewise very quickly would become involved in technicalities that would be meaningless, and therefore most boring, to you unless you possess the time, the interest, and, most of all, the determination to embark upon a somewhat extended course of reading and of training in mathematical forms of language that will enable you to go further and to understand. Of course, this takes time and effort. If the time and effort is spent, the person who does it is well repaid.

ANY general interest you may have in these matters has now been covered almost as completely as it is possible to do so. In so far as it is possible to pass those great fields of study that are involved in physics in review before you, we have about completed our attempt to do it, taking for granted you have no intention of ever becoming a physicist, and treating you accordingly, with that same respect for your personal wishes as we would hope you might have for ours had our positions been reversed.

There has been, however, one most serious omission. The study and analysis of light, as a grating or a prism spreads it out before us, has not yet been made. Ways and means of doing it have been gone into thoroughly. We have shown how the rainbow of color may be produced, by dispersion and by interference of radiant energy appearing in its rôle as the possessor of wave characteristics.

In our closing chapter we shall consider the rainbow itself.

CHAPTER 40

RADIATION AND ATOMIC STRUCTURE

WHEN Isaac Newton passed a beam of light through a prism, he performed the first experiment of a long series to be carried on by his successors, a series that was destined to bring to human beings more detailed information of the nature of the physical world than has resulted from any other single enterprise up to the present time. Indeed, it might be claimed by some that the study of spectra, directly or indirectly as these studies have confirmed and supported those in other fields, has been more fruitful than all of the others put together. We would hesitate to put forth such a claim even today when, in physical research, the study of spectra, i.e., the field of SPECTROSCOPY, certainly claims a large share of attention at meetings of, and discussions by, physicists.

In the other physical sciences, especially astronomy, and to a somewhat less degree in chemistry, spectroscopy is of signal importance. Independent as mathematics can be as the queen of sciences, it has frequently been true in the past that many of her most notable achievements have come as the result of following a path into which her footsteps were directed as a result of the need of a solution of some physical problem for the analytic expression of which no solution was known.

Modern developments of knowledge about spectra have likewise raised problems to be laid upon doorsteps over which only the most skilful and best-trained mathematicians might pass. Any person who would be so rash as to contemplate a career for himself in the field of physics at the present time recognizes very early in the pursuit of it that a skill in mathematics far beyond the ordinary garden varieties is of the utmost importance for him to acquire.

Contributions in current journals devoted to technical spectroscopy in many aspects resemble in appearance nothing so much as a purely mathematical discussion. Not only do equations bristle, in them; there is often nothing but equations there.

The significance of this for us at the present moment is that it indicates the depths of penetration of this subject of spectroscopy into the most fundamental aspects of the inorganic physical world, and the recondite character and complexity of the relationships that are being found therein. It may be that there will follow a period of some simplification; that when we know somewhat more about matters that are at present still rather obscure we shall acquire a better synthesis, though as yet we do not place very much hope in this. In the rather immediate past when we have had such hopes they have frequently been dashed by the discovery of additional confusion.

THE chief reason why the analysis of light is so important is that by means of a beam of light, in a figurative sense, we may travel with almost equal ease for a very great distance into the microcosmos of atoms, or into the macrocosmos of the stars. With respect to the latter, we really are going in both directions at the same time, for the radiation we receive from stars comes from the *atoms* in the *stars*, from a microcosmos within a macrocosmos.

It will be noted, as we proceed in this discussion, that we refer to astronomical sources quite as much or more than we do to terrestrial ones. Indeed, many of the more simple and fundamental relationships in spectra were originally found in the study of the spectra of stars.

IF WE put a poker in a fire and withdraw it rather soon, we can recognize that it is hot without touching it, or without having heated it to the slightest degree of incandescence. We can *feel* the *radiant heat* that it throws out as electromagnetic radiation in wave-lengths that are too long to affect the optic nerve, but still short enough to be readily absorbed by our skin

RADIATION AND ATOMIC STRUCTURE

where their incident energy is converted to molecular kinetic energy and a rise of temperature is experienced.

Further heating of the poker increases the intensity of the invisible heat that it sends out, but it does something more. A deep-red glow soon becomes visible; shorter waves are being sent forth and in sufficient intensity now to be above the threshold of vision. The heated rod now affects simultaneously two senses—sight, as well as feeling. Is the rod any different from what it was before? No, except with respect to the degree of agitation of its molecules it is quite the same. (Cf. drawing, p. 111.)

As these molecules, however, acquire more and more energy of motion, more energy in the radiant form is sent out into the surrounding region. The total amount of this energy is found by experimental measures to increase with the *fourth power* of the temperature of the rod when the latter is reckoned on the natural Absolute scale of temperature. This is the RADIATION LAW OF STEFAN.

AS THE rod gets hotter, it gets whiter in color, which means that the green and blue colors increase at greater rate than does the red. We cannot heat objects in the laboratory more than a few thousand degrees, and so they never reach that degree of "blue heat," as contrasted to red heat, that we find in the hotter stars, the temperatures of whose outer atmospheres, whence the light we see originates, may be as high as 20,000°–25,000° C. Of the more familiar stars, Rigel (β Orionis)* ranks among the highest in temperature, 16,000° Absolute, and is a conspicuously blue star. Vega (α Lyrae), another bluish star, rates 11,000° of temperature. Toward the other end of the scale of stellar temperature lie Antares (α Scorpii) and Betelgeuse (α Orionis) at about 3,000°, Ras Algethi (α Herculis) at 2,500° and Mira (o Ceti), variable between 1,500° and 2,500°, approximately.

Over the entire range of temperature thus made available we find this common property of larger and larger relative amounts of radiant energy in the blue-violet region being spread out in the spectrum of the radiation as the temperature of the source increases. There is another law that precisely expresses the connection between the wave-length that sends out the maximum energy and the temperature. WIEN'S DISPLACEMENT LAW says that

---

* Stars in any given constellation are frequently referred to by the letters of the Greek alphabet, α, β, γ, ϵ, etc., approximately in their order of brightness followed by the name of the constellation. Thus Rigel is indicated by its alternate name as being the second brightest star in the constellation of Orion. To find it in the winter time, consult any good star map.

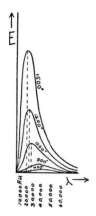

Distribution of energy in radiation with increasing temperature

the product of the wave-length containing maximum energy by the absolute temperature is a constant quantity, and that the energy in this wave-length increases with the *fifth* power of the temperature.

Because we have abundant independent means of establishing the truth of these relations theoretically and have fully verified them experimentally as far as possible on earth, these laws are one of the most widely used means of determining stellar temperatures.

The light from an incandescent solid or liquid, then, when analyzed by prism or grating into its spectrum, carries thus a thermometer, as it were, for measuring the temperature of the source—quite a long-range thermometer we must admit, when starlight is in question.

THE continuous band of color of which we have been speaking is characteristic of the light that comes from *any* incandescent solid or liquid substance. The relative intensities of different portions of it enable us to determine with considerable precision the temperature of the source. The sun's surface emits a CONTINUOUS spectrum, the total energy of which, and the distribution of energy in which, indicate a temperature of 6,300° C. for the center of the disk, where we look directly down into and through the atmosphere of the sun. At the edge of the sun (the "limb" as it is called) our line of sight passes at grazing incidence through higher regions of the solar atmosphere, and the light from this region is quite distinctly fainter in intensity. It is yellower in color also, indicating that the temperature of these higher regions is only about 5,000° C.

WHEN gases or vapors are heated to incandescence or excited to luminosity by electrical or other means, their light, when analyzed by a prism or a grating, gives another type of spectrum altogether. The radiation from gases does not contain all wave-lengths but only certain isolated ones. Consequently, the light, when sorted out, consists only of these detached items. A neon sign looks very red and one of mercury vapor, blue; but the colors in these cases have no significance with respect to the temperatures of the two different gases in the tubes. The electrical constituents, the outer electrons especially, in the neon atom are, as we have seen, of definite number, and make up a characteristic structure. When disturbed by the relatively infrequent jostlings that occur because of their comparative isolation in the gaseous state, this ar-

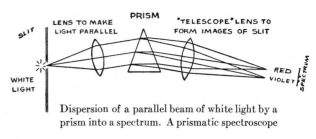

Dispersion of a parallel beam of white light by a prism into a spectrum. A prismatic spectroscope

rangement oscillates in a characteristic way and sends out radiation that is characteristic also of that particular atom. This radiation from neon consists chiefly of a number of intense rays of various shades of red and yellow.

In passing the light through a slit, then on through a prism or from the surface of a grating to disperse it and then through a lens to form an image that we may look at, the image, of course, is that of the slit of the instrument. If the light is but one color of red, there is but one image of the slit, one of that particular color. If however, there are several different shades of red—say fifteen different separate wave-lengths in the radiation—then there will be fifteen different images of the slit—one for each of these. If several of the colors are only very slightly different, i.e., of nearly the same wavelengths, the images corresponding to them may not be clearly separated or RESOLVED. Now the object of the slit becomes apparent. Two colors, even if quite different, would be seen only as a blend if the slit were absent or very wide. As the slit is narrowed, the images narrow. Thus one obtains, in the case of a gaseous source, a spectrum of isolated fine narrow lines, slit images, one for each characteristic color radiated by the atoms in the source.

The spectrum of neon thus consists chiefly of quite a number of red and yellow slit images or LINES; mercury vapor (Plate 95), on the other hand, has two intense radiations close together in the yellow, a very intense one in the green, and several strong blue and violet wave-lengths. The color difference between these two sources is so obvious, of course, that any but a colorblind person could instantly distinguish neon from mercury vapor.

Helium gas has an intense long yellow radiation that is its earmark, or wave mark. Sodium vapor has two yellow radiations *not sensibly different in color* from helium, so that the unaided eye alone could not detect any difference in the yellow light of these two gases. The quantitative separation by grating or prism, however, shows them clearly as distinct and separate slit images, very like in color, but dispersed or diffracted to different amounts and occupying on the wave-length scale quantitatively different places.

The spectra from incandescent gases are thus known either as LINE or as BAND SPECTRA. This distinction is due to the fact that the light from *molecules* in the gaseous condition consists of so many lines that to the eye they may often group themselves more or less in bands, or associated groups. Often these superficial appearances are of no significance. More often, however, the clustering of numbers of lines into band arrangements is of profound significance with respect to the nature of the radiating material. Line spectra of unbanded character are characteristic of the radiation from ATOMS in the gaseous state; banded spectra are characteristic of the MOLECULES, groups of atoms in close association and radiating as a unit rather than as individuals. Thus line and band spectra are frequently classified as ATOMIC and MOLECULAR SPECTRA, Plate 96B.

OF ALL atomic spectra, probably that of hydrogen (Plate 96), the lightest element, has been of the most importance. Certainly it is the prototype of all other spectra that originate in atoms. In the laboratory, under the usual conditions of excitation, where an exhausted tube is filled with hydrogen gas, H_2, at low pressure, and a glow discharge sent through it in the manner described in pp. 294–95, a very complicated spectrum is observed, practically all of which comes from the hydrogen molecules. There is, however, a brilliant red, a strong blue, and some fainter violet lines that can be shown to be due to the atom, Plate 96A.

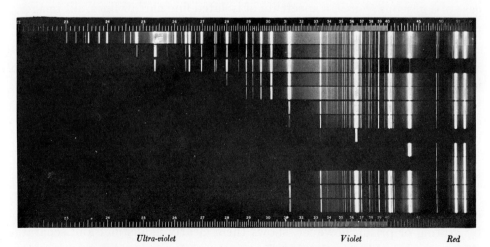

PLATE 95 LINE SPECTRUM OF MERCURY. The spectrum of a mercury vacuum arc photographed through quartz (top) and through various different kinds of glass.

RADIATION AND ATOMIC STRUCTURE

These same atomic lines form a very conspicuous part—indeed, the major part—of the spectra of all those stars whose temperatures range around 10,000° C. Furthermore, stars much hotter and much cooler show this same spectrum as well as many others. It is conspicuous in the sun. As the spectrum of hydrogen appears in the stars, it is a much more elaborately developed series of radiations than we usually obtain from the gas in the laboratory. The reasons for this are well understood, but we shall not go into them. This illustrates the value of studying the spectra of elements in the stars where they may be observed under conditions we can never hope to attain on earth.

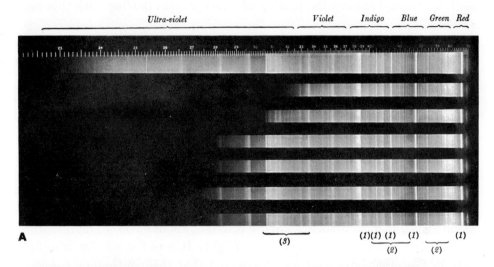

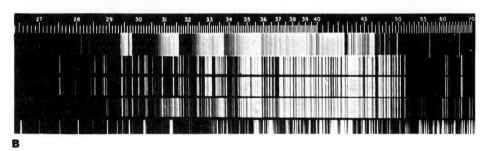

PLATE 96 (A) SPECTRA OF HYDROGEN ATOMS (1), HYDROGEN MOLECULES (2), AND OF WATER VAPOR (OH ONLY) (3). Top photograph through quartz prism and lenses entirely. All others with various kinds of glass in the path showing how glass absorbs the ultra-violet, but that some kinds are more transparent to it than are others.

(B) ATOMIC AND MOLECULAR SPECTRA. Photographed through a prism and lenses of quartz. Immediately below a scale of wave-lengths at the top is the band spectrum of a molecular system. The four lower spectra are line spectra of several metallic elements.

The chief point of interest for us is that the radiations form a definite SERIES, and their respective wave-lengths conform to a simple mathematical formula. By analogy it may be considered as something similar to the overtones of an organ pipe. There are definite relations between the frequencies that a pipe, a vibrating string, or a bell emits when sounded. These relations in the light that is given off from an atom are, in a figurative sense, the radiant tones and overtones that originate within its structure, and by their quality reveal much about the nature of their origin.

The subject is one that we can suggest merely. An adequate development of it would require the writing of another book dealing with this matter alone.

We have said that hydrogen was the prototype of all spectra. It is in this sense that this is so. The spectra of all the other atoms, when their radiation is excited while they are in the gaseous state, are arranged in series, as is that of hydrogen. From these atomic spectra and their detailed analysis we have learned about most of the details that we gave of the arrangements of electrons in the atoms, in that table which you did not understand any too well on page 310.

IF WE analyze the light of the sun by means of a spectroscope, we find a continuous band of color—light of every wave-length—which indicates, as we have said, an incandescent solid (or liquid) at about 6,000° C. temperature. But this is not all of the spectra of sunlight. It was Fraunhofer who first, early in the nineteenth century, discovered that this continuous spectrum of the sun was overlaid with innumerable fine DARK lines (Plate 97). These dark lines have been found to correspond very closely in position to the bright lines that are radiated by glowing gases. There must be some very close connection. An accurate check-up of the spectrum of sunlight shows the lines of hydrogen, calcium, magnesium, silicon, iron, titanium, and many other spectra that we know from terrestrial sources when we excite the vapors of these substances by heat or electric discharges. Can it be that there are outer regions of the sun that are gaseous, and contain these elements? If so, why do we not get bright lines? As a matter of fact, these lines *are* bright lines in the atmosphere of sun, if we look at that atmosphere alone without the dazzle of the incandescent lower layers.

We can do this at times of total eclipse, just at the instant the intercepting bulk of the moon passes over and cuts off the light of the sun's incandescent surface. Then for a few seconds only, the spectroscope shows a brilliant bright-line spectrum of the very same elements that we see as dark lines against the solar disk.

The same experiment may be performed with certain elements in the laboratory. A small chunk of sodium burned in a flame gives forth an intense yellow radiance. Two brilliant yellow lines constitute its entire visible spectrum. If we burn the sodium in an enclosed chimney, through the opposite sides

PLATE 97 ABSORPTION SPECTRUM OF THE SUN. Several different exposures to the ultra-violet part of the spectrum of sunlight. The fine absorption lines that overlie the continuous portion do not show except on the left-hand portion. All the light to the left of the center is invisible. It is that region of sunlight responsible for the photochemical and physiological effects, vitamin production, sunburn, and the destructive effects of light on living organisms.

of which two windows have been placed, and if we send through these windows the strong light from an incandescent lamp, and then look with a spectroscope at the light after it has thus pierced the incandescent sodium vapor, we will see two velvety black lines crossing the continuous spectrum of the lamp in the precise position where we see the two bright lines of sodium glowing when we quench the continuous source. The experiment leaves no doubt in our minds as to how the spectrum of our great luminary should be interpreted.

An incandescent radiating gas or vapor, when light from any other source shines through it, absorbs exactly those same wave-lengths that it normally emits. The effect of dark lines is largely one of contrast. Absorbing this light, the gaseous molecules acquire energy which they must later lose by reradiating the very same energy out again. The point is that the energy reradiated is sent out in all directions through the vapor; hence only a small amount of it is returned back to your eye as you look through the vapor in the line of the beam of radiation from the outside source. You perceive as dark, lines that are dark only by contrast.

This contrast effect occurs in other connections that have nothing to do with spectra. We see it in the case of sunspots. These look black as ink on the sun's surface; yet the temperature of these regions is around 5,000° C.;

they are far more brilliant than the brightest electric lights. They are dark only by contrast with the brighter surface of the sun around them.

Dark line, or ABSORPTION spectra, may be of the types of "line" or "band," atomic or molecular in origin, as described above. They indicate a glowing vapor between the observer and an incandescent liquid or solid. With respect to the sun, they indicate the presence of an atmosphere containing many substances familiar to earth-dwellers. Not merely are atoms there, but some molecules. Very stable ones they must be to resist dissociation at these high temperatures. In the spectra of the slightly cooler sunspots such compounds as an oxide of titanium and ammonia gas have been recognized.

The spectra of over 90 per cent of all the stars are dark-line absorption spectra. In this proportion the stars have, then, substantial interiors and luminous glowing atmospheres. In the spectra of the stars we recognize a great many of our well-known terrestrial atoms and a few of our familiar molecules. The stellar universe thus at once becomes to us a sort of familiar "homey" place. It is quite true that with respect to temperature we live in a very frigid spot, as we pointed out when we were speaking about temperature (p. 128). We could not survive an instant in any of these places to which our journey on a beam of light has transported us. We have evolved in, and are marvelously adapted to, our own abode. Life, as we know life, perhaps exists nowhere else, unlikely as such a supposition may seem. Certainly if organic processes are developing and evolving elsewhere in the universe, we shall probably never know about it. If the life is anything like that we know, we certainly shall not. Such life as that which breeds so richly on the surface of our tiny earth needs but little heat. It thrives in a place far too cool to send out radiant messages into the depths of space. But all the myriads of blazing worlds around us send us *their* radiance to read and to decode. In their incandescent atmospheres the stars show us our own familiar atoms often under temperatures that we can never hope to attain here.

How illuminating to us these spectacles may be is well illustrated in connection with certain stars of the hottest types and in certain of the gaseous nebulae that are generally associated with them.

These stars and their associated nebulae are included among the 10 per cent of celestial objects that exhibit bright-line spectra. Purely luminous gas they are. The radiation is seen to be largely from atoms of helium and hydrogen. But interspersed among the lines of these spectra, there were early recognized radiations that had no counterpart in those of any known terrestrial elements. The first hypotheses were that here we were privileged to see the evolution of matter itself taking place. Primitive or proto-elements were supposed responsible for these unknown spectrum lines. Such names as "nebulium," for unknown substances in nebulae, and "coronium," responsible for lines of unknown origin in the sun, seen only at times of total eclipse, were given. A faint green line in the spectrum of our own terrestrial aurora was assigned to a rare unknown atmospheric gas, called geo-coronium; and there were others.

Little by little, by the painstakingly slow but certain methods of exact science these matters have been checked up. Researches in spectroscopy and in other fields of physics revealed no support for such ideas. Work in X-ray spectra that originate from the innermost electrons in atoms, from those that find their places just outside the nuclei, gave indisputable evidence that our periodic table of the elements was quite complete. The integral character of atomic weight, our certainty of knowledge of atomic number indicated no great gaps in the list of elements. Theoretical studies assisted in experimental interpretations. New experiments supplied incentive for more careful theory. Only a few years ago the riddle of the unknown elements was solved.

In the transcendental conditions of temperature and pressure that exist in stars it appeared that radiations forbidden according to terrestrial rules, but not impossible in the broader sense of the word, were possible in these stars and nebulae. The atoms in question responsible for the unknown radiations were among our most common ones—oxygen, nitrogen, components of the air, and carbon, often known as coal. So many of these unknown spectra have now been thus identified that it would appear to be but a question of time until we will recognize nothing new under the sun or in all the universe of stars.

A curious by-product of this was the fact that in these studies it appeared that there were other lines of these familiar substances that we had never recognized or produced. New combinations of radiant harmony they were that might be brought out by a little more intensive effort right here within our own laboratories. These attempts were made, and sure enough the atom responded as predicted.

Some time you may be asked, "Why do the astronomers and physicists spend so much time and effort in the study of the light that comes from these far away places that no man shall ever visit save in his mind." Your questioner may not be one of those who would understand the answer that

Michelson invariably gave to queries of this nature, to the effect that such studies were so much fun. But you may answer that we study the light of stars and nebulae that we may learn more about air and coal.

NOT only do we learn about the structures, the chemical composition, the temperatures and pressures, in these far away regions and the extent to which ionization processes have proceeded in them. The light they send out reveals their motion also. To sit and wait to see if the stars shifted their positions in the sky, if we except an insignificant few near ones, requires centuries—indeed, millenniums—of time. The Doppler effect in light gives us, literally overnight, the answer with respect to that component of the star's velocity in the line of our sight as we look at it.

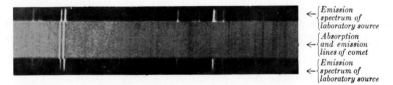

PLATE 98 DOPPLER EFFECT IN THE SPECTRUM OF A COMET. The central band shows the lines of sodium in the comet displaced to the left (toward the red), as compared with the lines of sodium taken at the same time from a laboratory source. This displacement indicates how fast the comet was receding from us at the time the photograph was taken.

Thanks to the precision of interference methods of wave-length measurement, motions considerably less than that of our earth in its orbit are readily detectable. Many stars have velocities measured in scores, and some in hundreds, of kilometers per second. The motions of the great spiral nebulae run into thousands. Such motions shorten the observed wave-lengths if the star approaches us and lengthen them if the object recedes, as most of the nebulae appear to do. The true wave-lengths we know with an accuracy of a few parts in a million. Hence velocities that bear to the velocity of light a ratio even as small as one in several thousand are measurable with precision.

Because of this message in the beam of radiation then, and because of the spectroscope that can decode it, the stellar universe, our own galaxy of stars, has ceased to be a static thing and has become a vast dynamic picture.

RADIATION AND ATOMIC STRUCTURE

Great clusters of stars—some widespread, others closely knit—are to be found traveling in company—stars of one sort streaming this way, others migrating in quite a different direction; there is plenty of random motion, but a great amount of organized motion is also to be seen.

OF COURSE, the very nature of the case in a realm so large prevents our getting a view of everything as it is now. Light takes time to go places. Even in our own little universe of stars, a thousand years is far from being enough to take a beam of light across it. As we look at regions more and more remote, we see things more and more in the past. We view our sun, not as it is at the moment, but as it was 8 min. 20 sec. ago. Of course it does not change much in this interval, so the point is somewhat academic.

EARLY in the year 1901 there appeared in the constellation of Perseus a new star, a brilliant star, where one had never been before. Brilliant outbursts of new stars, NOVAE they are called, are rather rare; but on the average about ten new stars appear every year, usually only telescopic objects. They are ephemeral and quickly fade, but not before their light has been recorded for study by the spectroscope. This bright one of 1901 had attendant phenomena that enabled us to place a date as to the approximate time of the cataclysm which occurred out there in space. It then came out that what astronomers were watching so eagerly in 1901 actually had taken place during the reign of James I of England.

When we look at stellar universes other than our own, the time that is taken by the light to get here really becomes quite respectable. For our nearest neighbor among other stellar systems, the great spiral in Andromeda, visible to the unaided eye on a clear September evening, the figure is almost a million years. Could we but travel thither and keep our tiny earth in sight what might we not see taking place as thus we rolled Time back through one geologic period after another! This, of course, can never be, and yet our geologists read much about it in the records of the rocks that the rest of us pass by, unseeing.

On the other hand, in these distant universes our telescopes see something, and our spectroscopes see more. From the pulsations of light from gigantic stars in these far places, we can estimate how far away from us they are. Our light waves can be made to plumb the distances that they have come. To go into the details of such matters we are sorely tempted, but space does not permit us even if our reader's patience would. For anyone who wishes to read more, these stories have been ably told elsewhere.*

IT IS a bit unfortunate that the term "cosmic" should have come to mean one particular brand of radiation that is not yet certainly known to come to us from the depths of space at all.

The rays of visible light that come to us from the stars are truly *cosmic* in the messages they bring, if we could but read and decipher all their story. Of distant time and space this ordinary visible light has much to tell. But there are other octaves whose waves are much shorter—shorter by thousands of times. These rays we never see; but if they are not too feeble, they may be photographed; and when so recorded, we discover that the terms "spectral lines," "series," "continuous," "bright line," and "absorption" apply to them as well. Spectra may be traced far beyond our visible octave in both directions. Infra-red spectra tell us many interesting facts about molecules. But it is the *far* ultra-violet that has told us the most about the innermost parts of atoms and has brought us face to face with that other aspect of radiation that we formerly never suspected. By "*far* ultra-violet" we now refer to those radiations usually called X-rays.

THE discovery of X-rays in 1895 by Roentgen was almost in the nature of an accident. A Crookes tube such as we described on page 295 was an important scientific plaything at that time. Roentgen found that one inclosed in a light-tight box gave off mysterious rays that rendered fluorescent materials outside the box luminous. In christening these rays he used the algebraic symbol for an unknown quantity.

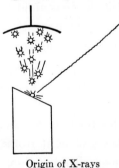

Origin of X-rays

After several false clues were followed, the origin of these rays was found to be the material on which the electrons shot out from the cathode impinged. Any sudden stoppage of these swiftly moving negative particles results in radiation being sent out—a kink in the lines of force, as we have so naïvely described it. (Cf. p. 388.) The more sudden the stop, the sharper the kink and the shorter the wave-length of the radiation. Of course, the faster the electrons go, the more

* Cf. Herbert Dingle, *Modern Astrophysics* (Collins, 1924).

acceleration there is when they are brought quickly to rest. Therefore, the stronger the fields, and thus the higher the potentials across the tubes, the shorter are the emergent X-rays.

THE penetration of these mysterious things was their first discovered and most obvious property. Indeed, in this respect they did not appear to resemble light at all. Unlike all streams of particles that then were known, or have been known until very recently, X-rays were strictly neutral as to charge. They did not transport charge to objects in their path, nor did they show the slightest deviation in the strongest magnetic fields. Therefore, the first hypotheses as to their nature were that they consisted of electrically neutral "pairs" of entities, one positive, the other negative in equal amount —a concept very like that of the recently discovered NEUTRON that has an origin altogether different.

THERE were, as usual, a few doubting Thomases who clung to an idea that these rays were really of the nature of light. This school tried reflecting and refracting X-rays, attempted to apply the crucial test of interference to them, but were compelled to conclude that if these things were waves they could not be at most even a thousandth part as large as the shortest waves of light that then were known. To imagine waves as small as this was asking too much.

Meditating on these matters, Laue, working in Germany, had a bright idea. Nature in the marvelous regularities of crystals in solid matter where atoms and molecular groups are spaced with utmost precision has provided natural gratings of a sort. They are three-dimensional arrangements, it is true; but all three-dimensional arrangements, if regular, may be so oriented that two-dimensional regularities of planes become conspicuous. X-rays, if truly of wave character, ought to show interference phenomena if sent through or reflected from crystals.

Inspired by Laue's idea, two of his associates, Friedrich and Knipping, tried the experiment. The very first attempt was successful. There appeared upon their photographic plate a beautiful arrangement of diffracted points, not only significant of the waves of X-rays and of their size but also significant of the "grating space," the distances between atoms, rows of atoms, or planes of atoms in the crystal used. So clearly is the significance of this crucial test universally recognized that the great school of "neutral-pair" adherents, led by the elder Bragg, at once made public their abandonment of their position, heaped praises on their former adversaries, and moved over into their camp. It was a most sporting performance, quite as inspiring to some of us as was the scientific contribution.

FOR the crystallographer an entirely new technique for his studies of crystalline arrangements was opened up at once.

Very soon two different types of this X-ray light were recognized. One was the so-called "white" X-rays, consisting of a continuous band containing waves of all lengths, at least down to a certain lower limit that depends entirely on the potential of the tube, decreasing as the latter increases and the energy of the electrons, before they are stopped, gets greater.

THE other spectrum of X-rays is superimposed upon the continuous spectrum and consists of a series of bright emission lines. There are not large numbers of lines in these series, only a few in each. Their wave-lengths have nothing to do with the operating conditions in the tube, but only with the composition of the target source. As we provide targets of one element and another in the periodic table, these series, characteristic of each element, shift very systematically in position and in the separation of their lines as we pass in sequence through the table. Calling by the atomic number the position of the elements in the periodic table, it turns out that "the square root of the frequency of any particular line in X-ray spectra increases directly as the atomic number of the element whence the spectrum originated." This law was discovered by Moseley, a young English physicist already famous when in his twenties he was shot through the head in the Dardanelles campaign of World War I.

BOTH the short-wave limit of the continuous spectrum and the Law of Moseley in the characteristic spectrum of X-rays betray the existence of definite LEVELS OF ENERGY in atoms—levels that are associated with the electrons contained therein. To follow the arguments that lead inevitably to this would take us too far afield.* We content ourselves with stating that to remove an electron from one of these levels takes a certain amount of energy (E) which must be imparted by the absorption of the kinetic energy

---

* For a brief account of the external electron structure of free atoms in a gas, reference is made to Part Two of the author's *What We Know and What We Don't Know about Magnetism*, published by the Museum of Science and Industry (Chicago, 1946).

RADIATION AND ATOMIC STRUCTURE

of a bombarding electron from outside, or by the absorption of the right amount of radiant energy from some extraneous source.

Normally, these various levels in an atom are filled with electrons. Any electron that is withdrawn or knocked out leaves a vacancy into which some other electron subsequently falls. In falling into place, this electron gives up some of the potential energy it had before it fell, and this energy, *as light of wave-length λ and frequency ν is then radiated away into space.*

In all cases the energy (E) required to separate an electron in this manner is definitely related to the light that another electron gives off as it falls into the empty seat. If the frequency of this light be ν, then it is true that

$$E = 6.55 \times 10^{-27} \times \nu = h\nu.$$

The constant factor between energy and frequency, 6.55×10^{-27} was found first theoretically by the German physicist Planck, and is called the PLANCK CONSTANT.

INNUMERABLE experiments in many fields other than spectroscopy bring additional evidence that, when energy falls on atoms, it is absorbed not continually but only in definite amounts, called QUANTA. Similarly, when a change in interior arrangements takes place within an atom, altering its energy content, the emission of energy by the atom is likewise in discrete units or quanta—"photons," we have previously called them.

A light wave may have any wave-length, of course, depending on its origin. Its energy content, however, is always fixed thereby and is its frequency times Planck's constant, h. This energy content associated with color or wave-length is that which remains unchanged in a beam of radiation, no matter how far it flies. It is the essence that in the photoelectric effect impinges on electrons in the metal and releases them. To do so, it must give up this energy; and when it does so, the electron it releases acquires all of it.

Herein lies the corpuscular aspect of radiation.

A BEAM of yellow light grows dim in intensity as one moves far away from its source, but yellow in color it remains. Fewer and fewer of its photons strike per second any given area. But every one that does has its full complement of energy, $h\nu$, or in terms of wave-length, hc/λ.

Unlike the individual packets of matter, atoms, or molecules, and unlike the individual packets of electricity, electrons, protons, alpha particles, etc., these photons, these individual packets of radiant energy, are not uniform in magnitude. Atoms of hydrogen are, of course, less massive than atoms of helium by four times; but in white light or white X-rays there are present photons that differ in magnitude by indefinitely small gradations, since all gradations of wave-lengths or of frequency are present. The amount of energy contained in each photon is exactly proportional to its frequency.

THUS we see finally why these corpuscular characteristics of radiant energy were first discovered in connection with X-rays. Having frequencies a thousand-fold those of visible light, their energy packets are a thousand-fold larger. When they give up their energy to atoms, electrons, or to any other entity, their punch is correspondingly greater and thus more readily observed.

We have been speaking of continuous spectra. In the case of line spectra, whether of X-rays or of visible light, the relation between the energy levels in the atoms and the quantum sizes of energy they absorb and re-emit as discontinuous spectra now begins to dawn upon us.

TAKE the simplest case, of hydrogen. Outside the proton nucleus as figuratively indicated in our diagram there is a region, represented by a solid line as if it were a sort of orbit where the lone electron is most likely to be found, when the atom is in its normal condition. If an electron from outside bumps into such a system, it rebounds with perfect elasticity; i.e., it imparts no energy at all *unless* its kinetic energy of motion *exceeds a certain minimum amount*, that of the first quantum that a hydrogen atom may absorb.

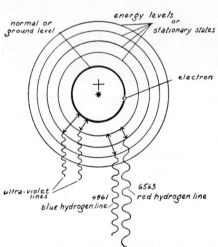

Simple Bohr picture of atomic structure

If, however, its energy is more than this, then the electron, after collision, will be found to possess less energy than it had before. It has collided inelastically, i.e., like a ball of lead against a ball of steel. The energy it has lost is just that of the first quantum that is characteristic of hydrogen. The electron of the hydrogen atom will now be

RADIATION AND ATOMIC STRUCTURE

found not where it formerly was but in a somewhat more remote region, the second "stationary state" or second "energy-level."

In this region it does not long remain. In about 10^{-15} sec. it "falls" back to its original level. As it falls back, radiant energy in the form of ultra-violet of a definite wave-length is sent out. The frequency of this light times the Planck constant is EXACTLY the energy lost by the disturbing interloper that collided with the atom and imparted the energy that shifted the atom's electron up one notch into the second stationary state.

A harder bump, from a swifter and hence more energetic particle from the outside, might have shifted the electron in the atom up two steps into the third energy-level that is characteristic of it. Subsequently this electron would have fallen back "downstairs," so to speak. It might have taken the two steps at once; and then another ultra-violet line shorter than the first, because of greater energy content, would have been radiated. It might, on the other hand, be bumped back from the third level into the second, and then dropped from there to the first ground-level. In this case two different radiations would have been sent out—one of much less energy than either of the others, a radiation of visible light of wave-length 6,563 A.U., and the other identical with the ultra-violet line we first mentioned.

We would not leave you with the impression that we can spot electrons in atoms with anything like the clarity that the drawings of orbits imply. Therefore, we include a very recent photograph of how atoms appear when "seen" by X-rays (Plate 99). In it all that we can resolve are respective regions, or shells, where the degree of haziness indicates the probability of finding an electron there, nothing more.

Thus we have found, beyond any question of doubt any longer,

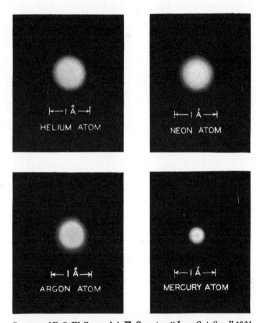

Courtesy of E. O. Wollan and A. H. Compton, "Jour. Opt. Soc.," 1934

PLATE 99 ATOMS AS SEEN BY X-RAYS

the origin of those series of spectrum lines that we said some pages back were characteristic of the spectra of all atoms.

With heavier atoms many of the lower stationary states are populated with one or, indeed, several electrons. Disturbances to such systems have been difficult to interpret; but that this has been done carefully, and done in such a manner, using all the facts of spectra about which all experts in this work are in complete agreement, was one of the supreme achievements of physics between 1915 and 1930.

These achievements cannot be popularized; they never make front-page copy for the press. To them we owe, however, the solution of the riddle of unknown elements in the stars referred to earlier and an increased knowledge about the nature of air and coal that might conceivably be of the utmost practical importance.

IN THE study of chemistry today he who runs may read how completely the chemist in his manifestly numerous and complex problems relies entirely, even for purposes of research as well as exposition, on these energy-levels for making clear to himself and to others the fundamental facts and principles of this great subject, the thoroughly practical value of which no one would question.

What we have been describing can be experimentally demonstrated not in one or two but in quite a number of independent ways. We will not discuss them here. You may read the details in other books.

ONE highly significant experiment, however, we must mention. Its fundamental character brought a Nobel Prize in 1927 to its discoverer. We refer to the COMPTON EFFECT.

A. H. Compton in 1923 reported the fact that when a "monochromatic" beam of X-rays (a single wave-length, i.e., stream of photons all alike, in that each has one and the same amount of energy), passes through and is "scattered" by light atoms such as carbon, there result *two* monochromatic beams of radiation. One is identical with the incident beam; the other is of greater wave-length, i.e., contains less energetic photons (photons of less energy). The atoms of carbon of course are ionized, electrons are vigorously expelled from them, with momentum and kinetic energy. These recoil electrons, whose existence and properties were first predicted by Compton from the implications of the arguments we have above recounted, were later found and observed to possess exactly their predicted momenta and energy. The calculations that apply are precisely those that would be used in our first chapters on mechanics, to compute, for example, the velocities, momenta, and energy of two billiard balls before and after a collision. Here, however, one of the "balls" is an aspect of a beam of electromagnetic energy, radiant energy. Direct proof is this, that radiation thus possesses in its corpuscular aspects all the properties that we usually associate with so-called "material" objects. But since material objects, after all, are defined only in terms of properties such as mass, inertia, momentum, energy, etc., and in these terms are described with respect to their effects upon one another, radiant energy found to possess all of these identical attributes and to affect and be affected by other material objects in just the same manner as they affect one another becomes admitted to exactly the same category. Radiant energy is thus a material object.

THE largest packets of energy that we know anything about are the famous "cosmic rays."* In knowing more about them today than we formerly did, we still have no certain knowledge of their ultimate source of origin. That their energy is not primarily electromagnetic in form we are certain. The primary cosmic rays are to a large extent indeed, if not entirely, composed of charged particles, in which the positive sign seems to predominate before their entry into our atmosphere.

Now in all of the work we have hitherto described, there has seemed to be very little difficulty in detecting the difference between charged particles and radiation, or in detecting the difference between different kinds of charged particles. Why should it be so difficult in the case of cosmic rays?

With respect to differentiating between different kinds of charged particles (protons and electrons, for example) there is, first of all, a very great difference in their masses, at least when the velocities of the particles are small compared to the velocity of light, as has always hitherto been the case.

In the case of cosmic rays, however, so great is the energy that if they are electrons their speed is very great, nearly that of light. Their masses, therefore, in accordance with demands of the theory of relativity, are enormously greater—greater by many hundred-fold—than their masses when at low velocities or at rest. If, on the other hand, these cosmic rays are protons, the characteristic mass of these likewise has been increased for the same reason but not to the same extent. Consequently cosmic ray electrons and

* Their detection by means of the gaseous ionization they produce has been described in chap. 29, pp. 281–82.

cosmic ray protons turn out to be so nearly equal in mass as we measure them that differentiation between them becomes very difficult.

THE deflection of a stream of charged particles in a magnetic field is another common test, not only for sign of charge but for the mass of the particle as well. Cosmic rays are excessively faint, for one thing; and we do not know how to concentrate them into bundles; direct experimentation on them is not as yet accomplished; indeed, the little that we know of their properties we know by observations made upon atomic wreckage caused by them, débris that is a by-product of their devastating energy. Laboratory magnets may be made to swing these by-products around in circular arcs, and thereby hangs a most interesting tale to which we will recur in a moment (pp. 438–39). No laboratory magnets have yet been designed to experiment upon the primary rays themselves.

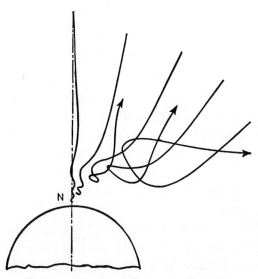

Various trajectories of electrons in the earth's magnetic field

The magnetic field of the earth has been appealed to. Our planet has been crisscrossed throughout equatorial and temperate zones by cosmic ray exploring expeditions. (Polar expeditions have not yet been quite so extensive.) These have been planned, directed, and not infrequently accompanied by R. A. Millikan; A. H. Compton; J. Clay, a Dutch physicist; and P. Auger, of France. For years the results were in great confusion. But order has gradually resulted, and one can now feel certain of several facts: (1) The origin of this phenomenon lies outside the earth's atmosphere. (2) The deviations of the rays in the earth's magnetic field show the majority, if not all, of the primary rays to consist of charged particles—not photons. (3) The absorption of the rays by atmosphere, by artificial shielding of the detecting apparatus, and by the sinking of the latter deep in high mountain lakes and far beneath the earth's surface in mines shows the existence of at least two components of different penetrating power, or, as we would say in X-ray terminology, of "hardness." (4) Guided by computations made by Störmer

RADIATION AND ATOMIC STRUCTURE

first, then by Lemaître and Vallarta, it has been shown that one can predict just what are the minimum energies with which charged particles at different latitudes and altitudes can penetrate this deflecting magnetic field of the earth. Since particles approaching the poles of the earth are moving parallel to the magnetic lines of force, even those with small energies can penetrate to the earth's surface. At the equator, particles with energies less than 20,000 million (20 billion) electron-volts are deflected away by their crossing the magnetic field. There are abundant cosmic rays at sea-level at the equator; hence a large part of these must have large energy, as, indeed, their penetrating power testifies.

The number of slower particles able to get through the field more and more as one proceeds north from the equator ceases to increase very much north of latitude 40°.

By making observations by means of free balloons at high altitudes at various latitudes, it is found that, as one ascends, the proportions of slow to fast increase from 1:2 at sea-level to 1:1 at about $2\frac{1}{2}$ miles to 5:1 at 9 miles up. Below ground or deep under water, however, the fast particles exceed the slow ones by 8 or 9 to 1, and this small residual of slow particles never falls to much below this ratio.

Now this seems most confusing, and, indeed, the best investigators found it so. In the struggle to interpret which came first, i.e., which is primary and which is secondary, one was reminded of the older arguments about the hen and the egg.

IN THE complex and increasing mixture of particles and photons into which the energy of the incident particle is dissipated as it is absorbed down through the atmosphere (finally to become ordinary thermal agitation), one highly penetrating and tremendously energetic component not only persists but becomes increasingly conspicuous at low altitudes. The interest in this component, whose ionizing and penetrating power cannot be figured out in terms of any particle previously discussed, has been very great. There are energies per particle here of 10^{10} (or even more) *million* electron-volts—energies that exceed those evolved in nuclear transmutations from fissioned uranium by as much as, or *more* than, the latter exceed ordinary chemical energies! Are we to expect soon the advent of a supernuclear age? What is the nature of these superenergetic particles arising out of cosmic ray phenomena?

Ordinary electrons endowed with energies as high even as a *billion* million (10^{15}) electron-volts (where theories about electrons begin to be uncertain) should not be able to penetrate more than a few centimeters of lead—but fast cosmic particles have been found experimentally by Auger and Ehrenfest to pass through *half a meter* of it. If they are not electrons, then what are they? No solution seemed tenable except on the basis of assuming that these exceedingly energetic particles must be endowed with more *mass* than ordinary electrons but not so much as that of protons, the lightest of atomic nuclei. So the word, MESOTRON or MESON, meaning a particle with "intermediate" mass, began to appear in the literature, to the confusion of all but the experts. "Heavy electrons of either + or − charge" is more completely descriptive, if, in addition, we think of "heavy" as meaning mass about two hundred times as great as electrons or one-tenth that of protons. In addition to this property of intermediate mass, it seems certain that the mesotron has a *kind of "radioactivity" associated with it,* in the sense that as a heavy particle it "lives" only a few millionths of a second after its birth! This property rules out any chance of its being a primary cosmic ray traversing great distances before it hits our atmosphere. Because of its great speed and energy, however, it can travel quite a respectable distance during even this short life. That upon its "death" it becomes just the simple garden variety of electron (− or +) seems assured. To this electron it passes on its energy. But how the mesotron is born in the uppermost reaches of our atmosphere or within the interstices of denser matter in our apparatus and by what manner or process it dies are still unclear. Of course, one must not imagine that abundant theorizing is not going on.* To the hypothetical neutrino, of no charge and negligible mass, appeal has again been made, in order to formulate an explanation. Since no independent experimental evidence has yet been adduced to testify to the neutrino's reality, we must leave the matter here.†

Before we dismiss this fascinating topic completely, however, comment should be made on the fact that it was as far back as the night of August 2, 1932, that the POSITRON was discovered by Anderson, of California Institute

---

* Or active experimenting either. Recent announcements from the General Electric Company announce the production of cosmic ray phenomena in the laboratory by means of artificial X-rays, now stepped up to requisite frequencies.

† For a much more detailed and excellent account of this whole subject, reference is made to Auger, *What Are Cosmic Rays?* trans. Shapiro (Chicago: University of Chicago Press, 1945).

RADIATION AND ATOMIC STRUCTURE

of Technology, in connection with his studies of cosmic rays as part of the program initiated by Robert A. Millikan at that institution.* Our knowledge of the symmetry of nature with respect to the existence of electrons of both + and − charge we thus owe to the careful and painstaking work of cosmic ray investigations.

IN THIS connection also the interchangeability of energy from mass into radiation and vice versa has received another experimental confirmation. An electron has a mass of approximately 10^{-27} gm. In energy terms from the relation $E = mc^2$; this means $E = 10^{-27} \times 9 \times 10^{20} = 9 \times 10^{-7}$ ergs. Since 1 erg = 6×10^5 mev,

$$E = 9 \times 10^{-7} \times 6\tfrac{1}{4} \times 10^5 = 56 \times 10^{-2} \text{ mev}$$
$$= \tfrac{1}{2} \text{ million electron-volts, approximately}.$$

A million electron-volts of energy are thus sufficient to supply the *energy equivalent of the mass of 2 electrons*. When radiation of this high energy,

$$E = h\nu = 1.8 \times 10^{-6} \text{ ergs},$$

and hence of frequency

$$\nu = \frac{1.8 \times 10^{-6}}{6.6 \times 10^{-27}}$$

(see p. 431) or 2.7×10^{20}, or about 300×10^{12} (300 trillion) megacycles, falls on matter, its energy may disappear from the radiant form (as photons) and reappear in a more "material" form as an "electron pair," a negative and a positive electron. If the incident radiation has an energy even greater than this, *the excess* over the million electron volts is exhibited in the *kinetic energy* with which these two electrons hasten from their place of birth. This phenomenon has been repeatedly observed; and energies, frequencies, and velocities, at least within the limits of experimental uncertainties, have been found to check with theoretical predictions.

Conversely, it has also been observed that when positron and negatron— an electron pair, as they are sometimes alternately called—encounter each other in a "head-on collision," ultra γ rays (a pair of photons [p. 334]) of frequency in excess of 300×10^{12} megacycles are radiated from the place of the "collision."

* A graphic and full account of this will be found in Millikan, *Electrons (+ and −) Protons, Photons, Neutrons, and Cosmic Rays* (Chicago: University of Chicago Press, 1935).

Again the *excess* of frequency over 300×10^{12} megacycles is found to correspond with the *combined kinetic energy* possessed by the electron pair at the instant of their collision.

In addition, if the recoil of an electron or a nucleus whose presence is essential in the vicinity of this cataclysm is taken into account, conservation of momentum likewise is exhibited. Thus, incidental to the study of cosmic rays, results of most fundamental character in our understanding of the relationships between radiation and atomic structure have been unearthed.

FINALLY to balance completely the discovery of Compton which gave direct experimental testimony to the corpuscular character of radiation (Compton Effect, p. 434), it was not long afterward, in 1927, that two men in the research laboratories of the American Telephone and Telegraph Company, Davisson and Germer, announced that *electrons, particles par excellence,* when shot through a crystal grating, exhibited *interference patterns, the crucial test for the possession of wave characteristics.* These men were not alone to be credited with the honor of this discovery. It had been made independently abroad by Professor G. P. Thomson in England, and by Kikuchi, a Japanese. Subsequently Professor Arthur J. Dempster, of the University of Chicago, found interference patterns from protons, hydrogen nuclei; and the presumption is strong that *any* material particle would give them, although the scale of the effect is too small for our techniques to observe as yet with the more massive particles.

Faced by the indubitable facts that everything we know about in the physical universe, every manifestation of energy, whether in the radiant form or in the form of atoms and their constituents, possesses dual characteristics —those of particles and those of waves—depending solely upon how we observe these manifestations, Theoretical Physics has girded its loins and taken a fresh start.

The names of De Broglie, Schroedinger, Heisenberg, and Dirac are a few, among many, that are responsible for the new wave-mechanics, a mathematical method of attempting to express the manifold laws, experimentally established, in a way that incorporates the dual aspects of these processes going on in the physical world, logically, coherently, and concisely.

Further than this it is not our present purpose to conduct you. We have already taken you, no doubt, at times somewhat beyond your depth. However, if you were at all intrigued by the range of subject matter suggested by our title, we have at least made good its promise. We have carried you from the first dawning of that spirit of approach without which scientific progress is impossible, that spirit first consistently achieved throughout a long lifetime by Galileo, to the full tide of the present moment, when scientific work is recognized, prized, and supported, at least to some extent, by all great nations.

Unfortunately, within the last five years, mankind has learned to unlock and, to some degree, to control the vast supplies of energy which hitherto had remained locked up in atomic nuclei. Having greatly accelerated his knowledge under the impact of fear of destruction in a global war, which, WITHOUT these recent discoveries, has destroyed millions of mankind and perhaps was only stopped, but not won, by the atomic bombs, will man be able in the future so to conduct his political, social, and economic life as to put an end to war and prevent for the future the much greater powers of destruction that soon *all nations* will possess?

Unless this can be accomplished, the darkest of all ages surely lies ahead; the rare joy that in the past has gone with the acquisition of knowledge will be known no more.

Quite characteristic is it of the acquisition of knowledge that as one learns a little, the horizon widens and what lies ahead to be learned is seen to cover a much vaster area. It is with our eyes on these horizons that we climb. What has been surmounted and overcome seems small and insignificant. How small and insignificant it is, compared to achievements that now must promptly be attained in other fields of human activity, is beginning to become apparent. Let us devoutly hope that in some way these problems may be met in time so that future generations again can get some joy from probing still more deeply into the heart of Nature.

INDEX

A

Abbreviations, 9
Absolute temperature, *see* Temperature
Absolute zero, 162
Acceleration: defined, 9; of gravity, 16, 18, 19; uniform, 16; units of, 10
Action: defined, 47; at a distance, 388; and reaction, 200, 243
Activity, defined, 90
Air: compressibility of, 101; made conducting by flames, 276; made conducting by other agents, 278; weight of, 92
Air resistance, 20, 21; effect on trajectory, 37; increase of, with speed, 21; *see also* Friction
Alpha particle tracks: lengths in air, 323; photography of, 322
Alpha particles: charge and mass of, 319; helium ions, 319, 321; helium nuclei, 327; properties of, 318; structure of, 327; *see also* Bombardment experiments
Alpha rays, *see* Alpha particles
Ammeter, 244
Ampere, A. M., 247
Ampere, practical unit of current, **228, 240**
Anderson, C. D., 437
Andromeda, great nebula in, 342, 343
Annihilation radiation, 439
Anode, 269
Archimedes' principle, 102, 103
Aristotle, 13, 17
Armature, 241
Arrhenius, Svante, 274
Arsinoe, 205

Astigmatism, 356
Astron, 313, 330m
Atmosphere: of Mars, 22; of Venus, 145; *see also* Air
Atmospheric pressure, *see* Pressure
Atom: nucleus of, 325; "seen" by X-rays, 433; structure of, 321; *see also* Bohr orbits; Electron
Atomic character of electricity, light, and matter, 334
Atomic dimensions, 171, 274
Atomic number, defined, 324
Audibility, limits of, 376
Auger, P., 436, 438

B

Bacon, Francis, 149
Bainbridge, K. T., 330m
Balance, analytical, 245
Barlow, 241
Barlow's wheel, 243
Barometer: aneroid, 99, 100; mercury, 97–99
Batteries, "A" and "B," 290
Beats, 380, 381; of light, 407; in stroboscopic illumination, 382; tuning by, 382
Becquerel, H., 317
Beebe, W., 96
Belanger, 57, 58
Beta particles: originate in nucleus, 329; properties of, 318; *see also* Cathode particles; Electrons
Beta rays, *see* Beta particles
Betatron, 330
Block and tackle, 86, 87
Bohr, Nils. 317, 326, 432

Bohr orbits, 326, 432
Boiling-point and atmospheric pressure, 113, 143, 167
Bolometers, 395
Bombardment experiments: with alpha particles, 328; with protons. 328; Rutherford's, 325
Boyle, 149, 150 158
Boyle's Law 159, 160
Bragg, W. H., 430
Broglie, de, 440
Brown, 170
Brownian movements, 17, 279
Buckley, H., 149
Buoyant force, 103

C

Caloric, 78, 150
Calorie, defined, 79
Canal rays, 311; *see also* Positive rays
Cathode, 269; glow, 295
Cathode particles: determination of ratio of charge to mass, 304–6; forces on, in magnetic field, 304; mass, 306; mass independent of origin, 307; speed, 306; *see also* Beta particles; Electrons
Cathode rays: appear at low pressures, 296; carry negative charges, 298; compared with photoelectric and thermionic "rays," 293, 299; deflected by magnetic field, 297, 298; energy of, 298; energy equation of, in magnetic field, 305; equilibrium equation in magnetic field, 304; momentum of, 298; photographs of, 297; travel in straight lines, 298; *see also* Beta particles; Electrons
Cathode-ray tube: appearance of discharge through, 294, 295; an X-ray tube, 299
Cavendish experiment, 28, 29, 216
Centigrade thermometer scale, 113
Centrifugal force, equations for, 303
Centripetal force, equations for, 303
Chadwick, 315, 317, 328, 329, 330i
Chain reaction, 317, 330f, 330q
Chamberlin, T. C., 50
Charles, 158, 161
Charles's Law, 159, 161
Clausius, Rudolf, 274
Clay, J., 436
Clouds: electrification of, 193; temperature in, 191; updraft in, 192
Coblentz, 122, 123
Cockroft, 330i
Collision, *see* Impact
Colors, primary, 359
Columbus, 211
Commutator, 242, 252

Components: defined, 41; of motion, 33, 34; of vectors, 61; of velocity of bullet, 34
Compton, A. H., 433, 434, 436, 440
Compton effect, 434
Condensation, 344
Conductors, 183; distribution of charge on, 184, 194, 195
Conservation of Energy, Law of, 81, 82, 223, 259
Conservation of Momentum, Law of, 48, 71, 72; applications of, 49–56
Coriolis, 58, 59, 64
Cornu, 401
Cosmic rays, 276, 280, 281, 282; energy of, 435; nature of, 435–39
Coulomb, 189
Coulomb, practical unit of quantity, 227, 240
Coulomb's Law, 189, 208
Crookes, Sir William, 295
Crookes dark space, 295, 296, 311
Crookes tube, 428
Crystal structure, 429, 430
Current: alternating, 232, 233, 268; analogy to flow in pipe, 267; direct, 232; electric, 228, 231; electromagnetic unit of, defined, 240
Current, induced: direction of, 248, 249; by moving magnetic fields, 248–50; strength of, 251
Cycloid, 345
Cyclotron, 330

D

Dalton, John, 151
Damped wave, 338, 339
Davisson, 440
Davy, Sir Humphrey, 150, 246
Democritus, 13, 151
Dempster, Arthur J., 313, 330m, 440
Density: of air, 22; defined, 93; of fresh water, 93; of sea water, 93
Descartes, 57, 401
Dew-point, 192
Diffraction grating, 407; crystal, 429, 430; equation of, 408
Diffraction of water ripples, 404
Diffusion, 154–56
Dingle, Herbert, 428
Dirac, 440
Disintegration, artificial, *see* Transmutation
Dispersion: defined, 358; by grating, 408; none in sound, 360; by prism, 359, 419; *see also* Spectra
Doppler effect: in comet spectrum, 426; in ripples, 362; in sound, 360–62; in sound, moving across an interference pattern, 380; in stellar spectra, 426

INDEX

Dugan, 50
Dunning, 330g
Dynamo, 252, 253
Dyne, 79, 189; defined, 11

E

Earth: a magnet, 206, 221; mass of, 28, 29; rigidity of, 126, 128; temperature of interior, 126; tides 127–29; wobble of axis, 126; see also Cavendish experiment
Edison, Thomas Alva, 288, 290
Edison effect, see Thermionic effect
Edison's "light within a bottle," 111
Efficiency, 88, 90
Ehrenfest, 438
Einstein, A., 287; theory of relativity, 330j; mass energy relationship, 330k, 412
Elastic, defined, 51
Electric: appliances, cost of operation, 229, 230; charge, dimensions of, 274; motors, 241–44; signs, glowing-gas type, 296; wind, 200–202
Electricity, atomic character, 266, 302; evidence for, 272, 273
Electricity: hypothesis of mobility of negative, 181; practical units of, 240
Electrification: by contact, 185; by friction, explanation of, 184, 185; one- and two-fluid theories of, 182; phenomena, 178; positive and negative, 178, 179
Electrodes, 199, 266, 269
Electrolysis: copper in, 273; of copper permanganate, 269; defined, 269; Faraday's laws of, 271, 272; hydrogen in, 272; iron in, 273; silver in, 272; transport of material in, 270
Electrolyte, 270
Electromagnetic induction, 247
Electromagnetic units, 240
Electromagnetic waves: contrasted with sound and water waves as to medium, 385, 386; kinds of, and their lengths, 391, 392; mechanical and corpuscular properties and aspects, 383, 384, 386, 387; methods of detection of, 393–97; transverse character of, 291; velocity, 389; wave aspects of, 387; see also Light; Radiation; Waves
Electromagnets, 241
Electrometer, 188, 286, 299
Electron: arrangement in atoms, 310; a cathode particle, 307; "heavy," 438; mass, 313; positive, 330; shells, 310; transitions, 432, 433; see also Beta particles; Cathode particles
Electronegative, 308
Electrons: atoms of electricity, 190; pair production, 439; planetary, 323, 324; planetary, and chemical properties, 325; planetary, and ionization, 325; recoil, 434
Electropositive, 308

Electroscope, gold-leaf, 179, 180, 185; in determination of amount and sign of charge, 187
Electrostatic generator: Kelvin's, 197; Van der Graff, 199; Wimshurst, 198
Electrostatic smoke precipitators, 200, 203
Electrostatic unit of charge, defined, 189, 190
Elements: transuranic, 330f, 330p; see also Periodic table
Elizabeth, Queen, 176, 178
Elster, 281
Energy: kinetic, 64, 67; measure of, 65; potential, 65, 66, 67; relations between forms of, 76; sources of, due to sun, 75; transformations in impact, 68, 71; transformations in pendulum, 68–70; units of, 59, 79; see also Conservation of energy; Heat; Nucleus; Quanta of energy; Work
Energy levels, 430; in hydrogen atom, 432
Erg, 79; defined, 59
Ether, luminiferous, 386
Ether-drift experiment, 410–12
Euclid, 60
Expansion: of bridges, 112; constant, of gases, 118; rates of, of materials, 114; variable, of liquids and solids, 117
Eye defects and their corrections, 356, 357

F

Fahrenheit, 117
Fahrenheit thermometer scale, 113
Falling bodies, free, and on inclined planes: early ideas, 13; equations for, 18; from great heights, 22; law of odd integers, 14; law of velocity increase, 16; numerical relations, 15, 16, 18; total distance, 15, 16
Falling water jet, 36, 37
Faraday, Michael, 187, 246, 247, 248, 250, 266, 271, 388
Faraday, the: defined, 273
Faraday cage, 187, 188, 194
Faraday dark space, 295
Faraday's laws, see Electrolysis
Fermi, E., 317
Field of force: defined, 214; electric, 217; gravitational, 214–16; magnetic, 220; represented by lines of force, 215, 216; represented by vectors, 215; strength of, 215, 216; work on charged object in electric, 224
Fizeau, 401
Fleming, 290
Florentine thermometer scale, 117
Fluid defined, 93
Fluorescence: of barium platino cyanide, 325; of glass, 295; of zinc sulphide, 296, 298, 325
Focus, 351

Force: on current in a magnetic field, 238, 239; definition, 11; between electric charges, 189, 190; between magnetic poles, 208; nuclear, 330*m*; *see also* Newton's laws of motion

Forces: balanced, 21, 39–45; on canal boat, 40; on falling feather, 21; polygon of, 42, 43

Foucault, 401, 402

Fraunhofer, 423

Fresnel interference experiments, 405

Friction: air, 21; coefficient of, 23; effects of, 77, 78; *see also* Air resistance

Friedrich, 429

G

Galileo, 3, 4, 6, 11, 13, 14, 16, 32, 57, 68, 71, 98, 116, 117, 118, 206, 336, 299, 401

Galileo's: gas thermometer, 115; method, 3; water clock, 14

Galvani, Luigi, 231, 232, 234

Galvanometer, 120, 121, 245; moving-coil, 240, 241; tangent, 240

Gamma rays, properties, 318; therapeutic uses, 397

Gases, properties of, 118, 153, 163; *see also* Kinetic theory

Gay-Lussac, 161

Geitel, 281

Geometrical sum, 33, 42

German long-range gun, 37, 38

Germer, 440

Gilbert, 176, 178, 206

Goldstein, 311

Gram molecular weight, defined, 273

Gratings, *see* Diffraction gratings

Gravitation, Newton's Law of, 27, 28, 29, 189; *see also* Cavendish experiment

Gravitational attraction of moon, 30

Gravitational constant, 28

Gravity: acceleration of, dependent on distance, 26; surface, of sun, moon, and planets, 31; *see also* Acceleration

Groups in periodic table, 308

H

Hahn, O., 330*f*

Half-period, defined, 320

Harmonic analyzers, 375

Harmonics in pipes, 379

Hartmann, 206

Heat: energy, 76; of evaporation, 146; a form of kinetic energy, 68; of fusion, 143; in impact, 73; mechanical equivalent of, 79–81; units, 79; of vaporization, 143

Heisenberg, W., 326, 440

Hertz, 290

Heterodyne receivers, 393

Heterodyning, 383

Hooke, 149

Horse-power: apparatus to measure, 85; defined, 84, 85; equivalent in watts, 228

Hotchkiss, 126

Humidity, 110

Huygens, 401, 402, 404

Huygens' principle, 405

Huygens' wave construction, 402, 404, 408

Hydraulic press, 95

Hydrogen: in electrolysis, 272; mass of atom, 307; nucleus, a portion, 314, 315; ratio of mass of atom to mass of electron, 307, 313; spectra of, 421, 422

Hydrometer, 104

Hydrosphere, 92

Hypo (sodium thiosulphate), 144

I

Ice-pail experiment, 187

Image, *see* Real image; Virtual image

Impact: apparatus, 53; elastic, equal masses, 51–53; elastic, unequal masses, 54, 55; energy transformations in, 68; inelastic, 53, 54; loss of kinetic energy in, 73; of molecules, 156, 157; of radiation and matter, 434, 435; *see also* Conservation of Momentum, Law of

Impulse, 54; defined, 58

Inclined plane, work on, 59–61; *see also* Falling bodies

Independence: of motions, 34; of velocities, 35

Induction, charging by, 186; *see also* Electromagnetic induction; Magnetic induction

Inertia: balance, 12; of car, 9; defined, 7, 8; principle of, 11; proportional to weight, 17; standard of, 12; *see also* Newton's First Law of Motion; Lines of force

Ingersoll, 126

Insulators, 183

Interference: bands, 406; conditions for, in light, 405; crucial test of wave character, 368; fringes, 409; of light from two apertures, 406, 407; pattern and beats, 382; pattern, moving, 381; patterns from X-rays, 429; of ripples, 368, 369; of waves in rope, 362; *see also* Diffraction, Newton's rings; Stationary waves

Interferometer, Michelson: diagram of, 409; proves wave character of light, 405; standard of length determined by, 410; stellar diameters measured with, 410; *see also* Ether-drift experiment

Ionization: by alpha, beta, and gamma rays, 318, 322; gaseous, 296; hypothesis of, in electrolytes, 274; hypothesis of, in gases, 277, 278; by ultra-violet light, 394; by X-rays, 396; *see also* Electrolysis

INDEX

447

Ionogen, 274

Ions: charge on gaseous, 281; defined, 274; mobility of, defined, 279; *see also* Millikan's oil-drop experiment; Electrolysis

Isobar, 330d

Isogonic lines, 221

Isotopes: defined, 314; history, 313; light elements, 330c; production, 330b, 330e; table of, 314

J

Jacobi, 242

Jacobson, E. G., ix

Joule, defined, 79

Joule, James Prescott, 79, 80, **81**, 117

Joule's apparatus, 76

K

Kelvin, Lord, 197

Kenotrons, 290

Kepler, 26, 325

Kerst, D., 317, 330

Kikuchi, 440

Kilogram, international, 12

Kilowatt hours, 228

Kinetic theory: calculation of gas pressure, 157, 158; of heat conduction, 164; of latent heat of evaporation, 164–67; of latent heat of fusion, 164; of quantity of heat, 162; of specific heats, 162, 163

Knipping, 429

Knowlton, 344

L

Latent heat, 142; *see also* Heat, of evaporation, of fusion; Kinetic theory

Laue, 429

Lawrence, E. O., 317, 330

Left-hand rule, 236; applications of, 390

Leibnitz, 64

Lemaitre, 437

Lens: concave, 355; convex, 355; cylindrical, 356; double convex, 355; positive, 355; prismatic, xi, xii

Levers: classes of, 86; work relations in, 88

Light: corpuscular theory of, 370; ionization by ultra-violet, 394; some energy characteristics of, not weakened with distance, 342; wave-length of, 392; wave theory of, 370; *see also* Electromagnetic waves; Radiation; Velocity of light

Lightning: places likely to be struck by, 196; reasons for occurrence of, 193

Lightning rods, 195, 198, 200

Liquid air, nitrogen, oxygen, 131–33

Liquid helium, hydrogen, 134

Lines of force: inertia of, 388–90; mutual repulsion of, 388; number per unit charge, 217; represent fields of force and field strength, 215, 216; tension in, 388

Liter, 12

Loops, 364, 367

M

Machines, 89, 90

MacMillan, W. D., 29

Magnetic: induction, 207; materials, 209; moment, defined, 208; phenomena compared with electrostatic, 207, 208; phenomena dependent on motion of electric charge, 212; properties of oxygen, 211

Magnetic compass: declination of, 206; dip of, 206

Magnetic field: accompanies moving charges, 233, 248; of bar magnet, 220; of current in a helix, 236; of current in a straight wire, 234, 235, 237; direction of, 236; of moving static charges, 238

Magnetic poles, 206; location in magnet, 208; unit defined, 208

Magnetism: effect of heat on, 209, 210; effect of jarring on, 209; a molecular phenomenon, 212

Magnetite, 205

Magneto, 251; energy conversion in, 252

Magnets, atomic, 237, 238

Marconi, 290

Marsden, 317, 328

Mass, 11; atomic unit of, 330i; energy equivalence, 330k; numbers, 330m; velocity dependence, 330j; *see also* Cavendish experiment; Inertia

Maxwell, 385, 388

Mechanical advantage, 86, 87, 88, 89

Mercator, 206

Mesotron, 438

Meson, *see* Mesotron

Metallurgical Laboratory, 330g

Meteor crater, 24

Meteorites, 24, 25, 73

Michelson, A. A., 128, 331, 401, 402, 403, 405, 409, 410, 411, 412, 413, 426; *see also* Earth, tides; Ether-drift experiment; Interferometer; Velocity of light

Millikan, R. A., 279, 281, 436, 439

Millikan's oil-drop experiment, 279–81

Mirror: concave, 351, 352; convex, 352, 353

Molecular: hypothesis, 153, 156; kinetic energy and pressure-volume product, 158; kinetic energy and temperature, 159; momentum, 156–58; motion of translation only, identified with temperature, 163; other motions, 163; numerical data, 171; velocities, 171, 172

Momentum: of car and track, 47; defined, 46; a vector, 48; *see also* Conservation of Momentum, Law of

Morley, 412, 413

Moseley's Law, 430

Motion, rectilinear, 7; unconstrained, 7; *see also* Newton's First and Second Laws of Motion; Speed; Velocity

Moulton, F. R., 50
Musical intervals, 382, 383

N

Neutrino, 330*d*
Neutron, 315, 329, 429
Newton, Sir Isaac, 4, 5, 11, 26, 57, 68, 358, 359, 370, 401, 415
Newton's First Law of Motion, 11, 39
Newton's law of cooling, 135
Newton's Law of Gravitation, *see* Gravitation
Newton's rings, 370, 401
Newton's Second Law of Motion, 11, 58, 243; units in 11
Newton's Third Law of Motion, 45, 46, 201, 240, 259
Nier, 330*m*
Nodes, 364, 367
Norman, 206
Novae, 427
Nucleon, 330*m*
Nucleus, 325; energy of, 76, 330*h*–330*q*; radioactive, 329; structure, 329

O

Octave, defined, 379
Odometer, 229
Oersted, Hans Christian, 234, 247
Ohm, Georg Simon, 268
Ohm's Law, 268
Oxygen, *see* Gases; Liquid; Magnetic

P

Parallax, xi
Particles, wave character of, 440; *see also* Interference
Pendulum: amplitude, 336; ballistic, 72; equation for period, 336; frequency, 336; period, 336
Peregrinus, 206
Periodic table, 308, 309
Phase difference, 337, 339
Phlogiston, 78
Photoelectric cells, 287
Photoelectric effect: in air, 283, 284; applications of, 287, 288; explanation of, 286; plate negative, 284; plate neutral, 285; plate positive, 284; unidirectional, 284; *in vacuo*, 285; *see also* Thermionic effect
Photographic emulsions, 287
Photographs in relief, xi, xii
Photography: photoelectric effect in, 287; stereoscopic, xiii
Photons, 334, 432, 434; relation to quanta, 431, 432

Pile-driver, 73, 74
Pipes, closed: lengths of, resonant to given note, 378; wave-lengths of notes emitted by, 378, 379
Pipes, open: lengths of, resonant to given note, 378; wave-lengths of notes emitted by, 379
Pipes, singing, 379
Pitch, 361; *see also* Doppler effect
Planck constant, 431, 433
Planets, diameters, masses, and surface gravities of, 31
Plate, 289, 290
Polarization: of electromagnetic waves, 391; of light, 338
Positive column, 295
Positive ray analysis, of atomic masses, 313; proves atomic weights integral, 314; *see also* Isotopes
Positive ray particles: charge of, 312; masses of, 312
Positive rays: charge and mass of, in hydrogen, 313; in electric and magnetic fields, 312; ions of residual gas, 312; in neon, 313
Positron, 330, 438
Potential difference: defined, 226, 227; unit of, 198, 227, 240
Power: defined, 83; electric, equation for, 227, 228; equations, 90; relations in machines, 89; units of, 84, 85, 227, 228
Preece, 290
Pressure: atmospheric, 96, 98; defined, 94; due to molecular impact, 156; beneath liquid surface, 94; variation of, in gases with temperature, 118; *see also* Kinetic theory
Primary, 250, 254, 255; mechanical force between secondary and, 259, 260
Prism, 358, 359, 419; *see also* Dispersion; Refraction
Projectiles, paths of, 32–38
Projection, simultaneous, horizontal, and vertical, 34
Protons, 313, 314; in atomic nuclei, 329; product of artificial atomic disintegration, 328
Prout, 314

Q

Quality of sound, relation to wave character, 374, 375
Quanta of energy, 431; relation to photons, 431
Quantity of electricity: electromagnetic unit of, 240; electrostatic unit of, 189
Quantity of heat: defined, 140; unit of, 141

R

Radar, 394
Radiation: corpuscular character of, 431, 435; kinds of, 386; laws of, 417, 418; magnitude of solar, 387; mechanism of, 431; penetrating, 436; relation of frequency of, to energy, 431

INDEX 449

Radioactive disintegration: products, 320; series, 320
Radioactive radiations, 318, 319
Radioactive substances, 319, 320
Radioactivity: discovery, 317; series, 320
Rain: electrification of drops, 193; formation of, 192; terminal velocity of falling drops, 23
Rainbow, 359
Rarefactions, 344
Rays of light, 346
Reaction, 45; defined, 47; see also Action
Real image: in concave mirror, 351, 352; in convex lens, 355; on retina, 352
Rebound, loss of energy in, 35
Recoil: of electrons, 434; of rails, 47
Reflection: independent of color, 357; law of, 350
Refraction: defined, 354; illustrated, 355; of sound over water, 360
Refrangibility, dependent on color, 358
Relativity theory: and atomic masses, $330j$; and electron masses, 306, $330j$
Resistance: the cathode-ray tube, 294, 295; change of, with temperature, 119, 268; defined, 268; specific, defined, 268; thermometer, see Thermometer
Resonance, 365; of air columns, 376–78; in detection of long electromagnetic waves, 393; in determination of wave-lengths, 378
Resultant: defined, 41; zero, 41, 42, 43
Retardation, 10; see also Acceleration
Roemer, Olaus, 399, 400, 401
Roentgen, 317, 428
Rowing on a river, 410, 411
Rowland, Henry A., 80, 238
Rumford, Count, 78
Russell, H. N., 50
Rutherford, Sir Ernest, 317, 321, 324; scattering, 325, 328, $330i$; see also Bombardment experiments

S

St. Elmo's fire, 195, 199, 223
Schroedinger, 440
Scientific method, 3, 56
Scientific theory, genealogy of a, 152
Scintillations of a luminous watch dial, 324
Secondary, 250, 254, 255; see also Primary
Seebeck effect, 121
Sensitive flame, 366, 367
Sensitivity of ear, diagram of, 376
Series in periodic table, 308
Short-wave detector and transmitter, 392
Shunts, 244
Sine, curves, 344; defined, 408

Smythe, H., $330p$
Soddy, 313
Solar system, origin of, 50
Solutions, conducting, 269; see also Electrolysis
Sound waves: amplitude and loudness, 374; complexity and quality, 374, 375; frequency and hearing, diagram, 375, 376; frequency and pitch, 361; longitudinal, 373; media of transmission, 373; pressure differences in, 344; see also Beats; Doppler effect; Waves
Specific gravity, defined, 104; see also Density
Specific heats: defined, 139; of gases, 163; of liquids, 139; of solids, 138; table of, 140; see also Kinetic theory
Specific resistance, see Resistance
Spectra: atomic, 420; band, 420; dark-line absorption, 422–24; of hydrogen, 420, 421; line, 419, 420; molecular, 420; normal, 408; orders of, 408; stellar, elements in, 425; see also Dispersion; Doppler effect
Spectroscope, 245
Spectroscopy, problems of, 415, 416
Spectrum: absorption, of sun, 422, 423; continuous, 418; grating, 407, 408; prism, 359, 419; sun's, elements in, 422
Speed, constant rectilinear, 7; defined, 7, 8; see also Velocity
Speedometers, 9
Spiral nebulae, 217, 342, 343, 426
States of matter, 153
Static machine, see Electrostatic generator
Stationary states, see Energy levels
Stationary waves: of audible frequency, 380; formation of, 363; in light, 409; in sound, 367; in stretched string, 364; from two sources, 368, 369; in water, 351, 352; see also Beats; Interference
Stefan's Radiation Law, 417
Stereopticon, 349
Stereoscopic photographs, x, 350
Stetson, 127
Stewart, 50
Stokes, Sir George, 23
Stokes's Law, 280
Stoney, Johnstone, 307
Stormer, 437
Strassman, $330f$
Stratosphere, 37, 100, 437, 438
Stroboscopic illumination, 349; photographs in, 350, 351, 352, 369
Sun: diameter, mass, surface gravity of, 31; energy from, 387; spectrum of, 422, 423; temperature of, 418; variable radiation of, 395
Surface tension, 165

T

Temperature: absolute, 159; measured by expansion, 112; rise on solidification, 144, 145; scales, 113, 117

Temperatures: graph of relation between Fahrenheit and Centigrade, 113; high, 111; low, 130–34; of stars, 417; with thermopile, 123, 124

Terminal velocity, 20–25; defined, 21; of heavy objects, 192; and pressure, 22; of rain, 23

Thales, 177

Thermionic effect: compared with photoelectric, 291; explanation of, 291, 292; increased with certain materials, 289; unidirectional, 290, 291; in vacuo, 288

Thermionic valve, 290

Thermocouple, 395

Thermojunction, 120

Thermometers: fixed points of, 112, 113; resistance, 119, 120; standard gas, 118

Thermopile, 122, 123

Thermostats, 115

Thompson, Benjamin, 78

Thomson, G. P., 440

Thomson, J. J., 263, 279, 313

Thunder, 194

Tides, 49, 75, 76, 127–29

Timbre, see Quality of sound

Time: of fall, 34; of flight of projectile, 33

Townsend, J. S., 279

Trajectories, 37, 38

Transformer, 254–57; energy relations in, 261; mechanical forces in, 259; power relations in, 255, 261; step-down, 255, 257; step-up, 258

Transmutation, 315; artificial, 328; see also Radioactive disintegration

Trigonometric functions, 61, 408

Tug-of-war, 43–46

Tuned receiving circuit, 393

U

Uranium: energy from, 330g; fission of, 330g; isotopes of, 330f; series, 320

Urey, H., 317

V

Vacuum arc, 296

Vallarta, 437

Van der Graff, 317

Vapor, saturated, defined, 165

Vapor pressure, 165

Vapor tension, 167; and boiling-point, 167; of water, table of variation with temperature, 167

Vector: addition, 41, 218, 219; defined, 40; diagram, 41; electric, 390; magnetic, 390; polygon, 42; quantities, 33; sum, 60; triangles, 60

Velocities: exchange of, in collision, 35, 52, 54; independence of, 33; line of sight, of stars, 426

Velocity: defined, 7; of secape, 27, 30; see also Speed; Terminal velocity

Velocity of light: demonstration of Michelson's method of determination, 403; methods of measurement of, 399–302

Verne, Jules, 143, 144

Virtual image: in concave lens, 355; in convex mirror, 352, 353; in plane mirror, 349, 351

Vis viva, 63, 64

Viscosity, 341; of air, 23

Volt, 198, 227, 240

Volta, Allessandro, 231, 234

Voltage, 226, 268

Voltmeter, 244

W

Walton, 330i

Water: maximum density of, 147; thermal properties of, 146, 147

Water waves: both longitudinal and transverse, 343; refraction and dispersion in, 359; relation between velocity and wave-length of, 347, 348

Watt, 228, 240; defined, 84

Wave: amplitude defined, 344; appearance of, 337, 343, 344; displacement equation for, 344; energy at a distance from source, 342; frequency, 348; frequency change with motion of observer, 360, 361; longitudinal, 339, 340, 344; motion of a form, 338; period of, 337, 344; transverse, 338, 344

Wave characteristics of electricity, light, and matter, 335

Wave-fronts: defined, 345; photographs of, in ripples, 350–52; relation to rays, 346, 347

Wave-length: change in, with motion of source, 361, 362; defined, 344, 345; interference method of measurement of, 407; in light, 409; relation to frequency and velocity, 367

Wave mechanics, 440

Wave motion: involves elasticity and inertia, 340; pendulum models of, 336–42

Wave number, defined, 348

Wave velocity: equation for, in stretched string, 363; relation to elasticity and density of medium, 340; relation to frequency and wave-length, 391; relation to period and wave-length, 344

Wedge, mechanical advantage of, 89

Weight, 17; apparent loss of, in fluids, 102, 103; equation for, 29, 60; see also Inertia; Newton's Second Law of Motion

Wheatstone, 401

INDEX

Wheeler, J., 330*g*
Whispering galleries, 360
Wien's Displacement Law, 417, 418
Wilson, C. T. R., 323
Wilson, H. A., 279
Work: defined, 59; equations, 63; equations in electric, gravitational, and magnetic fields, 225; against friction, 59, 62; in gravitational field of force, 222, 223; against gravity, and inertia, 62; and heat, 78 ff.; on inclined plane, 60; recoverable, 62, 64; units, 59, 79; *see also* Energy; Heat

X

X-rays: absorption of, 396; corpuscular properties of, 396; diffracted by crystals, 429; discovery of, 317; investigations as to nature of, 429; origin of, 396, 428; photographs, 397; scattered, 434, 435; spectra, 430; therapeutic uses of, 397; wave-lengths and potential difference of tube, 428; wave properties of, 396

Y

Young, Thomas, interference experiment of, 404, 405, 409

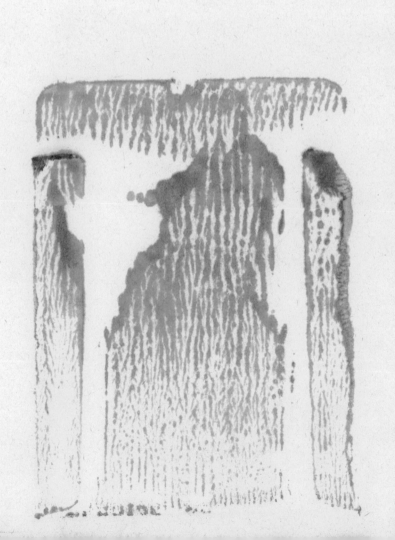